WIDENING MINDS

PROFESSOR TOM FRAME was appointed Director of the Public Leadership Research Group in July 2017 with responsibility for establishing the Howard Library. He served as a naval officer for 15 years and completed postgraduate studies in history (including the completion of a doctorate in history at UNSW Canberra in 1991), theology and sociology before being ordained to the Anglican ministry. He has been Bishop to the Australian Defence Force, patron of the Armed Forces Federation of Australia, a member of the Council of the Australian War Memorial, and judged the inaugural Prime Minister's Prize for Australian History. Professor Frame is the author or editor of 38 books, including *The Shores of Gallipoli: Naval aspects of the Anzac campaign*; *Living by the Sword? The ethics of armed intervention*; *On Ops: Lessons and challenges for the Australian Army since East Timor*; and *The Long Road: Australia's train, advise and assist missions*.

Dedicated to

Professor Jeffrey Guy Grey, 1959–2016
UNSW colleague and close friend

Mr Leonard Victor Hume, 1925–2016
Defence Force Academy Secretariat (1967–1984)
and former soldier

Mrs Trish Burgess
UNSW staff (1977–2000) and research assistant

'The nation needs people whose minds have been widened, whose perceptions of large defence issues have been quickened by an education which stressed inquiry, objectivity and innovative thinking [...] [the] self-reliant Australia about which we often speak needs more than technical training [...] it needs independent thinking about Australia's defence policy and its deployments and its force structure [...] thinking not dependent on borrowed ideas nor an inherited idea about strategy and organisation.'

Sir Arthur Tange, Secretary of the
Department of Defence,
Hearings on the Need for an Australian Defence Force Academy,
Parliamentary Standing Committee on Public Works, 1978

WIDENING MINDS

The University of New South Wales
and the education of Australia's defence leaders

TOM FRAME

UNSW PRESS

A UNSW Press book

Published by
NewSouth Publishing
University of New South Wales Press Ltd
University of New South Wales
Sydney NSW 2052
AUSTRALIA
newsouthpublishing.com

© Tom Frame 2017
First published 2017

10 9 8 7 6 5 4 3 2 1

This book is copyright. Apart from any fair dealing for the purpose of private study, research, criticism or review, as permitted under the *Copyright Act*, no part of this book may be reproduced by any process without written permission. Inquiries should be addressed to the publisher.

National Library of Australia
Cataloguing-in-Publication entry
Creator: Frame, T. R. (Thomas R.), 1962– author.
Title: Widening minds: the University of New South Wales and the education of Australia's defence leaders / Tom Frame.
ISBN: 9781742234397 (hardback)
   9781742234427 (paperback)
   9781742244068 (ebook)
   9781742248462 (ePDF)
Notes: Includes bibliographical references and index.
Subjects: University of New South Wales.
 Australian Defence Force Academy.
 Military education Australia – History.
 Soldiers – Education, Non-military – Australia – History.
 Australia – Armed Forces – Officers – Education (Higher) – History.

Design Josephine Pajor-Markus
Cover design Luke Causby, Blue Cork
Cover images FRONT (top, left to right) Graduation parade, 1988 (*UNSW Canberra*); Last graduation parade, UNSW Faculty of Military Studies, Duntroon, 1985 (*ADFA*); The Defence Academy, 1986 (*Dept of Defence*); Professor Sir Philip Baxter, UNSW Vice-Chancellor, 1968 (*UNSW Archives*); The Defence Academy, 2016 (*UNSW Canberra*). BOTTOM (left to right) Professor Rupert Myers, UNSW Vice Chancellor, 1971 (*UNSW Canberra*); Undergraduate Academy cadets during technical studies (*ADFA*); Undergraduate Academy cadets during classroom learning (*ADFA*); Her Majesty Queen Elizabeth II with Academy cadets, 1988 (*ADFA*); Royal Australian Naval College, HMAS *Creswell*, 1968 (*UNSW Archive*). BACK Images courtesy of Dept of Defence, UNSW Archives and UNSW Canberra picture archive.
Illustration (page xvi) Geoff Pryor, National Library of Australia, nla.obj-156314547

All reasonable efforts were taken to obtain permission to use copyright material reproduced in this book, but in some cases copyright could not be traced. The author welcomes information in this regard.

Contents

A note on titles and acronyms		vii
Preface		ix
Acknowledgments		xii
Dedication		xiv
Foreword		xvii
Introduction		1
1	The universities and service officer education, 1901–1966	14
2	Firm foundations and the faculty solution, 1964–1967	56
3	Establishing a tradition, 1967–1972	98
4	The University as a unifying agent in Defence, 1967–1975	123
5	The Academy commitment, 1974–1976	153
6	The joint educational enterprise, 1970–1980	178
7	From autonomy to uncertainty, 1975–1978	200
8	Parliamentary works and political will, 1978–1981	230
9	From affiliation to Academy, 1981–1986	260
10	Reviews, reforms and restructures, 1986–1995	305
11	Controversy and consolidation, 1996–1998	353
12	Commercial contracts and institutional partnerships, 1999–2004	385
13	Stability and sustainment, 2005–2017	418
14	Observations and conclusions	452
Notes		482
Index		499

A note on titles and acronyms

In 1949, an Act of the New South Wales Parliament created the 'New South Wales University of Technology'. Within a decade the University's title was changed to the 'University of New South Wales'. More recently the University has styled itself the 'University of NSW' and 'UNSW Australia'. The Canberra campus has also experienced several changes of name. Initially referred to as the 'University College, Australian Defence Force Academy', it is presently branded 'UNSW Canberra at the Australian Defence Force Academy' or simply 'UNSW Canberra'.

Unless specified, where UNSW appears I am referring to the University collectively or specifically to the staff and facilities located at the main campus at Kensington in Sydney. References to UNSW Canberra denote academics or activities at the Australian Defence Force Academy.

Where 'RANC' appears, the concern is with the Royal Australian Naval College (RANC) located at HMAS *Creswell*, Jervis Bay. Where 'RMC' or 'Duntroon' appears, it relates to the Royal Military College (RMC), Duntroon where the University's Faculty of Military Studies (FMS) was located from 1968–86. The Royal Australia Air Force Academy (RAAFA), located at Point Cook in Melbourne, also appears in the narrative. It will be referred to as the RAAF Academy or as 'Point Cook'.

Although there is mention of the Australian Defence Force (ADF) and the acronym 'ADF' is used throughout the book, the 'ADF' did not become a formal entity until officially promulgated in 1976. Before then, the armed forces of the Australian nation consisted of the three single Services. The acronym ADF does not usually include the Department of Defence, which is the other half of the Defence 'diarchy'. Hence, when speaking of 'Defence' I have in mind both the ADF and the Department.

Where possible I refer to *uniformed* leaders to refer collectively to officers of the Navy, the Army and the Air Force because the term *military* leader might imply the exclusion of the Navy and the Air Force.

Preface

This is a commissioned history of the relationship between the University of New South Wales (UNSW), the Australian Defence Force (ADF) and the Department of Defence. It begins with an agreement signed in July 1967 to provide for the delivery of undergraduate courses at the Royal Australian Naval College (RANC) at Jervis Bay and the formation of the Faculty of Military Studies (FMS) at the Royal Military College (RMC) Duntroon. This book does not deal in detail with any of UNSW's divisions, faculties, departments, schools or libraries at the main campus at Kensington, Jervis Bay, Duntroon or the Australian Defence Force Academy (ADFA). It does not dwell on the considerable individual achievements of the University's staff and students or the construction of buildings, the acquisition of equipment, the design of courses or the progress of research. The focus throughout is institutional interactions and the aims and objectives of UNSW and Defence in establishing and maintaining a relationship to educate the nation's uniformed and civilian leaders. There remains plenty of scope for others to examine the people, places and programs that I have deliberately overlooked and consciously omitted. Given the constraints of space, I could not deal adequately with every theme and topic worthy of extended treatment. I hope that former staff and students of the University will write their own accounts of the particular contributions they have made. It is a story characterised by many accomplishments.

It is the lot of the historian to assert his independence and his freedom in researching and writing a work commissioned by an institution – in this instance, UNSW Canberra – that would naturally prefer praise to criticism, especially when the invitation to write comes from the historian's employer. I am no different. In preparing the manuscript for this book I was not given a brief to which I was ever obliged to adhere nor did any officer of UNSW (or the ADF) suggest how I might treat any person or event described in the pages that follow. I was at liberty to take the narrative where the archival material and personal interviews led me and to make whatever judgments I felt were occasioned by the facts. This book is my own assessment of the 50 years in which UNSW and the ADF have maintained a close and continuing relationship to educate Australia's Defence leaders. It is not an official history in the sense of being endorsed by the University (or Defence) and I alone am responsible for everything in the pages that follow.

One further qualification: I made a conscious decision not to read Dr Ian Pfennigwerth's lively account of the first 25 years of the Defence Academy published in December 2012 before commencing research for this book in early 2014.[1] This decision was made on two grounds. First, I did not want to be influenced by the shape or substance of his narrative, which is much shorter than this one. Second, I wanted to avoid my book being a review of or a rejoinder to a book that was produced for another moment in time. Ian's book was produced for both the Academy Commandant and the Rector of UNSW Canberra. It focused on the totality of the Academy's life and described its military and academic dimensions. My interest in the cadet experience is their academic education. I make no mention of military training or sport. While I do not share some of Ian's contentions and would differ on some of his conclusions, this is not a rival account. To that end, this book does not include a running critique of Ian's ideas or insights. I would prefer that the two books stand side-by-side with readers left to make their own judgments of the decisions and events we both address.

Readers should also be aware that I am a product of the UNSW–Defence relationship. I was a naval officer for fifteen years between 1979 and 1994, graduated with a Bachelor of Arts (Honours) degree from UNSW in 1985 and completed my doctoral studies at UNSW Canberra in 1991. An uncharitable onlooker might think it inevitable that I would laud the benefits and dismiss the deficiencies of the education I received. To do otherwise, it might be thought, would be to diminish my own experience. But I was critical of the education I was receiving while an undergraduate at the Kensington campus and made no secret of my dislike for the Defence Academy's culture while a postgraduate. It is not in my nature to dissemble. While these disclaimers might seem excessive, I have tried to write this book conscious of my sympathies and allegiances so that the existence of bias and prejudice could be more readily detected and more aggressively addressed. I leave readers and reviewers to decide whether I have been successful in dealing with the bias and prejudice that every writer inevitably brings to their subject.

Acknowledgments

Many people have generously assisted me in the preparation of this history. They include former Vice-Chancellors of UNSW – Professor Sir Rupert Myers and Professors John Niland, Rory Hume, Mark Wainwright and Fred Hilmer; former Rectors of UNSW Canberra – Professors Geoff Wilson, Harry Heseltine, John Richards, Robert King, John Baird and the current Rector Michael Frater, and Deputy Rector Professor John Arnold; former Commandants Rear Admiral Peter Sinclair, Major General Peter Day, Rear Admirals Brian Adams and James Goldrick, former UNSW Kensington staff members Professor Jeremy Davis, Dr Jessica Milner-Davis, Associate Professor Jane Morrison; serving and former UNSW Canberra academics, administrators and students – Associate Professor Anthony Bergin, Professor John Burns, Dr Kelly Frame, Dr Bob Hall, Professor Wolfgang Kasper, Professor David Lovell, Professor Bill Maley, Laurie Olive, Sally Phillips, Associate Professor Hugh Smith, Dr John Sneddon, Trevor Short; at the UNSW archives, Karin Brennan and Katie Bird; at the Academy Library, Annette McGuiness, Rose Holley, Anna Papoulis, Christopher Dawkins, Paul Dalgliesh and Jenny Huntley; former Federal parliamentarians including Roger Price, Ian Sinclair and John Stone; members of the Commonwealth Public Service including Brendan Sargeant, Alan Capp, the late Tony Ayers, Norm Attwood, Rob Tonkin, Dennis O'Connor; former President of the UNSW Students' Association, Professor Ian Lowe; former Deputy Rector of UNSW Canberra and later

Acknowledgments

Vice-Chancellor of the ANU, Professor Ian Young; Professor Francis West, formerly of Deakin University; Emeritus Professor David Horner, inaugural Visiting Military Fellow at the Defence Academy; sadly, I could not interview Professor Ian Zimmer, who has a serious debilitating illness, but I received some information from his review colleague, Professor Bruce McKern; serving naval personnel including Captain Stephen Hussey, Commander (now Captain) Letitia Van Stralen and Lieutenant Commander David Jones of RANC HMAS *Creswell*; former naval officers Commodore Malcolm Baird, Admiral Chris Barrie, Rear Admiral David Campbell, Captain Andrew Craig, Rear Admiral Raydon Gates, Rear Admiral Tony Horton, Commodore Paul Kable, Captain Graham Wright, Captain Laurie Watson and Commander Gerry Purcell; former and serving Army personnel including Brigadier Peter Evans, Colonel Clive Badelow, Major Bruce Hughes; former and serving Air Force personnel including Group Captain Ken Given, Group Captain Callum Brown, Wing Commander Andrew Gilbert; Bob Hogan, Wayne Dalton and Jenny Oldfield of Defence Archives; Dr Roger Lee and Lieutenant Colonel Bill Houston of the Army History Unit; Dr Ben Wadham of Flinders University; and Mr Matthew McNeill at the Australian Civil Military Centre.

Dedication

This book is dedicated to Jeffrey Grey, Len Hume and Trish Burgess.

The idea for this book originated with my friend and colleague, the late Professor Jeffrey Grey, who suggested me to the Rector of UNSW as a possible author. As one of the few remaining UNSW academics at the Academy who had been part of the Faculty of Military Studies at Duntroon, he personally carried much of the corporate experience and had some forthright views on the strengths and weaknesses of the relationship between the University and Defence. Jeff died suddenly in July 2016 as this book was being written. I had managed to speak with him about a number of people and events in the draft manuscript but I was looking forward to his overall assessment of the text. Sadly, this was not to be. But I acknowledge my continuing personal and professional debt in this dedication.

For twenty years, Len Hume was the officer within the Department of Defence who served as secretary of the various inquiries, committees and councils charged with creating a tri-Service Academy. He worked closely with UNSW faculty and staff, gaining their admiration and respect for his attention to detail, grasp of the issues and ability to foresee difficulties. He was an eminent public servant whose administrative skills were crucial to the development of the Academy. Len also died while this book was being written. Although there are no memorials to Len Hume at the Academy, this book is a tribute to his memory.

Dedication

Finally, Mrs Trish Burgess worked for over twenty years at RMC Duntroon and at the Academy in a variety of roles. Not long after I started the research for this book Trish offered her assistance as someone familiar with archival research and having personal acquaintance with the people and events featured in this book. Trish was an indefatigable researcher and a source of energy and enthusiasm. Without her help, much of what will hopefully make this book more enjoyable would be missing. I am grateful for her interest, kindness and companionship and dedicate this book to her with deep personal thanks.

Tom Frame
UNSW Canberra
Australian Defence Force Academy

Foreword

Professor Sir Rupert Myers KBE AO
Pro Vice-Chancellor, University of New South Wales,
1961–1969
Vice-Chancellor, University of New South Wales, 1969–1981

As I read the draft manuscript for this book, I was instantly transported to another world. In the 1960s, the war in South Vietnam was escalating and the armed forces were conscious that they needed a university partner to meet their educational needs. I well recall the discussions of 1966–67 that led the Vice-Chancellor of the young University of New South Wales, Professor Sir Philip Baxter, to propose collaborative agreements with the Royal Australian Navy and the Australian Army. Sir Philip was a towering figure whose ability to see opportunities for the University to contribute to national life was unsurpassed. I was to share the task of negotiating with the Navy and the Army, and devising the agreements that led to the creation of the Faculty of Military Studies at Duntroon, and the delivery of UNSW undergraduate courses at Jervis Bay in 1968.

It is, perhaps, not surprising that some of my own colleagues could not see the value of what the University was attempting to do. In the context of a complex and controversial war that would not end for another five years, the need for uniformed officers to receive a balanced and liberal education seemed to me self-evident. Writing in this period, the British soldier-scholar General Sir John Hackett explained the importance of liberality in education:

> You may hear the question asked why an officer in a fighting service needs a liberal education. Why should he have any more than the requisite professional skills for his job with the minimum technical knowledge to support them? [...] In the profession of arms, what can be demanded of a man extends to life itself. It is this which does more than anything else to ennoble his profession and to set it apart among the true vocations. Would anyone seriously claim that the preparation of a man for such a calling can be limited to the acquiring of a few professional skills and the basic techniques of reading, writing and calculation necessary for their acquisition?[1]

These are wise words and they remain compelling. I certainly shared the Vice-Chancellor's view that the University ought to assist the nation during a time of national upheaval. Indeed, we had a duty to do so.

After my appointment as Vice-Chancellor in 1969, I continued to work closely with the Services and with the Department of Defence after the Commonwealth Government had decided to establish a tri-Service Academy in the national capital. As I re-read the letters I sent and received during the 1970s and the transcript of the Parliamentary Works Committee whose inquiry recommended against building the Academy, I am conscious of the formidable array of forces opposing this vital initiative. Although I had strongly supported the creation of a new university to educate young officers, I had no hesitation in accepting the Defence Minister's request for a continuing partnership with UNSW when the proposal to develop Casey University did not proceed. My last official duty before

Foreword

retiring as Vice-Chancellor in 1981 was signing the agreement to establish the University College that would operate within the Australian Defence Force Academy. It was a moment of immense pride and what I still consider as one of my most satisfying contributions.

The relationship between UNSW and Defence has come a very long way in 50 years. The former technical college located in Ultimo that became UNSW is now a Group of Eight University that includes a very substantial campus in Canberra with annual revenue now exceeding $100 million. Such an outcome was unimaginable in 1967 for either partner. Much more has been achieved than detractors thought possible, while the critics' worst fears have never materialised.

It is an honour to contribute this foreword and I am grateful for the opportunity to congratulate Professor Tom Frame warmly for his wonderfully detailed and very insightful recounting of an unusual collaboration between the Commonwealth Government and a university. The University's staff and students also deserve our gratitude for their achievements across the past half century. I will not be present for the centenary celebrations but I am confident they will be shaped by thanksgiving for what has been an incredibly productive relationship for both the University and Defence. Long may it continue.

Introduction

Western societies place great store and invest deep faith in what education can achieve. Education is considered the foundation of a civilised society and is expected to prevent or remedy all kinds of social evils and moral ills. A society that does not esteem education is accused of undermining its past and impoverishing its future. For good reasons, education is also tasked with preventing those mandated by the state to use lethal force in the pursuit of national interests from descending into barbarism. Uniformed leaders within the armed forces are provided with an education in the hope that right judgment and intellectual discretion will empower them to promote virtue and honesty in all their dealings.

The Australian nation invests substantially in education. In addition to training people for specific tasks, there is an insistence that uniformed officers in particular need to be educated people. There are plenty of reasons for this insistence. An abiding unease with the use of lethal force is partly ameliorated by an assurance that this force is deployed by educated thinkers rather than by ignorant thugs. If Australia is going to impose its will by using force against its adversaries, we prefer those tasked with this duty to be thoughtful and disciplined, creative and imaginative. The use of force must be accompanied by discipline lest it descend into mindless violence in which human life is needlessly lost and the benefits of using force are quickly outweighed by the costs. Australians believe that education will assist uniformed leaders to see when and where they can use conciliation to avoid conflict, and where and how to draw a line between acceptable and unacceptable conduct. Education will prompt a preference for the mind over muscle and the superiority of reason over coercion.

Conscious of these expectations and aspirations, this book considers three main questions. First, why did the University of New South Wales (UNSW) enter into a relationship with Defence to assist with the provision of officer education in July 1967? Second, how has this relationship influenced the University and Defence over the past fifty years? Third, what can be said about the education that has been provided? In answering the first question I am not concerned with officer training other than when it has influenced or shaped the educational programs offered by the University. I am not concerned with vocational programs developed and delivered to make uniformed officers sound navigators, capable soldiers or competent pilots. My interest is in their education and assessing the ways in which it has been shaped by judgments about what will make them a more able person and a better leader. In answering the second question, I will look at the ways in which the University has influenced both the uniformed and civilian components of Defence, and contributed through research and development programs to national security – programs that it would not have undertaken but for its relationship with Defence. The third question is much more subjective. Education means different things to different people in different contexts. The answer resides in the importance of the education offered by UNSW to Defence's leaders being 'balanced and liberal'.

Over the past half century, Defence has clearly considered UNSW to be its university of choice. While it is difficult to determine the precise extent of Defence's reliance on the University for external assistance, particularly in the fields of science and technology before 1967 when a formal relationship began, subsequent annual

Introduction

Defence reports give the strong impression that most requests for assistance were directed to UNSW in the first instance and that, in the majority of cases, the University was able to meet Defence's requirements. Conversely, when the University has sought assistance from Defence, the close relationship between the two institutions has meant the majority of requests have been met. Much of this work has not been the subject of sustained attention.

This book's remit goes well beyond UNSW's contributions to the Service colleges (the Royal Australian Navy (RAN) College at Jervis Bay and Royal Military College (RMC) Duntroon in Canberra) from 1967 to 1985 and the Australian Defence Force Academy (ADFA) since 1986. Much of what has been written about officer education has unsurprisingly taken the vantage point of the single Service colleges and the Academy, and made their evolution and achievements the main focus. Conversely, there is only passing reference to Defence in the commissioned histories of the three universities that have had some formal association with Defence, namely the Australian National University (ANU), the University of Melbourne and UNSW. Histories of UNSW have not made detailed mention of the Defence relationship, nor did the later commissioned history of the Kensington campus colleges (Basser, Philip Baxter and Goldstein).[1] In UNSW's 50th anniversary year (1999), Patrick O'Farrell was commissioned to produce a history of the University.[2] Obliged to observe a tight word limit while describing a large subject, O'Farrell's account of the UNSW–Defence relationship was brief. Had more space been available I am sure O'Farrell would have acknowledged that the University's relationship with Defence reflected the complex character of officer education and explained that UNSW was well placed to make a valuable contribution to the nation through its interactions with Defence.

WIDENING MINDS

Why is officer education complex? When each of the single-Service colleges opened in the first half of the 20th century, they had in mind a particular kind of youth and what they wanted him to become. (Women were not permitted to join the Services as general entry officers until the 1980s.) Joining the Service was probably the first job these young men secured and for many it would be their only job during a long working life. They did not volunteer for the Navy, Army or Air Force to gain an education but to enter a profession – although government inquiries conducted in 1944 and 1946 drew attention to the link between a desire for education and the quest for professionalism. By the late 1950s, indifference to tertiary education was no longer the prevailing view within Defence. There were at least five reasons for this change in perspective.

First, as the character of warfare changed and technology became even more of a decisive factor in battle, the demand for leaders with particular attributes and special expertise could only be met by extending and enhancing education. *Second*, within the Services, individuals or groups are able to capture influential positions and implement their ideas and opinions. Those persuaded by the benefits of tertiary education (as distinct from expanded professional training) were in the ascendance by 1960. *Third*, rather than being seen as a one-time, early-life experience, there was an emerging sense that education was a continuing process leading to intellectual development and vocational competence. An undergraduate degree was a 'licence to learn' and the prerequisite for highly attractive postgraduate qualifications. *Fourth*, the individuals who were attracted to peacetime uniformed service as a career had different expectations to those who had joined the armed forces on a short-term engagement in anticipation of war. *Fifth*, education was increasingly seen as a commodity with commercial value which could be used to attract talented young men to uniformed service who might not otherwise join at a time of economic prosperity and full employment.

The Services realised they could not continue to function as a separate social enclave because their collective learning needs were driven by an array of evolving demands. In addition to the acquisition of more technically advanced equipment, the rise of nationalist movements and the escalation of insurgency throughout Australia's immediate neighbourhood required a great deal more, both personally and professionally, from those who would exercise uniformed leadership. The contexts in which Australian forces operated remained highly fluid and unpredictable. So the enduring question was: what does the operational environment demand of the nation's uniformed officers in terms of public administration, resource management, financial acumen, ethical sensitivity, cultural awareness and so on? Neither Australian culture nor those with an adversarial attitude to its national interests would remain static. Educational needs were constantly changing and those who provided for those needs had to respond in a dynamic way.

There was also the persistent requirement to find the balance between education and training. Every report on officer education, from that produced for the Department of Defence by Professor Sir Leslie Martin in the late 1960s to that tabled in Federal Parliament by Roger Price MP in the mid-1990s, has addressed an imbalance between educational wants and training needs in the context of financial stringency. Politicians and the public sometimes see Defence as a drain on the public purse that ought to be minimised because such expenditure diverts resources away from necessary investment in infrastructure, industry, agriculture and healthcare. And in the context of constrained Defence funding, the provision for education is often blighted by parsimony. Rather than adopting a generous approach that reflects the importance of education, the attitude is usually one of minimal investment. The notable international exception to this approach was the United States during the 1930s, when the absence of armed conflict allowed the American services to invest heavily in education and training – investment that reaped rewards during the Second World War. It was then that the United States Naval War College developed 'War Plan Orange' and the 'Rainbow Plan' for use in the Pacific against Japan.

As Defence legislators and leaders have focused on 'results', the difficulty associated with being specific about the immediate benefits and tangible outcomes of education has led pragmatists to assert education is a 'nice to have' and not a 'need to have' commodity. When the Navy's ships are constantly in dry dock for repairs and maintenance, when the Army's tanks become obsolete and ineffective, and when the Air Force's fighters no longer deter would-be attackers, the perennial cry is to move money from academic education to capital acquisition. But is education a product that can be evaluated like any other 'resource input'? And is there a persistent mindset among uniformed people that is consistently and deliberately suspicious of education?

In his widely quoted critique of uniformed service published in 1976, *On the Psychology of Military Incompetence*, former British Army officer turned academic Norman Dixon appeared to contend that a tendency to disparage learning and deprecate education manifested as incompetence within military institutions.[3] In sum, he conjectured, the benefits of education are most readily apparent because they prevent the spread of attitudes and actions, such as prejudice and pettiness, that actually gain ground in the absence of education. Concluding that he was observing a standing structural feature of officer development programs, Dixon argued that:

Introduction

> the possibility of incompetence springs in large measure from the unfortunate if unavoidable side-effects of creating armies and navies. For the most part these tend to produce a leveling down of human capability, at once encouraging to the mediocre but cramping to the gifted. Viewed in this light, those who have performed brilliantly in the carrying of arms may be considered twice blessed, for they achieved success *despite* the stultifyingly bad features of the organisation to which they happen to belong.[4]

He continued:

> The view taken here is that those intellectual shortcomings which appear to underlie military incompetence may have nothing whatever to do with intelligence, but usually result from the effect upon native ability of two ancient and related traditions. The first of these, originally founded in fact, is that fighting depends more upon muscle than brain, the second that any show of education is not only bad form but likely to be positively incapacitating.[5]

According to Dixon, there was evidence of an enduring preoccupation with sporting prowess and adventure training. Writing in 1976, Dixon did not seem to have been persuaded that the mood had changed much over the previous 50 years. Indeed, he pointed to the 'deliberate cult of anti-intellectualism [that] has characterised the armed services'[6] and that its 'saddest feature' was its 'actual suppression of intellectual activity rather than any lack of ability'.[7]

Dixon was, of course, assessing the British experience, where both the Navy and the Army were well established in the collective psyche and thoroughly embedded in the nation's history. Hence, their resistance to change and the reluctance of governments to compel them to change. By way of contrast, Australia's armed forces shoulder a significant but smaller burden of history, notwithstanding the Anzac myths, making them more amenable to change. In Britain, the need to evolve quickly and to achieve efficiencies after the mid-1960s, when the United Kingdom's economy was slowing down and fell behind other mature economies, encouraged the armed forces to co-opt the universities' academic resources rather than building internal capacity. This explains, in part, why Dixon's critique has only limited relevance to Australian circumstances. Dixon was correct, however, in contending that the 'gap between the capabilities of the human mind and the intellectual demands of modern warfare continues that expansion which started in the eighteenth century'.[8] This gap was further widened by the diminished status of uniformed service in peacetime as other vocations rose in stature – particularly medicine and the law. There is an echo here of Alexis de Tocqueville's 19th-century observation about peacetime armies in democratic societies: 'When a military spirit forsakes a people, the profession of arms immediately ceases to be held in honour, and military men fall to the lowest rank of public servants; they are little esteemed and no longer understood'.[9]

Australian public leaders of all political persuasions have consistently shown great respect for those who volunteer to defend the nation and its interests. There has been acknowledgment that the profession of arms is different from the medical, legal and clerical professions, for instance, because of the demands imposed upon its members. This has created the need for highly specialised professional development that reflects the tone and tenor of national culture so that the armed forces are a true expression of the national character. Since 1901, Australia has sought to produce a particular kind of officer class – neither British nor American,

Introduction

not German or Russian – but Australian. Officers, more acutely than other ranks, are meant to embody the values and virtues of the national culture to the extent that these attributes are known and can be imparted during education and training. There is a parallel suspicion of an elitist officer class that defies the egalitarian national mood. Hollywood movies have also heightened fears of certain types of uniformed leaders, from the clichéd depictions of the heroic lieutenants Chard and Bromhead in the 1964 movie *Zulu* to the psychopathic colonels Kilgore and Kurtz in Francis Ford Coppola's *Apocalypse Now* which premiered in 1979. Australia seems to prefer a self-deprecating and pragmatic figure ever conscious of hubris and scornful of mindless jingoism. While Australian officers are afforded educational opportunities that are not routinely extended to those they lead, charges of elitism have never been levelled at UNSW or Defence on these grounds. That officers need to be educated to university degree level seems to be accepted and even expected by the parliament, the press and the people.

This acceptance may reflect a widely held view that educated people are more civilised and, consequently, more moral and that educated officers seized of the need to be moral will lead by example. By providing officers with an education shaped by the liberal arts as well as the rubrics of their own professional discipline, there is an implicit assumption that such officers will know instinctively how to distinguish between right and wrong, virtue and vice, altruism and nepotism, and usually choose the better part. They will see the 'bigger picture' and develop a more nuanced understanding of the context in which they are serving. Regrettably, there is no evidence that this is necessarily so. Educated people are still capable of barbarism, as the behaviour of British Imperial forces during the 1899–1902 Anglo-Boer War and the conduct of the German Army in Belgium in the opening months of the Great War of 1914–18 makes plain. But there is agreement that moral education and ethical training are necessary if those deployed to operational zones are not to suffer the kinds of unseen wounds that uniformed personnel have incurred since the end of the Cold War in 1990 and the beginning of protracted operations in the Middle East in 1991.

Less relevant to the Australian context is the conviction that a well-educated military is less likely to turn against civil authority and exploit its capacity to deploy physical force in the realm of domestic affairs. The ADF's aversion to the use of force in the domestic context, including Operation Sovereign Borders and the apprehension of 'home-grown' terror suspects, is probably stronger than in Britain and America. There is some suspicion of service officers who become politicians, especially if they are given parliamentary oversight of those in uniform. The only instance of uniformed resistance to civil authority in Australian history is the 1808 'Rum Rebellion', which led to the deposition of William Bligh as Governor of New South Wales. The whole episode is complicated and its significance muted by the fact that Bligh simultaneously held the rank of Governor and Post Captain (equivalent to Commodore in modern naval terminology) and exercised overall naval and military command in the colony. The point nonetheless remains. Part of the educative process is inculcating the view, some have called it a doctrine, that the military has no active role in domestic affairs and is completely subservient to civil authority.

Should the University have any association, then, with an organisation that relies upon indoctrination? Surely indoctrination is the antithesis of academic education and an affront to anyone committed to liberty? Although the word 'indoctrination' brings to mind visions of Soviet gulags and the inculcation of beliefs and commitments in a manner that bypasses reason and intellect, it is used by the armed forces to describe the initial induction of young

Introduction

uniformed officers into the customs and conventions of their profession. In military education there is no attempt to place prescribed beliefs beyond public assessment or rational critique. But there is a desire that young officers embrace certain ideals and habits of the mind and come to regard them as non-negotiable. These ideals include acknowledging the supremacy of civilian government, the importance of preserving the non-partisan political profile of the armed forces, the primacy of the rule of law at home and abroad, the exclusive role of the civilian police in securing domestic order, and the separation of the legislative, judicial and executive arms of government.

Given that Australia does not oblige its citizens to embrace a state-sponsored ideology or an established religion, participation in public life assumes that educators must 'indoctrinate' students about the value of reason and the importance of logic, the significance attached to freedom of speech and the right of free association, the need to challenge contentions and test assertions, and the expectation that personal disputes and public disagreements will be resolved without violence. As uniformed officers have access to both weapons and manpower, they must learn to respect the place of the ADF in the nation's body politic and the moral, legal and political dimensions of using force. Discharging the special responsibilities entrusted to the ADF requires a liberal democratic education that is cognisant of the complexities of modern politics and contemporary international affairs. This does not, however, require the University to associate itself with either the means or the ends of any indoctrination associated with preparing young women and men for uniformed service. Conversely, this book demonstrates the extent to which UNSW faculty and programs have exercised a civilising influence on the Defence community, brought depth and breadth to the professional performance of both uniformed people and departmental staff, and enhanced the provision and delivery of national security through critical engagement and innovative research. Conversely, the Defence relationship has helped the University to maintain a consistent focus on its undergraduate teaching and to preserve openness to social needs and national priorities.

The dynamics of the relationship between UNSW and Defence have been creative and constructive, leading me to argue that Australia has perhaps the best model for managing the remit of a university with the needs of a major public entity, in this case Defence, and that the Australian arrangements are superior to those in Britain, Canada and the United States. The relationship between the University and Defence has matured significantly over the years into one of empathy and trust. It has not only helped both the University and Defence to fulfil their legislated responsibilities; the relationship is a national asset and a measure of the country's self-sufficiency. This long-running relationship is another instance of the nation's coming of age since the 1960s, when it shed the last vestiges of British colonialism and Australian provincialism. That the education of Australian officers has endured far fewer upheavals than that experienced in Britain reflects better management of the intersection between the past, the present and the future. Tradition and innovation have been allowed to interact. The past does not dictate to the present, the present does not dismiss the past and the future does not dominate the past or diminish the present. This book reveals an achievement of national significance yet to be accorded its place in history – until now, that is.

In the chapters that follow, I trace the origins of officer education and the growth of Australian universities since Federation, focusing closely on the Navy, the Army, the Air Force and UNSW since 1949. Each chapter reveals the extent to which a tri-Service academy offering both an academic education and military training

Introduction

was the ultimate goal of the Commonwealth and the University. The arrangements agreed in 1967 between UNSW and Defence were always intended to be transitory. Why the shift from transitional arrangements to semi-permanent ones took nearly two decades is complicated. It need not have been so difficult and demanding but the weight of bureaucratic inertia and the power of vested interest was consistently underestimated. The opening of the tri-Service Academy in 1986 is, however, only half of the story. Once established, there was opposition to almost every aspect of the Academy's operation, with critics pointing to failure and doubters conceding shortcomings that opened the way to a succession of reviews canvassing the possibility of its closure. After 25 years, the Academy finally ended its formative phase, allowing full effort to be directed towards enhancing processes and improving outcomes. But questions of broad principle remain. These questions are addressed in the conclusion, where I critique the Academy's charter to provide a 'balanced and liberal' education. Despite becoming a mantra over the past thirty years, there is some confusion about what constitutes a balanced education and the benefits of a liberal education. Neither the University nor Defence has thought sufficiently long or hard about what the phrase 'balanced and liberal' actually means and why it is worth pursuing. Indeed, the phrase 'balanced and liberal' needs to be re-stated and its importance reasserted every decade. After 50 years of collaborative activity, the time is right for a new vision of education to emerge. Such a vision will help to bring clarity to the University's programs and conviction to Defence's planning. The relationship between the University and Defence has been positive and productive. This book shows why and how.

1

The universities and service officer education, 1901–1966

For the first hundred years of Australia's European history the small white community tried to transform the major venues of population located on the fringe of the vast continent into proud cities hosting institutions and a culture resembling the great centres of learning and inquiry that they had known or heard about in the United Kingdom. As the nineteenth century continued, each of the Australian colonies established schools and planned universities. When Australia became a Federal Commonwealth in 1901, the new nation hosted four universities: Sydney from 1850, Melbourne from 1853, Adelaide from 1874 and Tasmania from 1890. They were not large institutions by any measure. There were only 2652 students, less than 0.1 per cent of the population in 1901, with few staff teaching a small number of undergraduate courses. The sector developed slowly in the early years of the twentieth century, with the University of Queensland established in 1909 and the University of Western Australia two years later. By the time of the Great War, each state had its own university funded from state revenues with more than 3000 students enrolled in their courses.[1]

Although the Australian Vice-Chancellors Committee (the AVCC, which became known as 'Universities Australia' in 2006) was founded in 1920 to serve the collective interests of the six universities, the Commonwealth played no significant role in administrating or regulating what were considered elite institutions supported by the states. No new universities were established in the inter-war years, although the number of government and private schools expanded exponentially alongside trade and technical colleges which offered certificates and diplomas to the people they trained for a range of jobs, including teaching. By 1939, there were still only 14 236 students enrolled in the six universities (and two university colleges). Of that number, 10 354 were undertaking bachelors programs while nearly 4000 were completing non-degree programs. Incredibly, there were only 81 higher degree students in the entire country and none was engaged in doctoral research.

While the Second World War (1939–1945) exerted a downward pressure on enrolments, there was one notable national innovation in higher education. The Federal Minister for War Organisation of Industry, John Dedman, created the Universities Commission in 1942 with a view to extending the Commonwealth's contribution to academic training to support the war effort and managing anticipated post-war demands on the higher education sector.[2] The Commission was also tasked with regulating university enrolments and supporting the Commonwealth Reconstruction Training Scheme (CRTS) that began in March 1944. It aimed to provide education and vocational training opportunities for those who had rendered uniformed service during the war. More than 300 000 people had been assisted by the CRTS in the period between its commencement and mid-1951 when the scheme was closed.[3] The Commonwealth became a significant direct contributor to higher education when an act of Federal Parliament established the Australian National University (ANU) in 1946. It was to be a research-only institution that would focus on projects and programs to serve national needs, mainly in the humanities. Science was essentially an afterthought, with the Research School of Physical Sciences created in 1949.

For its part, the New South Wales Labor Government wanted the Sydney Technical College, founded in 1878 (and located in Ultimo after 1891), to raise its offerings in science, engineering and technology to degree level, and announced the creation of an Institute of Technology in 1946. The Second World War had brought together institutions of higher learning, business and the professions to show what enterprising collaboration could achieve. A Development Council was appointed in 1947 and chaired by the State Minister for Education, Bob Heffron. The Council resolved that the new foundation would instead be called the 'University of Technology', to attract better staff and to assert the Government's desire that it offer qualifications well beyond the diploma level. Its educational detractors claimed the new institution was not a 'real university'. The Liberal Opposition was critical of what they considered was a partisan plan designed to attract votes by a government that had been given a taste for greater control of tertiary education during the war years.

The mood of many at the nation's oldest tertiary establishment, Sydney University, was one of hostility and derision, if only because the proposed university was a government-controlled and financed institution that did not enjoy the scholarly independence that was thought to embody the essence of academic freedom. Sydney now had competition. The prospect of having its monopoly on tertiary education broken was the subject of enduring enmity towards 'the Institute' as many insisted on calling the new university. A Bill entitled 'The Technical Education and New South Wales University of Technology Act' was introduced into State Parliament on 21 March 1949 and passed the following month. The University was formally incorporated on 1 July 1949. A University Council was appointed, academic staff were recruited and students were enrolled. As a sign of what was to come, many of the inaugural intake were ex-Service or Citizen Militia men keen to acquire professional skills. The first classes were conducted at Ultimo while buildings were constructed on the new Kensington site adjacent to Anzac Parade. Many of the staff were recruited as public servants and were deemed employees of the State Government until 1954, when the *Public Service Act* no longer applied to academic staff.

As Patrick O'Farrell explains in his history of UNSW, the Council was clear about the University's role. It was the 'provision at university level of training and research in applied science and technology to provide scientists and technologists for the industrial and commercial development essential to State and nation'.[4] Part of the University's remit was to offer courses in regional centres, including the steel-making cities of Wollongong and Newcastle, and at the mining town of Broken Hill. Decentralisation was part of the University's culture and because many potential students had full-time jobs, there was a pressing need to deliver flexible learning, with evening classes being the norm in many disciplines. As some of the academic staff had experience of wartime production needs or were ex-Service personnel, there was nothing surprising in a close relationship developing between the new University and the armed services keen to see continuation of the close wartime collaboration with a range of technical disciplines. The defence industry also looked towards the University for assistance in meeting the production demands that accelerated dramatically when the Korean War began in mid-1950. Professor (later Sir) Philip Baxter became Director of the University on 1 January 1953 after the inaugural Director and head of the New South Wales Department of Technical Education, Arthur Denning, resigned in December 1952.

The senior academic staff recruited by Denning in 1950 and 1951 objected to working under public service regulations and resented having a public servant as their Director. Denning's strategy was to retain a close link between the new University and his Department to co-ordinate teaching and research, and to align University programs with industry needs. A campaign mounted within the Council to separate the University from the Department soon succeeded. Denning had no choice but to resign.

The Welsh-born, English-educated Baxter would become the first Vice-Chancellor of UNSW when the Director's title was changed in 1955.[5] Baxter was an academically well-qualified Professor of Chemistry but came to Australia with a bigger reputation in both the defence industry and the nuclear energy sector. He was also able to cultivate a close and co-operative relationship with state and federal governments. The University certainly benefitted from the tensions of the Cold War and the absence of any reticence on Baxter's part about the University making a contribution to national defence. He promoted partnerships with industry groups and later founded the research commercialisation unit that became known as 'Unisearch'.

Soon after the Coalition's electoral victory in December 1949, Prime Minister Robert Menzies had established the Mills Committee to examine university funding.[6] The Committee's Report led to the provision of block grants for previously state-funded universities alongside Commonwealth scholarships for undergraduate study in 1951 and postgraduate study in 1959. People aged over 25 were able to apply for 'Mature Age Scholarships' and education expenses became tax deductible. The Commonwealth now contributed one-quarter of the recurrent costs of the universities. They were no longer solely the responsibilities of state governments, whose declining revenues precluded necessary investment in facilities and infrastructure.

The universities and service officer education, 1901–1966

The Menzies Government created the Committee on Australian Universities in 1954 to deal with the Commonwealth's stake in higher education and two years later Sir Keith Murray, Chairman of the British University Grants Committee, was invited to lead a committee of inquiry into the entire higher education sector. Murray and his colleagues found the universities to be poorly funded, lacking necessary staff and infrastructure, facing high failure rates and offering inadequate honours and higher degree programs.

The Prime Minister described the tabling in Parliament of the Murray Report as 'a rather special night' in his political career as he outlined an expansive vision for Australian education in a statement to the House of Representatives on 28 November 1957:

> It is not yet adequately understood that a university education is not, and certainly should not be, the perquisite of a privileged few [...] We must, on a broad basis, become a more and more educated democracy if we are to raise our spiritual, intellectual and material living standards.[7]

The Government's response to the Report's recommendations included substantially increasing Commonwealth funding to the universities for capital works and operating expenditure in 1958, and the creation of the Australian Universities Commission as a statutory body in 1959.[8] The Victorian Government decided to establish Monash University as that state's second university in the wake of the Murray Report and was mindful of developments in New South Wales. In 1961, the eminent physicist and education policy advisor, Professor Sir Leslie Martin, was commissioned to produce a plan for the longer-term development of tertiary education in Australia, including the creation of colleges of advanced education as the sector grew rapidly.[9] There was a 13 per cent annual increase in enrolments in 1958, 1959 and 1960, with more than 55 000 students enrolled in ten universities by 1961. Within 15 years, the number of university students would increase to over 150 000 enrolled in 19 universities.

As a direct result of the Murray Report and the possibility of offering courses beyond science and technology, the University of Technology was renamed 'The University of New South Wales' on 7 October 1958 and began its transformation from a technology-focused institute to a generalist university with the foundation of Faculties of Arts and Medicine in 1960. Notwithstanding opposition to the change from those who thought the closest attention ought to remain on science and technology, Baxter could see that adding new faculties would help rather than hinder the existing schools. Expanding the University's remit would generate internal momentum, achieve organisational economies and demonstrate commitment to advancing the national interest. The new faculties would be created through increases in Commonwealth funding flowing from the Murray Report and would accommodate students unhappy with overcrowding at Sydney University. A series of 'Visiting Committees' was also established that linked faculties and schools with the government departments, industries and businesses that would recruit graduates. The committees would ensure that courses and subjects were relevant to professional demands and were consistent with workplace practice.

Baxter was especially keen for the University to establish a formal relationship with Defence. He was able to build on the platform that had been laid by Arthur Denning. With the introduction of a national service scheme following the outbreak of war on the Korean Peninsula, Denning wrote to the Secretary of the Department of Labor and National Service, Henry Bland, suggesting the creation of a militia unit for those young men at the University who were deferring the fulfilment of their national service training obligations for the duration of their studies. Denning thought these men might become part of an Army 'technical training unit' whose members would be inducted into the scientific aspects of modern warfare. The unit would be associated with the Electrical and Mechanical Engineers Corps (RAEME) which was then facing an acute shortage of technically trained commissioned officers. The Army was interested in such an arrangement. After meeting with senior militia officers, the University's Registrar, Godfrey Macauley, wrote to the Commander of the Second Military Division, Major General Victor Windeyer, suggesting they form a unit that would be clearly associated with the University. Macauley explained that around 200 students were interested in enlisting. Windeyer proposed the unit be known as 'The New South Wales University of Technology Workshop, RAEME' and that it specialise in the maintenance of anti-aircraft weapons. After further discussion, the Army hinted that the unit might have regimental status and recruit up to 400 men. Approval was granted in July 1951 for the creation of 'The New South Wales University of Technology Regiment, RAEME'. The Regimental Headquarters was located in Harris Street, Ultimo and was functional by February 1952. The first unit camp was held at nearby Moore Park in May 1952 with 86 men attending. By the end of its first year, the Regiment hoped to relocate with the rest of the University to the Kensington campus. In mid-1953, Baxter offered the Regiment the old Randwick Racecourse tote building as well as a hut located on High Street – the northern boundary of the campus site – while consideration was given to the Regiment becoming an Infantry unit (UNSWR) with a more general remit for officer training. Baxter went out of his way to create a tangible link between the University and the Army, and hoped that mature age undergraduates (many of whom had served in the Second World War) would combine their university studies with part-time military service.[10] By 1957, more than 600 men had enlisted in the unit. UNSWR was an integral part of campus life. In 1963, the Regimental Colours were consecrated at a major public ceremony held on the University's oval. There were no objections to the ceremony and no protests about the fact that the University maintained a formal link with the Army. Training depots were also being established near the University's campuses in Newcastle and Wollongong. It was not until the late 1960s that the Regiment's drill hall became the target of anti-Vietnam war slogans.

Although Baxter had a high regard for the Services, the thought that UNSW might become the nation's 'Defence university' was still a long way off. The Services were still pondering their own educational needs and how they might be best met. With more and more young Australians seeking a university education and the economy demanding a larger number of professionally trained people, the Navy, Army and Air Force were facing a challenge. They had to offer accredited tertiary education to their officers as a recruiting tool and better prepare them for the demands of uniformed service in a technologically advanced world with more intellectually complex challenges. The response of the three Services was uneven, resembling their historic provision of officer education.

When the Australian colonies were united in a Federal Commonwealth in 1901, there was little the newly created and cash-strapped national government could do until the cessation in 1911 of section 87 of the Constitution (the so-called 'Braddon Clause' or the 'Braddon blot', as it was called by its opponents, after the Tasmanian politician Ned Braddon who proposed it), which required the Commonwealth to hand three-quarters of all customs duties to the states.[11] National defence languished until just before the outbreak of the Great War in August 1914. While Imperial defence was the responsibility of the British Government, the Commonwealth was obliged to provide for local defence. As navies and armies needed leaders who were educated and trained, it was no surprise that the Government looked to the Royal Navy and the British Army for advice and assistance. But this did not obviate the need for the Commonwealth to offer education and training that was cognisant of Australian culture and attentive to the peculiarities of the Australian people. Section 29 of the *Defence Act* (1903) stated:

> The Governor General may establish an institution for the purposes of imparting education in the various branches of naval and military science, and in the subjects connected with the naval and military professions, and for qualifying persons for the naval and military service.[12]

Progress was slow because the Commonwealth could not afford to establish a college for either the naval or military forces. National leaders recognised that Australia could not provide for its own defence until it produced uniformed officers who were educated and trained. Notably, they distinguished between education and training but no description of what pertained to either was forthcoming. If officers needed an education, the obvious place to turn was the universities and the possibility of a partnership was canvassed. Meeting the Navy's needs was much more complicated than satisfying those of the Army.

The Royal Australian Navy (RAN) was more firmly wedded to its inherited traditions and strived consciously to be an antipodean clone of the Royal Navy. Because the RAN relied upon the Royal Navy for much of its training, its educational programs had to meet British requirements. Conversely, the Army was more flexible in its approach to education and tended to focus on content and delivery more than structure and administration. There were fewer debates within the Army about principles, with attention devoted mainly to practicalities. The Royal Australian Air Force (RAAF), which was not established until 1921, was able to look initially to the Army for its educational requirements because its leaders had been members of the Australian Flying Corps – a military unit. With its focus tightly on technology and tactics, it was not until the period after the Second World War that the RAAF began to concentrate on the specific educational needs of its officers. Hence, the balance of the following narrative reflects the complexity of naval education, the evolution of military education and the emergence of air education.

The Navy and officer education

As much of the Navy's early administration was based in Melbourne, the Council of the University of Melbourne expressed an interest in establishing a School of Naval Science in March 1906. As the Federal Parliament was also located in Melbourne, the Vice-Chancellor of the University of Melbourne, Professor Sir Henry Wrixon, sought the help of the Minister for Defence, Senator Thomas Playford, to encourage the Director of Commonwealth Naval Forces, Captain William Creswell, to take advice in England on the best way forward. Creswell conferred with Professor James Alfred Ewing, the Scottish physicist and engineer who had been appointed the Director of Naval Education at Greenwich in 1903, who did not think the courses the Australian university was proposing would achieve much of value. Creswell did suggest, however, that the construction of pre-fabricated destroyers for the soon-to-be established RAN would be a task with which the University could assist. For its part, the University wanted the Navy to fund a professorial position. The money was not available and a partnership with Melbourne never became a serious prospect. When established in 1913, the Royal Australian Naval College (RANC) was essentially a naval secondary school.[13]

The RAN followed the pattern of the Royal Navy in recruiting 13-year-old boys as the main source of its officers. Boys from around Australia were selected to enter a four-year program of academic study and naval training that resembled the syllabus at Britannia Royal Naval College (BRNC), Dartmouth and the usual range of secondary school subjects offered in Australian high schools. Shortly after its opening, a number of external assessors concluded that academic standards were impressive. In fact, the final-year program was judged to sit between the final year of secondary education and the first year of tertiary education, especially in the subjects relating to engineering, which had always been one of the College's curriculum strengths.[14] By 1922, every Australian university had accepted the College's 'Passing Out Examination' as equivalent to matriculation, with the sole exception of the University of Sydney.[15] Because post-matriculation academic education was conducted in Britain as a companion to professional naval training, there was no need at that time for the Navy to seek a partnership with any Australian university.

But the RAN College reached a critical juncture in 1920. Its future was in doubt because costs were high and student numbers were never more than a few dozen. The Naval College was smaller than the Royal Military College (RMC) that had been founded on the grazing property 'Duntroon' in Canberra; the Navy required fewer officers after the Great War as much of the Australian fleet was paid off into reserve. Put simply, officer education was a very expensive operation 'largely due to the isolation [of Jervis Bay] and consequent high cost of transport of all material and personnel'. Not surprisingly, the wisdom of the Commonwealth's investment in the Jervis

Bay site was being questioned. Was there a more cost-effective way of producing naval officers? A number of alternative schemes were considered, including abandoning local efforts to educate cadet-midshipmen and sending them all to England, or amalgamating the naval and military colleges with cadet-midshipmen sent to Duntroon to be educated alongside their Army equivalents. The Navy wanted to preserve the status quo but felt a responsibility for the young officers it could no longer employ.

On 27 June 1921, the Naval Secretary, George Macandie, wrote to Henry Barff, the Warden and Registrar of Sydney University, asking for a reconsideration of its refusal to accept that the academic standard achieved by fourth-year cadets at RANC was equivalent to the Matriculation Examination. The Naval Board hoped to find alternative career paths for those cadets facing involuntary discharge.[16] Boys from New South Wales had only one university open to them at that time: the University of Sydney. The Sydney University Professorial Board considered the matter on 25 July 1921 and a second time on 28 September 1921. Although Macandie wrote again on 23 December 1921 to point out that Queensland and Melbourne accepted the College's academic standards as sufficient for enrolment in their courses, Sydney held firm and would not change its attitude for another thirty years. This embittered many naval officers against the University of Sydney, thinking its decision mean-spirited and snobbish.

The Naval Board visited Jervis Bay in February 1922 together with Lieutenant General Brudenell White (representing the Military Board) to consider options for the future.[17] The consensus was clear: maintaining separate colleges was preferable but the cost was becoming prohibitive. The next month a small working party drawn from the Naval and Military Boards met to discuss the possibility of Duntroon becoming a 'Combined Services College' with academic staff drawn equally from the two establishments. The Navy opposed the proposal to raise the entry age for naval cadets from 13 years to 16 years and to lower the entry age for military cadets from 18 years to 16 years because it departed from the practice of the Royal Navy. The 13-year-old entry was non-negotiable for the Navy, which insisted on strict compliance with British customs and practices. When amalgamation proposals were considered formally by a joint conference of the Naval and Military Boards, the idea met with little enthusiasm and no support. Each service wanted to conduct its own education and training, notwithstanding the cost, while the Naval Board believed the strategic situation would soon change when the RAN would be inevitably expanded to deal with the rise of Japan as a naval and military power. A bigger Navy would create a need for more officers that would, in time, return the Naval College to viability.

The Minister for Defence, Eric Bowden, was persuaded that the cost savings flowing from amalgamation were slight and recommended no change to the arrangements. But Federal Cabinet decided in May 1923 that the cost of separate colleges could not be sustained and appointed a panel consisting of a senior naval officer, a senior military officer and the renowned Welsh Australian geologist and Antarctic explorer, Professor Sir Edgeworth David, to compile a list of cost-effective solutions. The panel's conclusions were stark: 'It would be a great advantage to the operations of the future if the officers of the three arms had been brought up together from youth, and had thus gained that personal knowledge of one another which is so essential to successful cooperation'.[18] But amalgamating the colleges to create an enlarged institution was not seen as a viable solution given the immediate need to reduce costs. Not even the admission of fee-paying civilians to the Service colleges or sending cadets to a civilian university would make much of a difference to the overheads. Further, the distinctive Australian character of the Navy and Army would be lost if cadets were educated in England. The savings would be minimal and the benefits marginal. The panel recommended that the Government do nothing until the 1924 Imperial Defence Conference, to be held in London, that would determine the nation's future security needs.

The conference did not lead to a larger RAN. The Admiralty was not in a position to financially support the Royal Australian Navy. The budgetary drain of the Australian colleges remained, with a succession of ministers unwilling to direct that RANC be closed down, given the College had only existed for a decade. The decision they made was to make no decision, until external circumstances forced one upon them. The Great Depression that blighted the nation's economy after October 1929 sealed the Naval College's fate. The Commonwealth Government needed to take drastic steps to deal with a rapidly escalating budget deficit. The Naval College was relocated to Flinders Naval Depot at Westernport on Victoria's Mornington Peninsula in 1930.[19] The College buildings at Jervis Bay were leased as a holiday resort. With the prospect of another world war looming throughout the 1930s, there was more emphasis on training than education and no attempt to offer post-matriculation study to young naval officers. In 1940, driven by wartime needs, the Navy introduced the 'Special Entry'. Young men who had attained the Leaving Certificate (or its equivalent) were given a six-month course limited to professional subjects before being sent to sea and active service.[20]

After 1945 and the return of peace, the RAN was obliged to rethink some of its attitudes to education and training as a consequence of changes abroad. The Royal Navy decided with some reluctance to end its 13-year-old entry in 1950 after the *Education Act 1944 (UK)* required all children to remain at school until the age of 15, while those at grammar schools were obliged to continue attending until they were 16. Although the RAN was not under similar pressure, it decided in 1951 to accept boys aged between 15½ and 16½ who had completed their Intermediate Certificate. The cadets of the 'Intermediate Entry' would complete a two-year program that was virtually identical to years three and four undertaken by the 13-year-old entry. National benchmarking of academic standards soon emerged as a challenge for the Navy. The College's annual report for 1954 noted:

> It had been found that the Intermediate examination was a most unreliable measuring stick, varying from state to state and often from school to school. Also, the fact that some candidates sat for external and others for internal examinations proved unsatisfactory for RANC requirements.[21]

When the Naval Board decided to end the 13-year-old entry in 1955 and to change the name of the 'Intermediate Entry' to the 'Normal Entry' in 1956, the majority of new-entry cadets would complete the final two years of their secondary schooling at the Naval College (which was still located at Westernport). Boys entering the College were now aged from 14½ to 16½, the range reflecting the existence of different state education policies. On finishing their studies, these officers were promoted from cadet-midshipman to midshipman and sent to the fleet or to shore establishments to continue their professional training.

In response to falling recruitment during the late 1950s, the Navy decided it would take a small number of young men (initially aged under 19½ years and later raised to 20 years) who had already matriculated and were seeking a naval career. This was termed the

'Matriculation Entry'. Joining the Naval College in January of each year, they would complete a course of professional subjects over two terms and graduate in August. This period was later deemed to be too short and was extended to three terms 'as there were academic weaknesses to be attended to despite the matriculation'. The study period was extended again to one year and seven months. The academic standard reached was 'designed to bring them to London University entrance standards'. This was, in part, because these midshipmen were then enrolled in professional courses in the United Kingdom that presumed this level of achievement.

Although the RAN had retained the 13-year-old entry for five years longer than the Royal Navy, it could not avoid being attentive to other developments in Britain. Australian officers were still being sent to Dartmouth, an institution sometimes referred to as a naval 'finishing school' for Dominion officers. The Admiralty's decision in April 1958 to commission Sir Keith Murray, Chairman of the British University Grants Committee, to chair the 'Dartmouth Review Committee' and to prepare a report on officer education would influence directly what happened at Jervis Bay. An agricultural economist by training before he became a government administrator, Murray noted that the Royal Navy had already moved substantially away from secondary schooling towards tertiary education, with degree programs being offered at the Royal Naval Engineering College in Plymouth as part of an affiliation agreement with the University of London. Moves were already underway at the Britannia Royal Naval College at Dartmouth to offer the first year of undergraduate degree programs to midshipmen seeking to become Seaman and Paymaster officers. Murray proposed making eligibility for admission to university a standard prerequisite for officer entry to the Royal Navy. He also recommended that the Royal Navy cease offering secondary education and accept only matriculation candidates. The Admiralty took this advice. Secondary education for young officers was ended and matriculation became mandatory for all new officer entrants to the Royal Navy.[22]

In response, the Naval Board in Melbourne asked whether graduation from Dartmouth 'involved the attainment of a standard equivalent to university graduation and that some, preferably well-known university, would accept this standard and confer degrees accordingly'.[23] The RAN Liaison Officer in London was directed to ascertain whether 'consideration could be given to the possibility of the attainment of university graduate status by naval officers in order to improve: i) the officer-man relationship and ii) the officer status in the community'. The answer was 'yes' but that Australian educational standards at that time did not meet London University 'A' Level exams.

It was clear that the RAN needed to augment the educational programs at Jervis Bay (the Naval College had returned to its former home in early 1958) if Australian midshipmen were to be enrolled in courses at Dartmouth. On 2 January 1959, the Commanding Officer of the Naval College, Captain Bill Dovers, produced a detailed document headed 'The Officer Training Structure Proposed by the Murray Committee Report'. Dovers, who was to have a very prominent role in the provision of officer education in the 1970s, noted the 'modern tendency seems to be toward a tertiary rather than a secondary education as the basis of training officers in all three services. As far as the RAN is concerned this would require a completely new approach to initial training'. The report noted five arguments in favour of providing a tertiary education.

> a) The future naval officer will require a higher scientific education than he now receives to be proficient; b) a university degree would provide a certain tone or standing in the future,

and will be the yardstick by which people are judged; c) if naval training offers education to a university degree we will attract more and better candidates; d) more candidates will be attracted to the service as they will have qualifications for civilian employment on retirement; and e) other services consider it necessary.[24]

These factors were weighed and evaluated. The principal question, in Dovers' view, was 'whether a university degree will result in a better naval officer and leader, and whether it is necessary for proficiency?' The answer was tentative: 'it has yet to be demonstrated that education to this standard is a pre-requisite of producing great leaders'. Notwithstanding growing Government interest in establishing a tri-Service academy and the expressed desire for its educational programs to have university recognition, Dovers did not think wholesale change was necessary. After all, he said, it was 'doubtful whether a degree conferred by such an Academy would carry much weight in civilian circles in the long term'. But what of the Navy's standing in the wider community? The report's conclusion was adamant: 'there is no justification for insisting that all naval officers should hold a degree until such time as a degree becomes the acceptable standard for executives throughout the country'. Achieving this standard would not be easy in any event:

> [A] great deal of expense and a great deal of reorganisation would be required to teach to a university degree standard at RANC. Due to the small numbers under training, it would be hard to justify such expense. There is the danger that RANC Jervis Bay would become redundant if such a scheme were introduced.[25]

The views that Dovers expressed in 1959 are significant in that he had repudiated practically all of them by 1975.

On 19 May 1959, Dovers and the civilian RANC Headmaster, Mr Q de Q Robin, met with the Second Naval Member of the Naval Board, Rear Admiral 'Arch' Harrington, at Navy Office in Melbourne to discuss Dovers' report and to consider changes to the RANC curriculum as a result of the recommendations made by the Dartmouth Review Committee. As the flag officer with responsibility for personnel, Harrington set the tone of the meeting when he remarked that:

> no-one should be a naval officer unless he possessed a degree as a hallmark of his ability to think. It did not matter if it was a degree in the humanities but the community itself demands this hallmark and unless one has it one will not be accepted.[26]

Harrington also indicated a preference for expanding the 'Matriculation Entry' but was advised the Junior Entry was needed to meet recruiting targets. One of the issues the Navy needed to consider was the possible advent of a 'combined Service college'. What would happen to the 'Normal Entry' (soon to be renamed the 'Junior Entry' with the 'Matriculation Entry' becoming the 'Senior Entry'), because it was a mode of entry without an Army or Air Force equivalent? The Navy was alone in providing secondary schooling. More work was needed on a comprehensive plan for naval education.

Taking its lead from the Admiralty in London, the Naval Board established a committee of inquiry headed by Jock Weeden, the Director of the Commonwealth Office of Education, with representatives of the secondary and tertiary education sectors from across the country. The 'Weeden Report' offered a comprehensive blueprint for the future of education throughout the RAN.

It recommended concentrating all academic study in Australia and proposed making degree courses available to those midshipmen who could handle tertiary studies. The committee recommended that all midshipmen complete the first year of degree studies to ensure a common standard. Those less interested in going to university or deemed unsuitable for undergraduate studies would undertake an alternative education program at the Naval College. The most controversial recommendation was educating engineer officers in Australia to degree level 'followed by one year of equipment training in the United Kingdom'. The Naval Board was not persuaded. This was too much change too soon and involved severing too many ties with the Royal Navy, to whom the RAN looked for the maintenance of its institutional standards.

The Weeden Report was nonetheless a turning point. The RAN would become more self-reliant. It would invest more heavily in tertiary education as it operated more technologically advanced equipment requiring higher levels of professional competence. The Naval College would no longer be a self-accrediting institution that developed its own curriculum. It would look to the expertise of the universities and consider their advice on what junior naval officers needed to learn, albeit with close attention to workplace applications. At a time when most other professions were looking to the universities for assistance in shaping the content and the conduct of programs for their practitioners, naval officers could not afford to be left behind. Notably, the Naval Board was far from persuaded of the need to offer degrees to increase recruiting or to lift the social standing of its officers. A Standing Academic Advisory Committee chaired by the Commanding Officer of the Naval College was given the task of implementing Weeden's recommendations.[27]

At its first meeting in February 1963, and emboldened by the Weeden Report's expansive vision, the Academic Advisory

Committee first canvassed the possibility of the Navy abandoning secondary studies and concentrating entirely on post-matriculation study. While Junior Entry cadet-midshipmen had their matriculation studies directed towards possible enrolment in a degree program, the Senior Entry midshipman undertook a year of academic study closely resembling the initial year of an undergraduate degree. In 1964, an agreement between the Naval College and the University of New England (UNE) allowed five cadets to study first-year Arts degree subjects (History I and English I) by correspondence and to have these results credited towards a UNE degree. One of the cadets, Gerry Purcell, later recalled having to 'read three Shakespeare plays as a part of the curriculum, which was a challenge, and having to work late into the night to keep up [...] In short, it was not a successful exercise and I believe it was abandoned after that first year'.[28] The chances of a midshipman building on these subject completions were, of course, minimal because much of the following year was devoted to naval training and eleven weeks at sea in the training ship. Very few naval officers ever received a UNE degree. The Captain of the Naval College explained that the need for extra tuition meant curtailing leave, starting each day at an earlier hour and reducing mandatory afternoon sport from three afternoons to two. He also lamented the RAN's reliance on the Royal Navy at a time when the Admiralty was constantly changing its educational requirements.[29]

Looking further ahead, the Academic Advisory Committee also recognised that the advent of a tri-Service academy would have a profound effect on whatever happened at Jervis Bay. To prepare the Naval College for this eventual transition, the Committee took the bold step of recommending the abolition of the Junior Entry (which had been the mainstay of officer recruitment) and the introduction of university-aligned post-matriculation courses at Jervis Bay.

The Naval Board decided that it was 'unable to abandon schoolboy recruitment' but accepted the need for the secondary school curriculum at Jervis Bay to be aligned with the New South Wales Matriculation Examinations (although the Naval College was physically located within the Australian Capital Territory).[30] This decision was not without some controversy, as the Navy would be relinquishing curriculum control to an outside authority for the first time in its history. The Board eventually accepted the need for such a concession, as it would be much easier for midshipmen to gain entry to universities if they possessed the standard entrance qualification.

On 19 August 1964, the then Minister for the Navy and a decorated former RAAF pilot, Fred Chaney, approached the Minister for Defence, Senator Shane Paltridge, seeking his approval for degree courses to be offered at Jervis Bay in order to 'bring the Navy to the level of the other two Services'. The possibility of a partnership built on the connections the Naval College staff had already made with UNSW academics based at the Wollongong campus was quietly canvassed in 1965. Notably, the Army started similar discussions with UNSW at the same time (see 'The Army and officer education' on page 45) and one appeared to add momentum to the other. After a significant delay brought on by the slow progress of another national inquiry into tertiary education and the untimely death of Senator Paltridge in January 1966, the new Defence Minister, Allen Fairhall, advised the new Navy Minister (and former wartime RAAF officer), Don Chipp, in March 1966 that informal negotiations with UNSW could proceed. By the end of the year, Chipp sought Fairhall's approval for the Navy and UNSW to sign a formal agreement. On 6 April 1967, he was authorised to proceed on the same conditions that applied to a similar agreement the Army was negotiating with the University. A new era was set to begin, although the Navy still struggled to articulate a clear vision for education: why it was important and of what it would consist?

The Army and officer education

There were a number of parallels between the Navy's experience and that of the Australian Army. The University of Sydney was the only institution that expressed an interest in educating Army officers, with the University Senate establishing a committee to consider the assistance the university could provide in April 1905. By year's end the University agreed to establish a Department of Military Studies and asked the War Office in London to select a suitable departmental head. It chose an officer of the Royal Engineers, Colonel Henry Foster. Commencing his duties in early 1907, Foster arranged for subjects to be delivered in everything from engineering to administration, and history to science. Classes were held in the late afternoons and evenings to allow students in the University's mainstream courses to attend in the hope they would complete the Diploma of Military History and Science and be eligible for appointment to a commission in both the Imperial and Commonwealth Military Forces. Serving officers provided specialist technical advice and other areas of the University offered the expertise of their disciplines, such as the history and engineering departments. The number of enrolments in the first year (63) was impressive, although the figure was nearly halved in the second year. The demand was limited and seems to have been met in the initial year. It would appear that those enrolling were motivated by personal interest in the particular subjects being offered rather than from vocational aspirations that would have been reflected if they had completed the whole program.[31]

WIDENING MINDS

When Lord (Herbert Horatio) Kitchener of Khartoum was invited to Australia by the Commonwealth Government to report on the young dominion's defences in late 1909, he was far from impressed by what he found. In addition to making a series of recommendations on command organisation and force structure, he spoke of the need for Australian leaders to have a 'complete military education', which he thought was best delivered by an Australian college based on the structure of the United States Military Academy at West Point. The cadets who achieved the best results would serve in the permanent forces, while those not selected for permanent service or who sought a civilian vocation would be required to serve for twelve years in the citizen forces (the part-time militia) in return for their military training. Kitchener suggested Colonel William Bridges, the Australian representative at the Imperial General Staff in London, as an officer well-equipped to serve as the first college commandant. The Minister for Defence, Joseph Cook, offered Bridges the post in January 1910. Acting on Kitchener's advice, Bridges was told to model the Australian college on West Point, an establishment he could visit on the way home from London. After initially declining the appointment, Bridges was persuaded to accept and visited the British military colleges at Woolwich and Sandhurst before returning via Canada, where he visited the Royal Military College, Kingston and then West Point in the United States.

The Royal Military College (RMC), located on the former Campbell family pastoral property of 'Duntroon' within the Federal Capital Territory, was opened in June 1911.[32] It offered a four-year program of education and training. 'Civil subjects' dominated the first two years; the final two were devoted to 'military subjects'. The former were externally validated by academic faculty at the University of Sydney, later supplemented by assistance from the Universities of Melbourne and Queensland. On completing their studies in

Australia, graduates would gain practical experience within a British or Indian Army regiment before receiving training specific to their corps allocation. Shortly after the 1914–18 Great War, when education and training programs were no longer circumscribed to meet operational demands, the Army managed to secure external recognition for the academic studies its cadets completed. It achieved this breakthrough in the discipline of engineering during 1920, with the University of Melbourne and the University of Sydney granting two years of credit for studies completed at Duntroon and admitting RMC graduates into the third year of their respective engineering degree courses after 1920. The numbers were very small but the precedent had been established. The Army was delivering educational programs at the undergraduate level and had achieved external recognition for its students.

The decision by many of the universities in September 1921 to recognise the first year of the RMC course as equivalent to matriculation made a difference to all RMC graduates. By then, however, the cost of maintaining the College had become the subject of adverse parliamentary and press commentary. There were numerous suggestions for making the College more cost-effective, including a proposal to make Duntroon the nucleus of a 'Federal University' which would be designed to support Australia's future development as an industrial and agricultural power.[33] As previously mentioned, the Army resisted amalgamating Duntroon and Jervis Bay but it did accept a small number of RAAF cadets. By the end of the decade and with the Great Depression seriously disrupting the economy, the College was relocated to Victoria Barracks in Sydney and the 1931 entry was cancelled. There was no further conversation with any university about accreditation. The future was uncertain and the number of cadets to be educated could not be predicted. The academic staff was reduced from five to three. Although the cadets were not far geographically from Sydney University, there was no attempt to enrol them in undergraduate courses. It was soon apparent to the Army and to the Commonwealth Government that Victoria Barracks was not, in fact, a suitable site for a military college. When the economic stringencies of the Great Depression had passed and the likelihood of another world war went from being possible to probable, the College returned to Canberra. RMC had been away from Canberra for six years. Only two classes graduated without an experience of Duntroon.

With the declaration of hostilities in September 1939, the RMC course was understandably shortened to just two years. The program was shaped by wartime needs and the constant demand for junior officers to lead a rapidly expanding Army. As the Second World War continued, a number of senior officers canvassed the possibility that the RMC academic program be raised to university level and a suitable degree awarded to graduates. The influx of men into the Army with tertiary qualifications and the close relationship that had developed between the Army and a number of universities in relation to research projects had created goodwill and an appetite for a continuing relationship after the war. In late 1944, General Sir Thomas Blamey, the Commander-in-Chief of the Australian Military Forces, commissioned a review of RMC and its future. The review committee included Major Generals Alan Vasey and Horace Robertson, and the RMC Commandant, Brigadier Bertrand Combes. The committee examined the future of RMC, particularly its entry requirements and the affiliation of students with civilian universities. The latter part of their brief was prompted by a proposal for Duntroon to host cadets destined for the Citizens Militia, RAAF students and cadets from other Commonwealth countries. Vasey's committee also examined the curriculum and, echoing Lord Kitchener's remarks from 30 years earlier, recommended modelling the RMC program on that offered at West Point. This reform would require raising the entry standard to matriculation and offering the course over four years to allow graduates to obtain 'a civilian degree granted by an Australian university, or a definite portion of the course'.[34] To ensure the RMC program was delivered at the required standard, Vasey's committee proposed including university academics in the curriculum revision and maintaining a clear separation between civil and military subjects with provision for a 'liberal arts and law' stream of study. The emphasis of the RMC program would remain on engineering and science but the offerings would be broadened to attract candidates with other interests.

Although he did not specify which university would be approached to accredit the program, Vasey assumed there would be no impediment to the award of degrees or diplomas based on the study completed at Duntroon. The recommendation was intended to increase educational standards at the College, to acknowledge cadets' achievements and to make a military career more attractive to young men in post-war Australia at a time when the best and brightest would be sought by other professional groups. Vasey and his colleagues could also remember the depletion of the Army after the Great War and were determined to provide attractive employment for those who were willing to serve in the peacetime Army. With the Army's attention focused principally on defeating Germany and Japan, the report was received in January 1945 but not discussed. Its content was important but not urgent.

General Blamey, who had proposed the inquiry and commissioned the report, took no immediate action. After retiring in January 1946, Blamey used the report to criticise the College and its educational program.[35] As his own initial military training predated RMC's establishment, Blamey had no particular affection for the

College. He claimed in May 1946 that its original charter was 'an insurance policy against any military caste ever being developed in Australia' and he lamented that the current program (which included post-matriculation academic study offered at a time when few young men went to university) had produced a Staff Corps 'segregated from the thoughts and minds of the body of an army which must necessarily remain a citizen army'.[36] It was not an impressive argument and few were attracted to it. An editorial in the *Sydney Morning Herald* appeared to side with the spirit of Vasey's reforms and saw university recognition of the RMC educational program having a much wider remit:

> If Duntroon were to produce not only permanent Army
> and Air Force officers, but militia officer graduates whose
> technical qualifications would stand comparison with those of
> the universities, not only would its influence be more widely
> felt but an important step would have been taken towards
> ridding the Australian Army of the discord between the
> professional and civilian officers which has long been its bane.

The public controversy Blamey's criticisms provoked reached the office of the Minister for the Army, Frank Forde, who asked the Military Board to assess the now retired General's comments. A committee chaired by Lieutenant General Sydney Rowell, the Vice-Chief of the General Staff, re-examined the RMC curriculum and rejected Blamey's criticisms as unfounded. Rowell's committee concluded that Duntroon met the nation's needs in a manner that was consistent with Australia's outlook on armed conflict and military service. The committee also reaffirmed the wisdom of Vasey's advocacy of a West Point-style program and the need for the RMC curriculum to restore the balance between academic and military subjects while providing cadets with a grounding in the liberal arts for the first time. He suggested that many of the science and technology subjects that had dominated the curriculum could be shifted to post-commissioning corps-specific training. The committee's report, which was released in November 1946, stressed the Army's need for enhanced educational standards:

> If the military profession is to retain its status, its members must be brought up to what is now the accepted standard for all professions in the Australian community, i.e., a Bachelor's degree. This is also necessary if the profession is to attract recruits. But above all else, the Committee is convinced that graduates of the College will not be fitted to take their place and assume their proper responsibilities in the Australian nation unless they attain that status.[37]

As the Australian university sector was still state-based and Duntroon was a Commonwealth establishment located in the Australian Capital Territory, the opening of the federally funded Australian National University (ANU) at Canberra in August 1946 appeared to be a most opportune development for the Army. The *Australian National University Act* had made specific provision for the University to collaborate with government agencies. It was, in a sense, to be the Commonwealth's university as a well as a nationally significant research institution. As Duntroon was too small to become a university on its own, a partnership with an existing university was considered the best way to allow cadets to study for a degree within a disciplined environment. After basic military training in year one, under this proposal cadets would be accommodated at RMC but travel daily to the ANU for lectures as part of a modified degree that would reflect the Army's particular needs. Engineering students would continue to be seconded to the Universities of Melbourne or Sydney because the ANU was without an engineering school. Oversight of the academic program would be vested in a standing committee consisting of a representative from the Army and the ANU, and the Director of the Commonwealth Office of Education.

The Army was being a little too hopeful. The provision of undergraduate education at the ANU had yet to be finalised and the programs to be offered were far from settled. Nothing practical could happen until it made more progress with the erection of buildings at Acton Peninsula above the Molonglo River where the new university was to be located. But there were continuing efforts to negotiate accreditation for RMC subjects with a number of universities. The most productive negotiations continued to be with the Universities of Sydney and Melbourne in relation to engineering courses. The University of Melbourne offered just one year of accreditation to Arts and Science students because its statutes included a two-year campus attendance requirement. The granting of a single year of credit made it virtually impossible for an officer who was not posted to the Melbourne area or who was managing a demanding posting to complete an Arts or Science degree.

There were few other options for external accreditation of RMC courses at that time. The founding legislation for most of the universities left them little room to manoeuvre, whatever their attitude to the Army's needs. The tertiary education sector in this period was marked by inflexibility, despite rapidly growing community demand for study programs. In response, the Army worked on the second- and third-year curriculums to expand the range of subjects and to enhance the quality of teaching. There was continuing attention to the possibility that RMC might secure degree-granting status but opinions differed within the Army as to the need and the timing. There were those who believed the course was already worthy of degree status and that awarding degrees to RMC graduates was a viable quality assurance strategy. Others, such as the College Commandant from 1954 to 1957, Major General Ian Campbell, felt that a 'Duntroon degree' would not be respected in the wider community while the entire Army officer corps would be unhelpfully divided between those graduating with degrees and those without them. Campbell did not personally believe that Army officers needed a degree and thought the effort involved in securing one was too great in the context of other competing demands. He felt the emphasis at Duntroon needed to be on developing military leadership and acquiring practical skills.

By 1958 the Army was able to secure credit for its general science program with the Royal Melbourne Technical College (later to become the Royal Melbourne Institute of Technology (RMIT)) while its engineering graduates were excelling at the University of Melbourne and at UNSW. The latter had started to accept third- and fourth-year cadets. RMC also had a new Commandant, Major General John Wilton, whose views contrasted sharply with those of his predecessor. Wilton was deeply concerned about the social status and the professional standing of military officers. He believed they needed to be the equal of those leading the way in public administration, industry, commerce and the law. It was Wilton who effectively initiated the negotiations that led to Duntroon being affiliated with UNSW a decade later.

The Air Force and officer education

The Royal Australian Air Force (RAAF) was established on 31 March 1921. Its officers were initially educated at RMC Duntroon because the number required was so small. Many of the subjects taught at the Royal Air Force College at Cranwell in Britain were not covered, however, in the RMC curriculum, leading the Professor of Physics at Duntroon, Dr Richard Hosking, to propose creating an Aeronautical Engineering Department and recruiting a suitable professor to take charge. There were plans to establish a RAAF Wing at Duntroon, with cadets destined for service in the Air Force undertaking a common course with Army cadets in year one, studying some RAAF subjects in year two and completing all RAAF subjects in year three. After graduation from Duntroon, the newly commissioned RAAF officers would proceed to flying training. These plans were still under discussion when the Air Force started to accept young men on short-service commissions who started flying training immediately. The financial stringencies of the Great Depression and the onset of the Second World War a decade later precluded serious discussion of the RAAF's long-term need for a distinct program and a separate college. For the first 25 years of its history, the RAAF was able to rely on the Army to educate its officers but the arrangement was not one that the Air Force's leadership wanted to continue indefinitely. It was fearful of remaining dependent on the Army, which would naturally respond to its own needs first.

With the end of the Second World War in August 1945 and the closure of the highly successful Empire Air Training Scheme, attention turned to the RAAF's peacetime officer development needs. The Director of Training, Group Captain Paddy Heffernan, proposed the development of a RAAF Cadet College offering a three-year academic program that would include flying training in the second and third years. Heffernan thought the Cadet College should be located in Canberra to make use of existing academic staff at Duntroon. He also suggested sponsoring a small number of selected cadets who would benefit from a university education at a civilian campus. These cadets would be obliged to remain in the RAAF for a period of ten years following their graduation. There was some sympathy for this view. The Deputy Chief of Air Staff thought 'the ultimate aim should be degree standard in Arts or Science or second year in Engineering' although he noted the Army's difficulty in having its courses recognised by the universities of Sydney and Melbourne. Nevertheless, he believed the Air Force should work towards founding its own college with degree-granting status in time. The Minister for Air, Arthur Drakeford, was not personally persuaded that cadets needed to spend two years undertaking academic studies. Consensus on how to proceed was proving elusive.

Ministerial approval to establish the RAAF College was given on 9 July 1947. The College would provide tertiary education and professional training for RAAF officers on permanent commissions. The College would be located temporarily at Point Cook until facilities could be built in Canberra. The initial offering was a four-year course covering science, mathematics and the humanities in addition to RAAF subjects and flying training. The following year an Air Force committee was set up to examine three sites in the Canberra area that might be developed as an aerodrome for the RAAF College. The committee found that none was suitable. It looked as though the College would remain in Melbourne for some time. There were other equally pressing concerns. It appeared that the educational standard of the College's inaugural entry was largely unsatisfactory, with many cadets struggling to reach the required proficiency, especially in science subjects. This partly reflected the uneven character of secondary school education around the country and variable academic standards. To its credit, the Air Force did not consider lowering its standards but actively raised them to ensure its officers were not considered inferior to those of the Navy or Army. The discussion shifted to achieving the right balance in course content between academic education and technical training.

In September 1956 the member of the Air Board responsible for personnel, Air Vice-Marshal Frederick Scherger, ordered a review of the College syllabus with assistance from the faculty at the University of Melbourne, namely Professors Sir Leslie Martin and Wilfred Frederick. Scherger's personal objective was to see the four-year course reduced to three years, with years one and two devoted to academic subjects and the third focusing on flying training and operational acquaintance. The Chief of Air Staff, Air Marshal Sir John McCauley, made clear his view that 'the ultimate aim should be the attainment of a Bachelor Degree in Science'. The Committee reported in November 1957, recommending that Point Cook should be the permanent location for the College because of its proximity to a RAAF aerodrome and a major university. It proposed that General Duties cadets should be educated to Bachelor of Science standard, with Melbourne University granting degrees to graduates on the condition that Point Cook staff, facilities and research programs met its specifications. The rationale was plain. In 20–30 years 'when the first graduates of this new scheme would be reaching high rank, the Air Force will probably be primarily a "missile" Service'. Its leadership 'should have a very broad and advanced education' based on a Science degree 'in order to establish a widely recognised standard which will encourage boys with the high academic aptitude and education which is desirable'.[38]

As expected, cadet applicants were required to meet Melbourne University's matriculation requirements, thereby relieving the Air Force of the need to develop its own entry levels for education. The report was adamant that, in any event, the Air Force needed to raise its educational standards substantially to meet the technological and operational demands of the future. As it also needed to attract able young men, the offer of a degree would ensure the brightest and the best at least considered a RAAF career. But the new Minister for Air, Fred Osborne, a highly decorated former wartime naval officer, was concerned about the cost of such a scheme and asked his colleague, the Minister for Defence, Sir Philip McBride, to consider merging cadet tri-Service education and training to achieve cost savings. This thinking was a little far advanced for some in the Air Force, with the chairman of the review committee, Air Commodore Ian McLachlan, stressing the need for young Air Force officers to acquire first a loyalty to their own Service and to become professionally educated men. He contended that this could only happen at the RAAF College, where substantial spending on new infrastructure was needed.

McBride's successor as Defence Minister, Athol Townley, was hesitant about developing the RAAF College into an establishment that offered degrees. He referred the whole matter of tertiary education to the Defence Administration Committee in early 1959. After much discussion, Townley, a wartime naval officer who had commanded a number of small vessels in the Pacific campaign, concluded that amalgamating the Service colleges had more disadvantages than advantages and subsequently approved investment in upgraded facilities at Point Cook. On 16 August 1960, Osborne wrote to Professor Sir George Paton, the Vice-Chancellor of Melbourne University, seeking concurrence with the changes that were being made at Point Cook and confirmation that the University would recognise the attainments of its graduates by awarding Melbourne degrees. With the goodwill of Melbourne University, the College was reconstituted at the beginning of 1961 as the RAAF Academy and was formally affiliated with Melbourne University.[39] The affiliation statute explained that 'the Academy shall be an educational establishment affiliated with the University of Melbourne for the purpose of instruction of candidates for the various degrees of the Faculty of Science and the degree of doctor of philosophy of the University'. The course for General Duties officers (the non-pilot cohort) was four years duration. The first three years were essentially a Bachelor of Science degree with a major in physics. There was no choice in subjects studied, which were spread over four 'divisions' – Military Studies, Pure Science, Aeronautical Science and Humanities. The fourth year of study was designed 'to build a bridge between the pure studies and their application to the Air Force'.

Cadets who had achieved the necessary results in first year could decide to become engineers. They would transfer to the University of Sydney, become residents of a campus college, and commence second-year studies leading to the award of a Bachelor of Engineering degree. The number of cadets who chose this pathway was always small, ranging between two and five during the 1960s. The academic staff at Point Cook were members of the University of Melbourne and civilians from the Department of Air. Most impressive were the number of doctoral candidates (around 20) and the continuation studies being undertaken by the Academy staff from both the University and the Department. Notably, the affiliation required the provision of adequate research facilities for teaching staff and for students working on higher degrees. There were also civilian higher degree students at the Academy conducting research projects related to the RAAF and aviation-related topics (the source of most PhD enrolments). The profile of aviation at the University of Melbourne was assisted by the Department of Air offering research grants to projects of interest.

The affiliation was co-ordinated by a 'Dean of University Studies' appointed from the professorial ranks of the University of

Melbourne, who was responsible to the Faculty of Science for teaching and research at Point Cook. The Dean worked with the Warden of Point Cook (a public service appointment) and the Assistant Commandant (a RAAF appointment) in relation to local administration. Those students selected to complete a science degree lived at the RAAF Base and commuted to Melbourne University by bus until new buildings were completed at Point Cook. The new science block was not opened until August 1963. In his history of the Academy, Air Vice-Marshal Roy Frost commented that:

> In many ways, exposure of the cadets to the less regimented environment of the lecture room and the campus, contact with the broad spectrum of the university population and the requirement to compete with their civilian contemporaries on their home ground as well as establishing contacts for the future might, if managed properly, have well proved beneficial to the RAAF in the longer term.[40]

He went on to note that:

> contamination by universities has little effect on retention, or indeed on promotion, despite fears which were often expressed to the contrary. Indeed, one cadet reported that he doubted that he could have stood the petty discipline and routine of the Academy and that his attendance at university was all that saw him through.

The small number of new entry cadets was supplemented by ex-cadets who were permitted to return to Point Cook to complete their Science degrees. This made the Academy atmosphere nearer to a university campus than a military college. There was also a very commendable commitment to postgraduate research and the need to recruit academics engaged in projects that would directly benefit the RAAF.

The first few years of the affiliation were challenging for both the Academy and the University, with a higher than expected failure rate and unexpected requests from cadets to withdraw from the degree program. The combined attrition rate would later rise to 60 per cent. The University was also proposing to extend the science program to four years (with the inclusion of additional pure science and non-science subjects) and wanted more control over the subjects it was being asked to recognise by the Air Force. By July 1964, there was deep disquiet within the Air Board. It had received a paper concluding that 'the aim of the Academy is not being attained':

> The motivation of cadets to the degree studies suffers because they do not see in the course an obvious application to Air Force requirements. As a result, their performance on the degree studies falls below what they are capable of. There is a tendency among cadets who have completed the degree component part of the course to feel that the training which they have undergone will not be utilised within the RAAF. As a result they look around for civilian careers which will make more use of their training.[41]

The Minister for Air and former wartime Fleet Air Arm pilot, Peter Howson, wrote to the Vice-Chancellor of Melbourne University, Professor Sir David Derham, proposing that it develop an Applied Science degree that would more closely meet Air Force needs. While the Air Force was keen to preserve the link with Melbourne, the university was now less than generous in its co-operation. Below the surface there were fundamental problems that the Air Force needed to address. Foremost was vocational frustration. Most cadets joined the Academy with a strong interest in learning to fly but spent most of their time studying pure science subjects that appeared irrelevant to routine service in the RAAF. There was also an apparent preference for research over teaching among lecturers, leading to the accusation that they did not take undergraduate learning seriously. The Air Board believed it had to regain control of the curriculum at Point Cook or look to the creation of 'a Defence university' as a means of meeting its educational needs. With the Navy and the Army both reaching out to UNSW and noting its technology focus, would it be a better partner for the RAAF than the University of Melbourne? Or was the best solution a tri-Service academy that granted its own degrees?

A critical juncture

In the absence of a shared conversation between the Departments of Navy, Army and Air, all three Services realised independently that they had reached a critical juncture in the provision of junior officer education by 1966. They might have come to this view earlier had they been able to transcend their reluctance to co-operate. Each Service needed to improve the quality of its educational programs and believed that the nation's universities (rather than comparable institutions in the United Kingdom or the United States) were a vehicle for both raising and maintaining acceptable standards. The Services also needed to offer their cadets a chance to gain university qualifications as a recruiting tool and as a means of intellectually enriching the contribution uniformed personnel could make to security planning and higher defence decision-making. Warfare was changing and strategic considerations were becoming more nuanced.

Competing ideologies also made it more difficult to explain why deployments were necessary and what the use of force could reasonably achieve in terms of security and diplomacy.

The universities were not especially generous in the options they offered Defence. None of the vice-chancellors ever approached the Services with an offer of assistance, nor did they devise a strategy designed to meet the educational needs of young officers. While several universities were prepared to make some minor allowances for uniformed students, expressions of goodwill were substantially on terms that would not require the universities to make substantive concessions. The governance and administrative processes of the universities were not sufficiently supple to deal with the complexities of educating service officers. University administrators were not prepared to seek even special amendments to their institution's legislation to provide sufficient latitude to assist the Services. Few senior academics understood the need for academic education to be delivered alongside professional military training. Although the universities were not completely without empathy, indeed many senior people in the higher education sector had served in uniform during the Second World War and retained an affection for the Services, they were not encouraged by the Government to exude an entrepreneurial spirit that promoted originality and creativity. The universities tended to take a 'one size fits all' approach to delivering undergraduate education, emphasising traditional degree structures that allowed little variation in content or subject combinations. They were comfortable with rules and regulations that made their processes and procedures effective and efficient. As the demand for undergraduate places was growing exponentially, the universities were not desperate for new students. In the 1960s, the expansion of the universities was about buildings and equipment rather than vision and opportunity. The Services needed a champion in an institution that did not equate innovation with compromise. That champion proved to be Professor Sir Philip Baxter and that institution was UNSW.

2

Firm foundations and the faculty solution, 1964–1967

Australia's armed forces are among the nation's oldest public institutions and uniformed personnel among its most revered citizens. The political management of the armed forces has been, however, anything but consistent and creditable. In the four decades after Federation, the Navy, the Army and later the Air Force experienced both rapid expansion and swift decline because of emerging and ebbing security threats, with men and a small number of women hastily recruited to serve in times of crisis and then just as quickly discharged. The possibility of being deemed excess to requirement undermined the standing of uniformed service and militated against creating a professional officer class that could combine experience with expertise in tendering advice and guidance to government. The education of officers, crucial to their continuing intellectual development, was considered a peacetime undertaking that could be suspended during wartime, when an immediate need for leaders prompted an exclusive emphasis on training to achieve technical superiority and tactical competence. In addition to relegating education to a list of second-order priorities, investment in education suffered from the desire of successive governments to reduce Commonwealth outlays as soon as the prevailing strategic situation appeared benign. Equipment acquisition and operational demands usually took precedence over education, with governments of all political persuasions falling victim to short-term thinking and even shorter-term planning.

When the Second World War began in September 1939, none of the armed forces had a comprehensive education policy, and nothing resembling a tri-Service education policy existed. Education was a single-Service activity, with each leadership group pursuing their own objectives and outcomes. The Navy, Army and Air Force essentially operated independently when it came to preparing junior officers for a career in uniform. In one sense, the absence of a cogent and co-ordinated policy in 1939 hardly mattered as pre-war education programs were suspended for the duration of hostilities. Prior to that, the universities were unconcerned with the education of officers as the Service colleges were small and the numbers of students were few. The universities were not encouraged to see the education of Service officers as a pedagogical priority or an administrative challenge to which they might respond. The universities were State entities and national defence was a Commonwealth responsibility. But the reliance of the armed forces on the experience and expertise of academic faculties around the country during the war brought the defence community and the universities into close and, in some instances, continuing conversations.

The end of hostilities in Europe in May 1945 and in the Pacific three months later marked yet another new beginning for Australia's armed forces and the re-emergence of many pre-war challenges, such as the need for an education policy. Although thousands of 'hostilities only' personnel were demobilised, fears that the Soviet Union would forcefully impose its will on international affairs meant the Services would not suffer the fate they endured after the Great War, when there were massive reductions in both personnel and equipment. The less than benign strategic environment created by the onset of the Cold War meant that Australia required a standing peacetime force for the first time in its history. While the young officers needed to lead the armed forces could be trained with the help of Australia's allies, principally the United Kingdom, each of the Services attempted to provide post-matriculation education that resembled university study in both content and delivery, without the benefit of a university that actually acknowledged or accredited their programs. While different opinions continued to exist within and beyond the armed forces over whether officers really needed a university degree, the context within which national security was managed was changing. There was greater civilian participation in both policy development and administrative oversight, while the social status of uniformed service was declining. The National Service scheme that had been introduced in 1951 was discontinued for the Navy and the Air Force in 1957. Registration remained compulsory but the Army's intake was cut to almost a third (12 000 trainees) and 'call-up' was subject to a ballot. In November 1959, Federal Cabinet decided that National Service 'call-ups' would end and arrangements for the January 1960 intake were cancelled.

As fewer young men were experiencing uniformed service, it was apparent that the needs of a growing and diversifying economy, coupled with raised personal and vocational aspirations, would make new and enlarged demands on the tertiary education sector. The workforce was changing in response to a series of influences. With near full employment, it was possible to leave one job in the hope of quickly finding another. As the risk of being unemployed for an extended period was low, many young Australians decided to make the kinds of pronounced career shifts their parents would not have contemplated. With increased private affluence in the second half of the 1950s, there was less pressure on teenagers to conclude their education at age 14 or 15 and enter the workforce to support their families. Young people could remain at school, complete their matriculation and then enter a technical college or university in the hope of a 'professional' career that would include formal qualifications.

These shifts accompanied changing attitudes within the armed forces. There was growing recognition that the rising generation of young officers required a better education than that provided to their predecessors, while those considering a career in uniform wanted recognition of their attainments, including a university degree, as a mark of their enhanced social standing. As a degree was usually required of those seeking admission to a professional community, many senior commanders realised the profession of arms needed to demonstrate that the education it was providing at the Service colleges was equivalent to, if not better than, that being offered in the nation's universities. The content might be different but the quality needed to be the same. Because a university qualification was fast becoming the accepted path to vocational recognition and professional advancement, fewer young Australians were likely to consider a career in uniform. In the longer term, if the armed forces could not attract the nation's brightest and best, it would steadily erode their community standing, affect their funding and diminish their operational capability.

Major General John Wilton recognised the need for external validation and formal accreditation of RMC's educational programs when he was appointed Commandant in March 1957.[1] He was a man of considerable intellectual depth, able to think laterally and expansively. With Wilton as the driving force, the RMC Standing Committee on Curriculum recommended in 1959 that 'a formal approach should be made to the Canberra University College to negotiate credits for Arts and Science Degree Courses; the first two years of which would be done at RMC and the third and final year at Canberra University College'. The Canberra University College (CUC) had been established on 19 December 1929.[2] It was associated with the University of Melbourne and existed to provide undergraduate education in the fast-growing national capital, with the first students commencing their studies in 1930. The College offered courses in Botany, Chemistry, Classics, English, Geology, History, Law, Mathematics, Modern Languages, Oriental Studies, Pacific Studies, Philosophy, Physics, Political Science, Psychology, Statistics and Zoology. Wilton lost no time in beginning a conversation with the Vice-Chancellor of the Australian National University (ANU), Sir Leonard Huxley, and the CUC Principal, Dr Joe Burton, about a relationship with RMC.[3] Huxley had served as a civilian academic in the British Telecommunications Research Establishment during the Second World War. Burton had no previous military connections.

On first inspection, Huxley considered the RMC course to be too focused on military subjects. Were the course to be diversified, he explained, the ANU would consider granting cadets two years' advanced standing, with the third year completed on campus. While a viable option had been placed on the table, no firm plans could be made as the ANU and the CUC (which was still affiliated with the University of Melbourne) were moving towards amalgamating and admitting the ANU's first undergraduates. As the negotiations were complicated, neither institution wanted an affiliation agreement with the Army to hinder the final arrangements. The door had been opened but Wilton was frustrated by the ANU's inability to proceed. The number of students would never be more than 600 and the proposed affiliation was anything but complicated. Duntroon had its own academic faculty; the existing RMC course was highly regarded; and the ANU was just three miles away. The minutes of

Firm foundations and the faculty solution, 1964–1967

the ANU Council meeting held on 18 July 1960 recorded that 'the Vice Chancellor had indicated to the Commandant that this matter had better be deferred until the School of General Studies had been established within the University'. By this time, Wilton had been posted from Duntroon to SEATO headquarters in Bangkok. The School of General Studies within the ANU was established in late 1960 but there appears to have been no follow-up contact between the new RMC Commandant, Major General Robert Knights, and the ANU during the next twelve months. The hiatus continued.

A new Commandant, Major General Basil Finlay, arrived at Duntroon on 15 January 1962 and was told by the Chief of the General Staff, Lieutenant General Sir Reginald Pollard, that his main priority needed to be establishing a degree course at RMC. Finlay needed no persuading because he thought that at Duntroon 'the situation could not be described as adequate or satisfactory'. Finlay believed implicitly in the importance of tertiary education and acknowledged that his own lack of formal qualifications had hindered his performance as a member of the Military Board before he was posted to Duntroon. Finlay hosted a series of meetings in early 1962 that ultimately recommended upgrading programs to 'a post-matriculation course of clearly tertiary status'. This meant imposing a new entrance requirement – matriculation. But this new standard could not be introduced until 1964, as recruiting had already begun for the 1963 intake of cadets.

Having resolved that it would press ahead with improving educational standards, the Army established the RMC Advisory Board on Academic Studies in October 1963 to assess course content and the quality of tuition. Its membership was drawn from seven Australian universities. The Board's main remit was to design post-matriculation tertiary courses in Arts, Engineering and Applied Science. Forty-one academics from across Australia, including Professors Rex Vowels and Albert Willis from UNSW, were involved as members of the Board or as external examiners.[4] The courses were first offered in 1964, with external assessors subsequently judging them to be too demanding for first-year undergraduates. The Army was not particularly concerned as it was easier to lower the bar than to lift it. Having delivered courses at the requisite level, the Army sought a partner university to develop the curriculum further and possibly to accredit the entire RMC program.[5]

An arrangement between the Army and the ANU still made sense. Defence was a Commonwealth responsibility and the ANU had a national remit. In practical and legal terms, the campus was a short distance from Duntroon and both institutions were in the Australian Capital Territory (ACT). By 1962, the ANU had existed for sixteen years and was reasonably settled with its four central research schools: the John Curtin School of Medical Research (JCSMR), the Research School of Physical Sciences (RSPhysS), the Research School of Social Sciences (RSSS) and the Research School of Pacific Studies (RSPacS). The newly formed School of General Studies contained the Faculties of Arts, Economics, Law and Science. Joe Burton, Professor of Economic History, was head of the new school.[6] Burton was a genial man who had excelled during his early years at university academically and in sport, particularly Rugby Union. He was also a Rhodes Scholar. While Burton was willing to consider an affiliation with Duntroon, his colleagues were of a different mind.

Although it was a relatively new institution, the ANU had already endured several highly publicised controversies over allegedly politically motivated research projects and the ideological affiliations of its faculty. There was very strong resistance to, if not resentment of, anything that might have suggested the University was under the control of the Commonwealth Government. As a number of ANU staff members were members of the Australian Communist Party or were deemed to have communist sympathies (an accusation levelled against many Australian academics at the time), preserving academic freedom and the entitlement to criticise government policy were pressing concerns. If academic autonomy and intellectual independence from government needed to be asserted anywhere, according to its staff, it was at the ANU. Nonetheless, Finlay met with Huxley and Burton who were:

> both very supportive of some kind of affiliation with RMC […] [but] the whole proposal was given the complete kybosh by Professor Manning Clark, Dean of the Faculty of Arts, who ridiculed the suggestion that cadets under a regimented disciplinary system could possibly study under conventional university conditions […] it was he who killed the ANU thing, just like that.[7]

Finlay was mistaken about one thing. Clark was not the Dean of Arts. He was, however, an enormously influential figure at the ANU. Clark had joined the ANU as Professor of History in 1949 and his stature had grown to the extent that he was among the University's most widely known academics.[8] Strong opposition to an affiliation might also have come from the celebrated poet AD Hope, who was the foundation Professor of English. Neither the nature of Hope's objection nor the source of its vehemence can be ascertained. Several sources mention Hope as being opposed but his reasons were not disclosed. Neither Hope nor Clark had any military experience; they had never lectured at RMC and seemed generally hostile to Government instrumentalities. Clark's politics were certainly on the political Left. He was already among those Australians contending that the Viet Minh was a nationalist movement within South Vietnam that the international community ought to recognise rather than a communist front that it should oppose. Clark had called on the Australian Government not to embrace American hostility towards opponents of the 'Saigon regime'. There may have been others at the ANU who were opposed to an affiliation agreement; only Clark and Hope are known to have expressed their opinions publicly. Of course, political disputes among its senior staff were not something that either Huxley or Burton wanted to see proliferate at this stage in the ANU's evolution. They decided the Army would need to look elsewhere, principally to a university that operated an engineering school, something the ANU still lacked.

A number of options were considered. The preferred option was to reconstitute Duntroon as an autonomous degree-conferring institution. This might have required some arrangement with an existing university or, if the Commonwealth were willing, creating a new category of institution whose operations would be supervised by a Federal rather than a State instrumentality. But Finlay did not think a direct move to autonomy was possible. He was right. Prime Minister Menzies, who as a long-term Canberra resident took a personal interest in Duntroon, intervened and indicated no decision on the granting of degree courses at RMC would be made until the Martin Committee on the 'Future of Tertiary Education in Australia' had completed its inquiry.[9] Menzies told the Navy Minister, Fred Chaney, on 10 December 1964 that he had discussed with Professor Sir Leslie Martin the possibility that 'the three Services [might] send undergraduate personnel to a single university or to establish such a university type institution for this purpose'.

The final instalment of the three-volume Martin Committee report was tabled in Federal Parliament on 25 March 1965. The Committee recommended the formation of a Commonwealth

Firm foundations and the faculty solution, 1964–1967

Institute of Colleges and proposed including the RAN College and RMC in its membership. The Government rejected this particular recommendation on 28 April 1965 – the night before Prime Minister Menzies announced that Australia would contribute combat forces to the war in South Vietnam. After another hiatus of nearly five months, the Minister for Defence, Senator Paltridge, advised the Service ministers that he was interested in exploring the prospect of a tri-Service academy. The Army Minister and decorated former wartime Army officer, Dr Jim Forbes, told Paltridge that he supported the proposal personally and went so far as to say it was 'essential for the training of those Service officers who in due course rise to the more senior ranks', particularly as they would advise on national security policy.

RMC would not become an autonomous degree-granting institution in the nearer future. There would be a halfway or interim solution. A number of possibilities were explored and this took even more time. Forbes and the Secretary of the Department of the Army, Bruce White, attended a meeting of the RMC Standing Committee on Academic Curriculum on 27 August 1965 and explained that officers in the department had concluded that academic independence for Duntroon was neither possible in the short term nor wise in the longer term. He suggested seeking assistance from one of the universities. The ANU was unwilling to participate, so would the attitude of the others be any different? Given the prevalence of growing left-wing, anti-military sentiment on Australian campuses and the rising tide of student activism across the country, especially after Australia became involved in the Vietnam War as a combatant, would any vice-chancellor be willing to face the concerted political opposition that would be generated by their university entering into a relationship with the Services? Although the Army was a servant of Government, public objection to Australian involvement in South Vietnam would soon be vented on individual officers and soldiers, who endured taunts and abuse. To many in the anti-Vietnam protest movement, the Army and those who served within it were personally complicit in a war they regarded as illegal and immoral.

As discussion of a university partnership continued at the Standing Committee meeting, Professor Rex Vowels excused himself and contacted Kensington. After a brief telephone conversation with the Vice-Chancellor, Professor Baxter, Vowels reported that UNSW would be interested in offering its support. Baxter believed some form of affiliation was 'in the national interest'. He was prepared to support the negotiations if the ultimate objective was Duntroon becoming an autonomous degree-granting institution. His attitude to Defence was not without precedent. Baxter had responded with similar generosity when the University of Melbourne decided against hosting the National Institute for Dramatic Arts (NIDA) in 1958. UNSW also had the biggest Engineering school in the country. This attracted the Army and impressed the Navy. There was one more positive element: the University had played the role of midwife when its Newcastle campus had become a stand-alone university on 1 January 1965. The Army appeared to have a very positive solution. The minutes of the Standing Committee meeting noted the University's offer and confirmed that affiliation might be 'an interim stage towards the ultimate establishment of the College as an independent degree-granting institution'. The subsequent discussions between the Army and the University were headed by the Pro Vice-Chancellor, Professor Rupert Myers, who was to become one of the most significant figures in the University's relationship with Defence.

He was not without personal acquaintance with uniformed service. While completing a Science degree at the University of Melbourne, Myers had joined the Melbourne University Regiment

Firm foundations and the faculty solution, 1964–1967

(MUR) and commenced military training in 1941. Several months later, the Army Signals Directorate learned that Myers was a metallurgist who could make himself useful by working with tungsten and tantalum as part of a research project related to radar. The MUR commanding officer instructed Myers to report to the Professor of Metallurgy at the University. Although his military service was essentially over, Myers retained a great affection for the armed forces. He thought that those who wanted education to remain entirely within the Service colleges and out of the universities 'immediately revealed an inadequate understanding of the situation, because not only were they prepared to sell somebody else's soul for their own misconceptions [...] they didn't conceive of the level of skill that needed to be generated in serving officers'. Myers thought the ANU should have accepted responsibility for officer education in the 1960s and later remarked: 'I give that body no credit for its narrow-minded, ill-conceived approach to the whole proposition. I make that comment in parenthesis but with full emphasis because I think it was a shame'. Myers was generally critical of the ANU for failing to have a well-rounded commitment to preparing professional people for their careers.[10]

With the UNSW offer on the table, the three Service ministers met with the Acting Defence Minister on 17 November 1965. (The Minister, Senator Shane Paltridge, was gravely ill and died on 21 January 1966.) They agreed that the overarching objective was to establish a tri-Service academy that would 'operate with separate wings at Duntroon and Point Cook with a headquarters at Duntroon'. The Naval Board later questioned the need for two sites, stressing that naval students would only be educated in Canberra and would not be sent to Point Cook. Early in 1966, the Minister-in-Charge of Commonwealth Activities in Education and Research (and former Navy Minister), Senator John Gorton, explained that there was no point approaching the ANU about an affiliation agreement given its recent attitude and that he 'favoured negotiations with the University of New South Wales on behalf of each College'. The new Defence Minister, Allen Fairhall, then advised the newly appointed Navy and Army Ministers, Don Chipp and Malcolm Fraser, to conduct 'preliminary negotiations with UNSW'.

After months of private conversations, the Minister for the Army, Malcolm Fraser, wrote to Baxter with a formal proposition on 1 April 1966. He began by explaining the context of his request:

> Today's senior commander in war must be not only a military tactician but must also be prepared, and have the intellectual capacity, to deal on equal terms with allied commanders, diplomats, politicians as well as leaders of commerce and industry. The impact of scientific advances in recent years has brought to the Services an acute awareness of the need for higher intellectual attainment so that their members may deal with sophisticated and complex equipment and techniques now available and being developed [...] there is no way whereby students who successfully complete their courses can have their achievements recognised, other than as a graduate of RMC Duntroon.

Fraser wanted to know whether UNSW 'would be prepared to consider these courses with a view to according to them a form of affiliated recognition'. As a formality, Baxter told Fraser that he would consider the proposal. Significantly, the initial request was for *affiliation* with UNSW. There was no mention of autonomy in either Fraser's letter dated 1 April 1966 or Baxter's reply of 12 April 1966 – the date that the proposal was considered and approved-in-principle by the University's Professorial Board.

Firm foundations and the faculty solution, 1964–1967

On the day he replied to Fraser, Baxter sought advice on whether any legal impediments existed to an affiliation with RMC, as the University's Act stated that its Council 'may establish and maintain branches, departments or colleges of the University at Newcastle, Wollongong, Broken Hill or such other place *in the State* as the Council deems fit' (emphasis added).[11] To ensure that legislation would allow UNSW to operate within the Australian Capital Territory (ACT), Baxter wrote to Charles Cutler, the Deputy Premier of New South Wales, who also held the Education and Science portfolio, asking him to add another amendment to those then being proposed to the University's Act. Amending the Act to include the ACT would remove any doubt that the University could operate beyond New South Wales.

But what of the separate negotiations then underway with the RAN College at Jervis Bay involving UNSW academics based at the Wollongong campus? Would the proposed arrangement with RMC help or hinder an agreement with the Navy? Baxter explained to Cutler in the same letter that:

> The Navy is concerned primarily with the training of engineers, though applied scientists or arts graduates may also be of interest to them. The suggestion at the moment is that the Navy will provide at Jervis Bay, facilities and teachers to train their officer cadets to sit for the first year examinations of UNSW […] after which students will come to Kensington for the remaining years of the course, during which they will of course be indistinguishable from other undergraduates. This pattern fits perfectly well within our rules and creates no uncertainties.[12]

The University could negotiate with both Services without any added complication or unexpected contradiction because the land on which the Naval College stood was also part of the ACT, having been resumed by the Commonwealth in 1911 as a possible port for the national capital, Canberra. Baxter ensured that Gorton was aware of these developments but only as a matter of courtesy. This was to be an initiative managed entirely within the Defence portfolio. The Department of Education was a mere observer. There was never any suggestion that the affiliation would be funded from the Commonwealth's Education budget; the Navy and the Army would be paying UNSW from their own financial allocations. Defence did not want its educational programs managed by another Commonwealth department. Baxter then sent a memo to the Professorial Board outlining Fraser's request and his own views:

> I think this is a service we should be prepared to undertake as anything we can do to effect an improvement in the standard of education of senior military personnel would be a service to this country. The proposal would not involve the University in any direct cost.

The Professorial Board gave its in-principle support within a month and suggested the appointment of a 'Professorial Board Standing Committee on RMC Duntroon'. The creation of this new entity was part of the proposal that Baxter took to the University Council on 9 May 1966. The Council resolved:

> that the University approve in principle the forming of an association with RMC Duntroon which could lead to the awarding of certain of the University's degrees to students of the College who have successfully completed approved

Firm foundations and the faculty solution, 1964–1967

courses in the College, subject to the University and the College reaching agreement on the details of courses, staffing, examining and related matters and to the representation of the University on the relevant College committees [these arrangements would be made under the provisions of Section 23(2) of the *Technical Education and University of NSW Act, 1949–61*].[13]

The first step was to deem RMC an 'affiliated college' of UNSW; the second was dealing with internal dissent in both institutions.

An exploratory meeting of representatives from UNSW and RMC met on 12–13 May 1966 to focus on 'affiliation' arrangements.[14] The meeting considered staff appointments, curriculum approval and campus time for cadets at Kensington. Professor Myers began by advising the meeting that the University had in mind a fixed rather than an open-ended commitment. His colleague, Rex Vowels, suggested that:

> any association would be for a fixed period, say ten years, after which it might be expected that RMC would get full autonomy. It was important that the mechanics of association should be developed in such a way that RMC's ultimate autonomy would be fully justified.

As the Army was seeking autonomy and realised it would take time to achieve, the University's suggested timetable made perfect sense. Although staffing and curriculum issues would need some sorting out, the matter of campus time for cadets was a problem for the Army. General Finlay explained that the four-year program at RMC made it impossible for the cadets to spend one year at Kensington, a period the University had suggested. He explained that:

a cadet was under constant observation throughout his four years with regard to the development of his qualities of leadership [...] to take cadets away from the College for any long period would unavoidably interfere with activities which contributed significantly to their military education.

Professor Paul George noted that a strong body of opinion existed within the University that 'a period on campus for cadets should be a necessary condition for any association' between UNSW and RMC. Professor Fred Ayscough dissented from this view, saying he was personally 'not convinced that the campus at the University could offer cadets any more than they gained at Duntroon [...] [and] could not see any real advantage in taking cadets to the University, particularly in view of the effort that would be involved'. His colleagues added that whether or not it was beneficial in an educational sense, 'it would go a long way to meet any objections their colleagues might raise if the College were prepared to consider some compromise in the matter'. It was agreed that a subcommittee would be appointed to consider the matter in detail.

Campus time appeared to be the 'deal breaker'. The prospect of a shorter time, perhaps a semester, was canvassed and left undefined. The Staff Association Representative in the University Council, Professor Jeremiah Hirschhorn of the Engineering School, was not content. He wrote to Vowels on 24 May 1966 insisting that the courses and subjects taught at Duntroon should be under the complete control of the Kensington faculty and that RMC cadets needed to spend at least one year at the main campus. With respect to the latter, he said that if 'the Army does not want to let the cadets out of its control during the later stages of their course [...] this reason should be completely unacceptable to academics. Under no circumstances must we be party to forcing students into an intellectual straightjacket'.[15] Ayscough was still neither persuaded nor concerned about the environment in which academic studies would be conducted in Canberra:

> In my view, a satisfactory atmosphere for the development of the enquiring mind could be developed at Duntroon. The cadets come from all over Australasia and from a wide variety of social environments. Discussions inside and outside the lecture rooms are rigorous. Because the cadets are expected to be leaders, they are largely thrown on their own resources in organising their academic and general college life.[16]

At the first meeting of the Professorial Board's Interim Committee on Affiliation held on 7 June 1966, Vowels explained that while the University might try to impose its standards and values on Duntroon, the College had a life of its own and faculty had to recognise and respect 'the type of compromise institution that the proposed affiliation was endeavouring to establish'. A joint academic-military oversight body was needed. In a letter to Fraser dated 24 June 1966, Myers mentioned the need for 'the establishment of an Interim Council as an autonomous governing body for the College'. Fraser replied:

> For reasons of governmental, military and departmental policy the ultimate control of the College in all matters must remain in the hands of the Military Board. However, within the broad framework of academic policy agreed by the Board (upon which matter the Interim Council should be the advisory panel), it is proposed that the Council should have autonomy in matters of academic application.[17]

While the University was content to proceed and the Army had come to the view that there was 'no real difficulty which could endanger a satisfactory working agreement', the Commonwealth Public Service Board opposed the extension of academic conditions and salaries to any new staff because it might lead to 'demands for extension of such privileges to other branches of the Commonwealth Service'. It proposed that the University employ all academic staff at RMC. UNSW would then invoice the Army for its services. The Army and University accepted this principle. Existing academic staff would be given the chance to join the University or remain Commonwealth public servants. The only sticking point was organising security clearances for faculty. After deliberation the University accepted that applicants for new roles would be subjected to the usual ASIO checks for anyone commencing government employment.

The Professorial Board met on 9 August 1966 to consider a report from the Interim Committee. The report proposed a governance structure, staffing arrangements and a suite of new degree programs designed to meet the Army's needs. It also recommended that UNSW support plans for Duntroon to become an independent degree-granting institution, noting that this development had been endorsed by academics from the Universities of Sydney and Melbourne, and the ANU, who were members of the RMC Standing Academic Committee. The report explained:

> Throughout these discussions it was recognised that these arrangements would obtain for an expected period of ten years and that initiation of arrangements for assumption by RMC for autonomy in degree granting status should be expected in six years.

Vowels opened the discussion as chair of the Interim Committee:

> Many views have been expressed ranging on the one hand from the urgent need to advise and assist the Duntroon College towards autonomy over a ten year developmental period and, on the other hand, to the point of view that the affiliation of the College within a University is completely incompatible. In addition, there are some who argue strongly that such an affiliation could be acceptable only if at least one academic year is spent at the University of NSW.

He observed that 'present world trends demand that Australia should have a College of Defence Science operating at degree status' but explained that whether there should be three colleges or a tri-Service academy was not a matter for the University to decide. After commenting on the broad principles undergirding the affiliation, Vowels noted that the agreement 'would remain in force for a period of ten years and it is anticipated that before this date the College would seek full autonomy as a degree granting institution'.

Myers later explained that Vowels was crucial to establishing the Faculty from the University's perspective because of his personal standing as an academic and his professional standing as Chairman of the Professorial Board. Myers said he was a 'stickler for academic standards and in almost everything he did he had in mind whether or not excellence was being sought and being achieved'. With Vowels involved 'the more lily-livered or disinclined to be involved with anything military among the staff' were confident that academic standards would not be sacrificed or subdued.

The two main opponents of affiliation – and their objections related to matters of both principle and procedure – were Professors

Gordon Hammer and John Wood. Hammer, a psychologist who had come to UNSW from the Sydney Teachers College, was against the affiliation on three grounds. He claimed the tabled report 'was inadequate; it was biased towards advocacy of the proposal; it gave no statement of the arguments for the opposed view; and included little of the data upon which a reasonable discussion might be based.' He went on to assert that there had not been a proper consideration of academic values:

> The University, by its very nature, was committed to seeking truth as an over-riding consideration. It was difficult to see the Duntroon Military College becoming a university in the sense that it would hold fast to academic principles no matter what political or other pressures were brought upon it. The importance of these principles could not be conveyed to students merely by requiring them to spend a term or two on the Kensington campus.

Hammer's second objection flowed from the first. What would other universities make of UNSW if it entered into an affiliation that he was convinced would adversely affect the University's image and reputation? He explained: 'the action proposed could give the impression that the University was not concerned with maintaining the importance of truth'. Furthermore, he could see no evidence that the opinion of other universities had been consulted nor had any approach been made to the staff association.

His third objection was the University's inability to maintain academic standards across a prolonged period of time in an institution more than 100 miles (161 kilometres) from Kensington. He feared that UNSW would have 'increasingly less power to determine standards, staffing, curriculum and other important matters'.

Firm foundations and the faculty solution, 1964–1967

Hammer disputed the report's claim that the affiliation would lead to an increase in the number and quality of applicants for the Army, dismissing it as a 'doubtful premise'. He discounted the argument that there were international precedents for these affiliations, claiming they were 'evidence of how academic malaise spreads, and it was the duty of this university to resist the trend'.

Finally, he dismissed the claim that the proposal had more merit because it came from Duntroon and not the Federal Government. Hammer thought it was 'possible to have a university dealing with military affairs which would train high-level military personnel' but he objected to it being a 'government-controlled institution'. He also argued that inasmuch as Duntroon was preparing personnel who had taken an oath:

> it was clear where their commitment would be in the last resort, and that was the vital issue. It was true that some universities were organised as religious institutions, and many of these existing did good work, but the ideal concept of a university was not compatible with the idea that it was based on any religion. Where there existed a commitment to something else, a departure had been made from the general nature and spirit of a university.

He received some support on the latter point from Professor Kit Milner, who was concerned about the teaching of a specialist group within the University. Milner thought that clause 44 of the University's *Act of Incorporation* prohibiting a religious test might have been relevant to an affiliation with Duntroon because admission to RMC was restricted to cadets who had taken an oath of allegiance. He thought 'the spirit of this clause should be considered in relation to the present proposal' because it had parallel aspects.

Professor John Wood of the School of Mechanical Engineering shared Hammer's objections. He thought the committee ought to have developed a series of proposals rather than settling quickly on one option and asserting its superiority. This was a thinly veiled criticism of Baxter's authoritarian style. Wood was unwilling to accept assurances that the University's processes would apply to courses offered at Duntroon (he did not elaborate on why) and argued that it was necessary for cadets attending the University for the final year of their studies to 'obtain a degree more worthy of this University'. Replying to these objections and criticisms, Professor Paul George explained that:

> the University had been asked to sponsor the emergence of Duntroon as a fully independent and autonomous institution within ten years. The University had the option of refusing the task as being academically unsound or it could take a more positive view and attempt to supply what was needed to establish courses and degrees which would be generally acceptable.

He felt the criticism of academic standards was pre-judging the issue. In a conciliatory gesture, Ayscough conceded that a more detailed proposal might have been given to faculty when their views were canvassed. He went on:

> in this era of advanced technology the Army, to be successful, needed personnel trained in basic science and engineering in order to develop its operations in line with modern scientific concepts. No army in these times could become an efficient fighting force unless it established and maintained a close liaison with scientific developments. However, a highly

Firm foundations and the faculty solution, 1964–1967

trained staff was necessary to effect this, and it was considered worthwhile that the University should give every assistance to that end.[18]

He did not accept that graduates would have 'a narrow outlook in comparison to the average university man', contending that military officers 'were trained to be leaders of large bodies of troops incorporating every shade of thought and opinion. It was known that they subscribed to a very wide range of political journals and freely discussed them'. Professor Crawford Munro, foundation Professor of Civil Engineering, rejected the criticism that the curriculum was too narrow, noting that 'there were many institutions offering courses at tertiary level which were provided solely for select specialist groups'. The notion of the vocational degree was gaining ground, he said. The award of 'special degrees' that closely met military needs, such as the Bachelor of Arts (Military) and Bachelor of Science (Military), was a good solution that helped to ease tensions among those at UNSW who did not directly control the courses at RMC. Degrees in Canberra would not be of a lower standard, just different. The Deans had already suggested to Baxter that the RMC faculty would need to teach the same courses as those offered at Kensington (other than specialist military subjects) because Duntroon had so few students and developing new courses would be too time-consuming, while having a consistent curriculum was a means of generating support for eventual autonomy. These were sound arguments of principle with important practical outcomes.

Professor Albert Willis (known widely as 'Al'), the Dean of Engineering and soon to become Pro Vice-Chancellor with special responsibility for RMC and RANC,[19] thought some of the questions posed by the Professorial Board were 'simply bizarre'.[20] He recalled that one member asked whether RMC students could join the

Communist Party. Willis thought they could if they wished but believed it most unlikely that they would. He personally thought the existing Duntroon courses 'were rather narrow. The Engineering courses were thin, they weren't recognised [...] [and] you couldn't call yourself an engineer unless you have qualifications acceptable to the Institution of Engineers Australia'. The Army needed the University and the University was in a position to help. Why not, he asked. While accepting the 'devil was in the detail', he appealed to his colleagues to approach the affiliation with 'the right spirit'.

In terms of supporting the proposal, Willis later recalled that:

> the number one voice in all of this was our Vice-Chancellor, very much a man of affairs, very much a man of vision, that is Sir Philip Baxter. Followed by Rex Vowels, Chairman of the Professorial Board, a key factor, in a leading academic position at the University.

Behind Baxter were the Faculty Deans who completely supported the proposal. UNSW was different from other universities at that time in that Deans were appointed and not elected by their colleagues, and were charged with supervising academic entities such as faculties and research institutes. They were Baxter's most important allies. The University Council was also firmly behind the Vice-Chancellor. The Navy and the Army were not dealing with a fractured organisation, despite the existence of critics. The resolve at Kensington to proceed was strong.

The question of campus time remained unresolved. Willis cited the University of London as an institution that awarded degrees to students who had never set a foot on its campus. In the end, the University did not press the matter and the critics simply conceded this point. During the first few years of the affiliation, groups of cadets spent a term on campus and stayed in motels because the Kensington campus colleges were full. When it proved to be impractical and too expensive, according to Willis 'it just petered out' and no-one at UNSW made much of a fuss.

In terms of organisation and governance, rather than replicating Kensington's various disciplinary schools in Canberra, Vowels floated with Baxter the idea of forming a single 'Faculty of Military Studies' at Duntroon. This was the ideal solution. Baxter responded in a personal letter two days later:

> The great merit of your scheme, I feel, is that it would take out of the hands of the existing faculties the responsibilities for the operations at Duntroon, and since the degree awarded would not be those awarded by our present faculties, there should be no difficulties over relative standards.[21]

In essence, Baxter didn't want teaching in Canberra disrupted by academics at Kensington who did not understand or who refused to empathise with the Army's needs. He wanted to establish an organisation that helped rather than hindered an easy shift from short-term affiliation to eventual autonomy. Myers explained that the Faculty of Military Studies 'was a sort of multi-disciplined faculty because it was a bit different from "the Faculty" as we understand it of Engineering or Arts or Science. It was a more broadly-based faculty [...] so that it could be brought into the ordinary consideration of the University's academic programs'.

In terms of staff appointments, Baxter was particularly sensitive about engaging highly regarded people after an article published in the *Bulletin* alleged that 'the new university' (which had been in existence for fifteen years) would attract only second-rate staff. The idea of securing the widely respected chairman of the

Australian Universities Commission, Professor Sir Leslie Martin, as the inaugural Dean of the Faculty of Military Studies was, Myers commented, 'one of those typical Baxter inspirations'. Martin was a Founding Fellow of the Australian Academy of Science and played a central role in affiliating the Royal Australian Air Force Academy, Point Cook with the University of Melbourne during 1961. Myers was convinced that Martin's standing 'lent immeasurably to the success of the operation, and it meant that the academic staff at the College knew right from the beginning that the leadership of the Faculty would be in the hands of somebody whose integrity could not be questioned'. Although Martin was conservative and cautious, he brought considerable 'lustre to this embryonic body'. Three foundation professors with immense personal standing would anchor the Faculty's academic standing. Alongside Sir Leslie Martin were Len Turner in History and Arthur Corbett in Engineering.[22] John Burns was a slightly later appointment but arrived with an excellent reputation in Mathematics.

In closing the Professorial Board's discussion, Baxter noted that the 'whole basis of the proposal' was establishing Duntroon as an autonomous degree-conferring institution. Willis recalled him saying: 'Let them go, take a broad look, control them in a broad sense, and let them go. They are going to be off our hands in ten years; we are doing a midwifery job'. The report of the Interim Committee, he reminded his colleagues, 'did not envisage a permanent relationship'. He stressed:

> this University is being invited to assist an institution raise its standards and develop its courses to gain for itself university status, but if it were believed that no university would emerge at the end of the affiliation period then the proposal would be a waste of time.

Baxter chided the detractors among his colleagues, noting that the breadth of support for the proposal beyond the University 'should be gratifying to all who had the well-being of this university at heart'.

The Vice-Chancellor had won the day, although the anxiety of his opponents persisted. The Professorial Board recommended that affiliation should proceed and that Duntroon cadets should spend one term at Kensington. The Interim Committee's proposal that students could take elective subjects at the ANU was not accepted. Hammer and Wood dissented from the meeting's resolution, which adopted the report of the Interim Committee that could serve as the basis of negotiations between the University and Duntroon. The RMC Interim Committee became the Interim Council and would consist of 23 members, five of whom would be University nominees. The extent of the University's goodwill had effectively imposed a burden of reciprocity on the Army, with General Finlay stressing that 'if the scheme for affiliation should not succeed, the consequences could be very serious for the College'. As the scheme had already received considerable publicity, 'the "public image" of its failure could be damaging to the present and future status of the College'.[23]

It looked as though a deal had been struck. The Chief of the General Staff, Lieutenant General Tom Daly, wrote to Baxter on 1 December 1966: 'Creating something new and unusual in a bureaucratic environment is never simple and we have been exceedingly fortunate in having the able support of people like Rex Vowels and Rupert Myers and, indeed, Bruce White [the Secretary of the Department of the Army]'. Once it was known that RMC would be affiliated with UNSW and practical preparations began, some of the longer-serving Duntroon academics felt that Kensington was indifferent to their interests and sought too much authority, including the right to make appointments. Several RMC academics also loathed Bruce White, believing he was unsympathetic to their concerns about employment and superannuation. The RMC Director of Studies and former wartime RAAF officer, Professor Traill Sutherland, 'quite definitely believed that he alone could create the appropriate resource. He was resentful from the start of all outside interference'. He wanted external recognition of existing courses, rather than new courses imposed by UNSW.[24] This was not an arrangement acceptable to the University given it was being asked to grant UNSW testamurs to graduates. Finlay thought that after 1966:

> relations between the military and academic staff in a way deteriorated as the negotiations with UNSW proceeded. It became apparent from the very first professorial statement that the RMC academic staff were not as highly regarded by UNSW as they themselves regarded their own status. Our professors were only to be associate professors. Some of our lecturers only had diploma level qualifications and were not directly denigrated but given quite definitely lower status.

These were genuine difficulties and the University tried to be generous in allowing a reasonable transition phase, making provision for lecturers without the requisite qualifications to gain these qualifications as part of their employment conditions.

With staffing and curriculum issues gradually being resolved (superannuation arrangements proved to be the biggest and most enduring challenge), a slightly discordant note entered the continuing negotiations. The Minister for the Army (Fraser) explained to Baxter on 11 April 1967 that the nomination of ten years as the duration of the agreement was now:

considered necessary in order that, if the Government decides in the interim to proceed with its intention to develop an Armed Forces Academy to meet the requirements of all three services, our affiliation might be amalgamated into the broader concept [...] I do not see that this should pose any serious problem.

Then, as later, Fraser underestimated the complexity of the issues and the extent of 'the problem' posed by establishing a tri-Service academy at a time when single-Service loyalties remained strong. Fraser was to learn the hard way that amalgamation was, indeed, a 'serious problem' that would ultimately be resolved by the exercise of raw political will and that he would be its principal source.

After the University's leadership had made its intentions clear and announced that an affiliation agreement was proceeding, the more politically active students at UNSW considered their reaction. With Australian involvement in the war in South Vietnam escalating rapidly in 1966 and into 1967 with the dispatch of additional troops and the deployment of Navy and Air Force units, the response of UNSW students and, indeed, undergraduate students across the country, to the RMC affiliation was remarkably mute. The leadership at UNSW had feared much worse. Writing anonymously in an article entitled 'The University Student – 67' published in the *Current Affairs Bulletin*, Richard Walsh spoke of the 'cultural revolution' that was breaking out on Australian campuses. Rather than fighting for a new socialist society, the New Left focused on 'opposing basic elements of the old society – liberal humanist ideas, established religious beliefs, traditional morality, conservative social institutions' which included, of course, the Australian Army. Disparaging the Services served to antagonise senior leaders in many Australian universities who had wartime military experience.[25]

Although the University's origins were more in the sphere of technology and engineering than humanities and social sciences, student activism was no less voluble at UNSW than at the 'other' university – the University of Sydney.

The day after the meeting of the UNSW Council that had given in-principle support to the affiliation (10 May 1966), Ian Lowe, the President of the UNSW Students Union (later emeritus professor in the School of Science at Griffith University, adjunct professor at two Australian universities and president of the Australian Conservation Foundation), wrote to Baxter. Lowe later explained the nature and extent of student concern, which focused on the wisdom of Australia following the United States into the Vietnam War 'as well as the ethics of conscripting men too young to vote into military service to provide the necessary foot soldiers [...] for what we regarded as an ill-judged and legally dubious intervention'.[26] The activists conveniently overlooked the alternatives to military service in South Vietnam, such as part-time service in Australia. Members of the Students Union felt it 'particularly unlikely that those in charge of preparing army officers would want to expose them to any criticism of the military campaign that they would be expected to prosecute with unquestioning vigour', although RMC cadets naturally read the newspapers, watched television news and discussed the war with family and friends whose views on the conflict were no doubt mixed. Another source of anxiety was the training of military officers to 'follow a particular approach' in countries like China and Indonesia 'where the military were politically influential'. The students were concerned that Australia might 'take the same line'. That China was a communist state and Indonesia a military dictatorship tended to diminish the persuasiveness of the comparison. The objections lacked detail and sounded like slogans that RMC cadets would have found demeaning if not offensive.

Firm foundations and the faculty solution, 1964–1967

A member of the 1965 intake, Bob Hall, later recalled that 'it was compulsory to listen to the ABC radio news each morning – the junior class had to recite it to the senior classmen over breakfast [...] Each class had its own recreation room and daily newspapers were provided in each. It was hard to miss the public debate'.[27] He confirmed that 'a major influence was our families [...] my sister was in her final year at Sydney University and later worked as a teacher. She was strongly anti-war and particularly when I was at home on leave when we had our own "vigorous debates" about the justice or otherwise of the war'. Several cadets are believed to have resigned on the grounds of conscientious objection or political opposition while the war was in progress. In the years to come, a number of RMC graduates from the period 1964–1970 who saw active service in South Vietnam became leading scholars and vehement critics of Australian policy in Asia, including Dr John Mordike, Dr Greg Lockhart, Dr Graeme Cheeseman and an official historian of the Vietnam War, Ian McNeill. The suggestion that all cadets were unthinking drones or were unwittingly imbibing propaganda was baseless.

The Students Union considered a resolution opposing the affiliation. It was passed without 'serious opposition'. Lowe then conveyed the resolution to Baxter:

> Having regard to the traditions of University education and its responsibility to protect the interests of its members, [the Students Union] strongly condemns the decision of the University Council to accept RMC Duntroon as a degree-granting affiliated college of this university without provision for periods of study on the University campus. Members of the [Student Union] Council feel particularly concerned that degrees from this University might be granted to students

who have neither to set foot on the campus or been afforded the opportunity for a liberal education which one associates with a university. I would appreciate an assurance that this is not the intention of the University Council.[28]

Baxter replied in his characteristically blunt style: 'it is a pity that your Council should have passed a resolution of this kind when it obviously could not have had before it a balanced statement of the case which it was discussing'. He told Lowe that the Council and the Professorial Board could be trusted to 'look after the interests of the University in a satisfactory manner'.[29] In an article published in the student newspaper *Tharunka* on 14 June 1966, Baxter observed that many UNSW students did not participate in campus life and that requiring the attendance of RMC cadets was inconsistent with the demands the University made of civilians.[30] The matter was closed.

Myers felt that student opposition was not widely felt and could not be described as 'vociferous':

> In our university we had one charismatic figure who turned out to have been a former Duntroon cadet [Graeme Dunstan]. He was by that time a very radical student in the university. He had changed his course; he was doing Arts and he knew everything about Duntroon and he was against anything and everything to do with the military. He was one of the leaders of the opposition to any involvement of the university.

Dunstan had been part of the academic engineering stream at Duntroon but had left in 1962. He continued his studies on the main campus at Kensington, was elected president of the Students Union in 1966 and was concurrently president of the UNSW Labor Club.

He twice edited *Tharunka* and was active in the anti-Vietnam movement, gaining prominence when he and other protesters attempted to obstruct a motorcade conveying United States President Lyndon Johnson to the centre of Sydney in October 1966. Many years later (2011), Dunstan was charged and convicted of causing malicious damage to an Army helicopter. He was placed on a three-year good behaviour bond.

In the same edition of *Tharunka* in which Baxter explained and defended the affiliation, Dunstan claimed it was 'the administration on the military side which has made an academic atmosphere impossible. Their demands are more immediate; it is much more important to have boots spit-polished for the morning inspection parade than it is to prepare for a maths tutorial that follows straight after'.[31] He alleged that the first-year intake at Duntroon were routinely bastardised and that, like 'all isolated training institutions, there will always be a form of group indoctrination'. These were serious indictments of RMC. But Myers was adamant that:

> we didn't ever feel that the venture was being threatened by that student opposition and, in fact, it was modest, not to say minuscule in its impact [...] I never felt for myself, and I don't think Sir Philip Baxter ever felt, afraid that we were doing the wrong thing and that we would not be able to bring about some proper program with RMC.

As the final form of the affiliation was being drafted, the National Affairs Officer of the National Union of Australian University Students, John Bannon (later Labor Premier of South Australia), also wrote to Baxter about the draft agreement. He relayed a resolution of his Union's recent annual conference expressing 'its strong condemnation of the proposals to affiliate the Royal Military College

(Duntroon) as a degree granting college of the University of NSW without adequate safeguards for the standing of University degrees'. Bannon told Baxter that his union:

> feels that the discipline, control and fundamental teachings of a military college (necessary perhaps for the training of an officer corps) are completely at odds with the concept of a university as formulated by the International Student Conference of which we are a member. Unless members of the RMC could spend a considerable amount of time at the UNSW campus, with no restriction on their participation in the many extra-curricular activities and discussions, to grant them university degrees would serve only to lower the standing and meaning of such degrees.[32]

Baxter responded a week later. He was obliged to Bannon, he wrote, for pointing out that his union 'disapproves of arrangements which are being made [...] to improve the education and training of those men who are to be officers of Australia's military forces and whose careers will be devoted to the protection of the lives of their fellow countrymen'. The response was classic Baxter.

Perhaps surprisingly, there was hardly a murmur from the ANU about the affiliation despite its proximity to RMC and other initiatives then underway 'across town'. Richard Walsh's article in the *Current Affairs Bulletin* observed that undergraduate life at the ANU 'is distinguished by its apathy, conservatism and indifference to national causes'. Walsh thought that given the ANU's location near Federal Parliament 'it might well have been expected to prove the most dynamic and involved campus in the country'. When Ross Garnaut (an Arts–Law undergraduate and later Professor of Economics and an advisor to successive Federal Governments) promoted a resolution committing the Student Representative Council (SRC) to oppose deploying Australian combat troops to South Vietnam in 1965, the motion was defeated on the grounds that the SRC was not competent to comment on student opinion. By the following year, however, the campus mood had changed and student opinion was fervent and politically engaged. Army recruiters and Duntroon itself were subject either to pranks or willful disruption, both of which served to convey the disdain with which many undergraduates viewed the Army and perhaps all forms of authority. But there were no protest rallies, no disruption to any classes and no questions asked in Parliament about the UNSW–RMC affiliation. It was not until late 1969 when revelations of the My Lai massacre involving American forces in South Vietnam first circulated and early 1970 when the Moratorium movement gathered momentum that the mood changed across the country. The countervailing influence was clear evidence from National Constituent Assembly and Presidential elections that the people of South Vietnam did not want to be part of a communist state and that the US-led military campaign against North Vietnam enjoyed majority support in the South.

Protesting against the RMC affiliation also seemed a bit incongruous given the ANU had established the Strategic and Defence Studies Centre (SDSC) in July 1966. It was located within the Research School of Pacific Studies and affiliated with the Department of International Relations. In addition to considering strategic issues in the regions surrounding the Australian continent, it would also consider military issues and provide a non-government forum for analysis. It would not, however, sponsor opinions of its own or seek to influence government policy in a particular direction. It was to be apolitical and non-aligned, although its instigators, Professors Tom Millar and Arthur Burns, were considered to be sympathetic to the political Right. The commissioned history of the ANU noted of these years that:

opponents suggested that the Centre was wrong in principle, arguing that strategists, who took for granted the existence of military force, suffered from intellectual blindness which inclined them towards a hawkish posture; that they were mere technicians, unconcerned with the morality of the situations under analysis; that they collaborated with the establishment instead of using their knowledge to argue on matters of conscience.[33]

That the opposition within the ANU to the SDSC involved a plethora of unexplored and unchallenged assumptions and assertions, prejudices and preferences, did not seem to matter. Being opposed to all things military was the default position, assuming a certain righteousness that looks smug, if not sanctimonious, with hindsight. The possibility that the Centre might be a collaborator with government to develop creative policy options was lost at a time when academic independence seemed predicated on a commitment to blanket criticism. Even Gough Whitlam, the Federal Opposition leader, had told the universities that whereas once their strength might have resided in isolation, it now resided in participation.

With the eventual resolution of issues associated with the superannuation entitlements of academics, only a small number of administrative matters remained to be finalised. There was an exchange of letters on 11 April 1967 relating to the security vetting of academic staff by ASIO. There was another round of correspondence on 19 May 1967 concerning the fate of faculty when the affiliation agreement ended and an understanding that the agreement would not commence until the University's Act was amended to allow operations beyond the state boundary. The final set of hurdles surprisingly appeared after a meeting of the RMC Academic Staff Association on 22 June 1967. The faculty highlighted the 'military

Firm foundations and the faculty solution, 1964–1967

nature of the College' and insisted that 'there should be no outside interference with matters of cadet discipline (including classroom discipline), physical education, compulsory sport and military training'. These points were made to ensure that academics applying for positions at RMC were aware that the College was different from a 'normal' university campus. This would be the last time such concerns emanated from academic staff, as those appointed by the University to serve in the new faculty were more concerned with the intrusion of military customs and expectations in the academic sphere. It was the integrity of the academic program that proved to need more protection than the conduct of military training. Those employed at RMC before the UNSW affiliation were left in no doubt the mood had changed; what would flow from the affiliation would be different.

With the endorsement of the Professorial Board, Baxter sent the final draft of the agreement to Fraser on 18 July 1967. The text was no more than 1000 words. The Minister and the Vice-Chancellor signed the document without fanfare. There were no media releases. No ceremonies were held. The University and the Army would work together but for no longer than was necessary. The Faculty of Military Studies created by the affiliation was intended to be a temporary arrangement. Baxter was resolute: the University was midwife, not mother or spouse. The affiliation was a short-term strategy designed to achieve a long-term goal. From the University's perspective the Faculty was merely a vehicle for Duntroon to become an autonomous degree-conferring institution. The agreement would run for ten years but both parties hoped its objectives would be achieved sooner. While the University was willing to grant its degrees to RMC graduates, its offer to assist was contingent on the transition to autonomy in the nearer future. The Army was hoping this would take no longer than five years.

Three days before the affiliation agreement was signed, the Defence Minister, Allen Fairhall, announced the formation of the Martin Committee. Chaired by the newly appointed Dean of the Faculty of Military Studies and veteran of numerous Commonwealth inquiries into higher education, Professor Sir Leslie Martin and his committee were tasked with making recommendations for the establishment of 'an Armed Forces Academy'. It was assumed that when the Academy opened, the RMC affiliation agreement with UNSW would end. By that time, the Army hoped, Duntroon would be an autonomous degree-granting institution and the other Service colleges would become its affiliates within the new Academy structure. Perhaps prompted by Fraser's mistaken perception of how straightforward its establishment would be, Baxter mentioned the 'establishment of a military university for all three services in the foreseeable future' in private correspondence during December 1967. When Len Turner wrote to Edwin Davis, the University Bursar at Kensington, about his superannuation entitlements on 22 February 1968, he was advised that 'when Duntroon became autonomous it would have the full status of a university and the appropriate title in which the word "university" would appear'.

By the end of 1967, UNSW and the Department of Navy had also signed an 'agreement of association' that would allow approved University first-year courses including Mathematics, Chemistry, Physics and Engineering to be offered at Jervis Bay. The University would approve additional courses as the need arose. On completion of their first-year studies, midshipmen would relocate to the main campus. Humanities students proceeded directly to Kensington as there was insufficient staff at Jervis Bay to teach the first year of a Bachelor of Arts degree. At the insistence of the University, the standard of facilities at RANC was raised to meet the requirements of tertiary teaching while qualified members of the Naval College

Firm foundations and the faculty solution, 1964–1967

staff were accredited as university lecturers. Those who were not qualified were restricted to secondary school teaching.

The faculty at Jervis Bay was not large. Dr Ewart Dykes, the Director of Studies and former Dean of Engineering at the Royal Navy's Engineering College at Plymouth, co-ordinated the four members of the Science Department, the four members of the Mathematics Department and the three members of the Humanities Department. Oversight was exercised by an Academic Standing Committee consisting of the Commanding Officer of HMAS *Creswell*, the Director of the Commonwealth Office of Education, a representative from UNSW, Professor Noel Dunbar from the Department of Physics at the ANU, the New South Wales State Director of Secondary Education and a number of naval personnel. The subordinate committees were the Joint Advisory Committee, 'whose duty is to foster the association between the [RAN] College and the University'; the Accreditation Committee, whose task was to determine whether 'staff were qualified and suitable for appointment'; and the Library Committee. The 'association' with UNSW led to a series of immediate changes at Jervis Bay. In terms of teaching and learning, the first semester of study for Junior Entry cadet midshipmen aged 15–16 years was devoted to bringing them to a standard consistent with the New South Wales matriculation syllabus, while the first semester of study for Senior Entry midshipmen was bringing them into line with UNSW degree requirements. The first UNSW classes at RANC were held in February 1968, ten years after the College had returned to Jervis Bay. It had been a tumultuous decade for the Navy.

The University had taken a flexible approach to the Army and the Navy, responding to their needs with an affiliation for RMC and an association for RANC. Both Services were prepared to make sacrifices to secure an adequate education for their junior officers.

They were not to know the extent of the sacrifices they would be obliged to make or the extent to which they had effectively entrusted the nurture of their cadets to civilian academics. Largely because of Baxter's sense of the national good and vocational need, UNSW was the only university willing to enter into such an agreement with the Services at that time. Putting aside legal matters that could be surmounted and distance issues which could be negotiated, there was no Australian university sufficiently concerned with the state of officer education to offer the manner of assistance needed, although seven of the nation's universities were then offering advice and oversight to RMC and RANC on curriculum matters. Conversely, the Air Force showed no interest in establishing any relationship with UNSW. The University of Melbourne was meeting its needs albeit with some difficulties.

In addition to the goodwill that UNSW had earned from the Commonwealth and the New South Wales Government as a consequence of its willingness to work with Defence, Baxter and Myers realised that collaborating with government served the University's own interests by raising its profile and expanding its alumni. There were clear benefits for the University. There was very little to be lost from the two Service college agreements and much to be gained from attracting undergraduate students who would be well disposed to postgraduate study with UNSW later in life. After a decade, the Faculty of Military Studies would no longer be a UNSW responsibility in any event. While there had been compromises, none had damaged the standing of the University nor had the potential to adversely affect its reputation. It was an instance of UNSW showing itself to be a better corporate citizen than the University of Sydney, and part of Baxter's mission to raise the status of the state's second university.

In their recent study of the 1967 agreement, Jason Andrews and

Firm foundations and the faculty solution, 1964–1967

James Connor note that 'absent throughout the process of affiliation was any suggestion that FMS might also assume a normative or "civilianising" pedagogical role with the RMC'.[34] They contend that the agreement 'established a compromise institution', that offered the Army 'the educational assistance it required without having to compromise or taper RMC's unique program of military socialisation', although it was 'now officially divided along organisational lines'. The enduring challenge for the University and the Army was dealing with 'an uneven, but progressive expansion of the normative horizons of the RMC's academic culture'. Although Baxter was not interested in changing RMC's institutional culture and the Army was adamant it did not want University-sponsored cultural change, the 1967 agreement profoundly affected the institutional dynamics at RMC and altered the experience of cadets in ways that produced tension between the military and academic staffs. This tension would become an enduring feature of the relationship between UNSW and Defence.

3

Establishing a tradition, 1967–1972

The UNSW Council gave its approval to establish the Faculty of Military Studies (FMS) at RMC Duntroon on 10 July 1967.[1] The resolution also created Faculty Chairs in English, History, Mathematics, Physics and Engineering. The agreement between UNSW and the Army signed eight days later provided for undergraduate teaching to begin in the first semester of 1968. But the University's senior leadership group faced an unexpected setback. To that point, planning for the new faculty had been the responsibility of Professor John Clark, one of the two Pro Vice-Chancellors at Kensington. Clark died suddenly on 4 June 1967. Recognising the increased demands on the leadership group, Baxter replaced Clark with two new Pro Vice-Chancellors, Professors Albert Willis and Rex Vowels. Professor Rupert Myers continued as the 'senior' Pro-Vice-Chancellor. He acted as a deputy to Baxter, with duties across the spectrum of University administration. Myers was also considered Baxter's likely successor. (He would become Vice-Chancellor in 1969.) Clark's portfolio of work was given to Willis although Vowels, who had been an inaugural appointee to the RMC Advisory Board on Academic Studies, remained personally interested in RMC and continued to contribute academic advice in curriculum matters. In contrast to the short-term posting cycle for uniformed personnel that saw a steady succession of

Service officers dealing with the University and trying to resolve complex institutional issues, Willis and Myers would be closely involved with Defence over the next decade, Willis until his retirement in 1978 and Myers until 1981, when he too retired.

The RMC Interim Council met for the first time on 14–15 September 1967. The meeting was chaired by the Commandant, Major General Finlay, and included representatives of the Army, civilian staff in the Department of the Army, the ANU, the Universities of Melbourne, Waikato and Sydney, and Monash University, the New Zealand Army, Jock Weeden of the Department of Science and Education, the Dean-Elect of the Faculty of Military Studies (Professor Sir Leslie Martin) and the Vice-Chancellor of UNSW, Professor Sir Philip Baxter. The Minister for the Army, Malcolm Fraser, was unable to attend. An opening statement from the Commandant set the scene:

> The Australian senior commander of today and of the future must be capable of dealing on equal terms with, of speaking the same language and moving in the same circles as the politicians, diplomats, industrial tycoons, and racial and religious leaders with whom modern crises bring him into close contact. Simple competence as a military tactician is no longer enough – if the soldier is unable to negotiate with the same finesse as his non-military colleagues, if he cannot demonstrate the same depth and breadth of intellectual capacity and flexibility of mind, his advice will not count in the councils of those responsible for national and strategic decisions.
>
> It is to meet this continuing and developing requirement for an appropriate educational and intellectual foundation in the

> future senior officer that the courses at this College have been developed along their present lines. Yet the courses must not be viewed simply in the light of their intellectual purposes [...] but also in the light of their specific orientation to the requirements of the service. They are clearly and essentially professional courses, aimed at setting a special qualification in the Australian profession of arms; they are distinctly different from conventional courses of similar name in usual university parlance.

The Interim Council was to be the 'lateral channel between the Military Board and the University'. This arrangement suited UNSW, whose staff appreciated the designation of a single point of contact with the Army.

There were, however, a series of administrative and operational issues to be resolved arising from what the agreement did not say. An early difficulty appeared to be the appointment of new academic faculty. Existing staff were members of the Commonwealth Public Service and were on better conditions than those in the tertiary education sector. Superannuation and the availability of Commonwealth-sponsored scholarships to support continuing study were the main points of concern. As the Commonwealth Superannuation Scheme was considered to be more attractive than the New South Wales State scheme, existing staff were later offered the option of accepting employment with the University 'while on leave without pay from the Public Service'.

The academic requirements for studying in the Faculty of Military Studies were also clarified at the inaugural meeting. Cadets had to meet the 'matriculation requirements of the University of NSW and the special requirements of the Faculty'. The University would generally accept the matriculation standards of each Australian state and territory, and New Zealand. Professor Baxter advised caution because 'of the difficulties inherent in defining what the term "matriculation" means'. Professor Martin explained that standards varied considerably and he preferred assessing each recruit on an individual basis when determining their academic standing. He felt that some cadets deemed to be 'of high officer potential' would have to be designated 'Diploma' students rather than 'Degree' students, although the Army's stated preference was for all students to complete a UNSW degree.

These requirements would influence recruiting for those entering Duntroon in 1968, as there was an expectation that these cadets would be eligible for the award of University degrees. The meeting was also advised of 'the possibility that cadets from the Royal Australian Navy and the Royal Australian Air Force might join the Arts Course at the Royal Military College in the near future; it was believed that this might necessitate modifications to the structure and content of the course'. (This possibility never materialised.) After recording that UNSW 'accepted the fact that military work was an integral part of each course, and its satisfactory completion would be part of the qualification for the degree', the meeting 'observed that it was too early' to give any consideration on the ultimate independence of RMC as a degree-conferring institution. A firm deadline of ten years was set for achieving autonomy. The remainder of the meeting was devoted to dealing with practical matters, principally the provision of buildings, equipment, staff and curriculum content.

By March 1968 and its second meeting, the Council was advised that appointments had been made to the chairs of Engineering, English, History and Physics, with the Mathematics chair to be re-advertised. It resolved to recommend the appointment of a Professor of Chemistry and the formation of a department embracing the 'fields of government, economics, administration and geography'. It was subsequently decided to separate Government from Economics, and to defer the appointment of an Economics professor. The Vice-Chancellor also pointed out that undergraduate teaching made it 'necessary for the College to have a small computer, or access to one [...] [and that] one IBM 1620 has been found adequate for teaching in Wollongong'. While accepting his point, the meeting felt 'there was no need to provide for major research computing at the College, considering the facilities available in Canberra'.

The Interim Council then suggested a series of 'Persons Prominent in Civil Life' for inclusion in its membership. The list of nine included Sir Robert Menzies, Sir Victor Windeyer and Sir John Overall (all of whom declined). Those who accepted the Minister's invitation were Mr Justice (Russell Walter) Fox of the ACT Supreme Court and a wartime Army officer, and Lewis Luxton, a former Olympic rower and member of the International Olympic Commission (IOC). They were joined in March 1969 by Major General Sir Jack Stevens, a decorated wartime leader and former Federal departmental head, whose tenure was brief. He died in May 1969.

The Faculty of Military Studies formally came into being on Monday 29 April 1968. It would 'operate in a manner prescribed in the by-laws made under the *Technical Education and University of NSW Act 1949*'. The Faculty's inaugural meeting was held on 20 June 1968 with the Dean, Professor Sir Leslie Martin, as Acting Chairman. Professor Len Turner was then elected Chairman. It was not a faculty that necessarily specialised in military studies, although particular expertise in defence and strategic-related areas would soon develop. There was also an emphasis on securing competitive research grant funding and building a vibrant postgraduate culture. The academic staff at Duntroon had not received any grant funding prior to the UNSW affiliation while its main effort was delivering undergraduate tuition. The UNSW Faculty insisted that quality research would both attract postgraduates and enhance undergraduate teaching.

Bruce White, the Secretary of the Army Department, attended the inaugural meeting and spoke on behalf of the Minister. Fraser was visiting military units in Queensland. He described the meeting as 'the end of a task commenced eight years earlier'. He spoke of UNSW as 'a university which occupied a proud and distinguished position in the world today'. Noting the requirement for 'highly competent men', he said the 'Army man needed to have a feeling for the ebb and flow of opinion and for the philosophies of the world he lived in [...] the need was for men with minds open to the wider sweep of the horizon'. The creation of the new faculty would 'improve the educational standards of the Royal Military College'. Representing the Vice-Chancellor, Professor Willis remarked that UNSW:

> knew that its role was the trying one of midwife and knew that one day the Faculty of Military Studies, Duntroon, would stand alone. The role of midwife was not a new one to the University which had helped in the establishment of the new University of Newcastle and was even now helping the University College at Wollongong to develop into a university in its own right. The University looked forward to performing this service for the College [...] [and he hoped] the University would always have a binding relationship with the College.

The inaugural Faculty was an impressive group. Both Professor Sir Leslie Martin, and the first Faculty chair, Professor Len Turner, were distinguished scholars with international reputations. The other foundation professors – John Burns, Arthur Corbett and Grahame Johnston – were leaders in their disciplines. Martin would retire in September 1970 after shedding the role of Dean in mid-1969, while John Burns soon replaced Len Turner as Faculty chair. Brian Beddie, the inaugural Professor of Government, took over from Martin as Dean of the Faculty. He, too, was highly regarded. John Burns would succeed Beddie in January 1973 after Beddie's health deteriorated. Grahame Johnston died in 1976 at the age of 47.

The Faculty gradually expanded to 12 departments offering degrees in Arts, Science and Engineering. As the Faculty settled into its own rhythm, the third meeting of the Interim Council held on 16 September 1968 considered a suggestion from one of its own members that 'cadets should spend some time on the University campus' at Kensington. While the Council members could see the point of broadening the cadet experience, they needed an opportunity to consider when campus time would be possible and how it might be arranged. But good progress had been made on devising the degree programs, with all of the academic components contained within the first three years of study in both Arts and Science. Handling Engineering was much more complicated.

Cadets undertaking the RMC Engineering course presumed they would be appointed to the Royal Australian Corps of Engineers on completion of their studies. In the first year of the engineering course, cadets were required to do a 'workshop practice' period of three hours each week. It was usually a Tuesday night between 6 p.m. and 9 p.m. at the Technical College, now known as the Reid Campus of the Canberra Institute of Technology. The cadets would consume an early meal before climbing into a green Army bus that would convey them to the mechanical workshops. One of the cadets, Peter Rose (later colonel), recalls: 'we would don navy overalls and eye protection, get ourselves covered in lathe oil and white cooling fluid for three hours, try to get our hands clean with solid detergent then mount the bus back to RMC for a very late arrival in order to be parade-ready for the following morning'.[2]

During their studies, the cadets were encouraged to apply for Student Membership of the Institution of Engineers. Those who passed their studies at the basic level were awarded a Diploma of Military Studies; those succeeding at the higher level graduated with a Diploma of Military Studies with Merit. Of those who graduated with Merit in 1967, two were able to continue studying at UNSW towards a Master of Engineering Science degree. One studied in the field of Structural Engineering and the other in Highway Engineering. Three members of the December 1968 graduating class who completed the Engineering course at the higher level submitted applications to continue at UNSW. They heard nothing from either the Army or the University. After completing their Engineer Officers' Survey Course at the School of Survey at Bonegilla and then commencing the Basic Engineer Officers' Course (BEOC) at the School of Military Engineering at Casula, they learned their applications to attend UNSW had been successful. They did not attend the Kensington campus. Instead, the new lieutenants started at the School of Highway and Traffic Engineering at the Randwick Campus and graduated with Masters of Engineering Science – Highway Engineering.

By this time negotiations were well underway with the Institution of Engineers Australia on the content, structure and recognition of the new RMC Engineering program. To gain civil accreditation, all future RMC Engineering students were required to study Engineering subjects in the fourth year of their degree. The more vexing question was whether external accreditation was desirable given that securing industry validation would probably hasten the departure of uniformed engineers for the civilian workforce. The six Engineering students who had graduated in December 1968 were raised from 'Student' to 'Graduate' membership of the Institute in mid-1969, well before they had completed their Masters degrees.[3] Full Membership of the Institution was awarded three years later. They had by then fulfilled the minimum period of professional experience for such recognition.

Other than the specific needs of engineers, the Interim Council was of one mind on the desirability of students completing their academic studies within three years so that the fourth year at Duntroon could be devoted entirely to military training.[4] As recent RMC graduates were leading troops in South Vietnam and some had been killed, the Faculty understood the pressing need for competent platoon commanders and the importance of military training.

As the new staff set about their work, the Dean noted 'the present deficiencies in academic facilities at the College. These occurred in all departments; the engineering laboratory was over-crowded and potentially dangerous [...] the science laboratories were quite inadequate [...] there were no proper lecture theatres or tutorial rooms'. The Council noted that the Army was obliged under the agreement with UNSW to provide adequate facilities which, Professor Martin commented, were integral to attracting the best staff to the Faculty. There was a suggestion that the Army's slow response reflected 'rumours' that a tri-Service academy might be established in the nearer future and that building work at Duntroon might not be needed. The Council urged the Military Board to approve building works at Duntroon as a matter of urgency.

By March 1969, the Council was discussing openly one of the objectives to be met ahead of the College becoming autonomous and to ensuring the long-term viability of the Faculty. Believing the Army would continue to increase in size and require many more officers, the projected student cohort was 'in the order of 600–700 cadets', leading the Council to assume that with this number of cadets 'the College would be a viable academic institution by the time it achieved autonomy'. This was the first occasion on which a material threshold for achieving autonomy had been disclosed. In effect, the Army was asked to make a ten-year investment based on an expanding student population that was itself predicated on the Army getting bigger, which presumed the continuation of the conflict in South Vietnam.

The minutes of the fourth Council meeting held in March 1969 recorded the Army's delight that a close and harmonious relationship had been established with the University and that no insurmountable hurdles had been encountered. Even the contentious matter of sending cadets to the Kensington campus had been resolved, with students in third-year Arts and Applied Science courses being sent to the main campus for three weeks in 1970. The Council had also resolved that the Arts course would contain 'a compulsory science subject involving a knowledge of mathematics, and it is therefore preferable that candidates should have successfully completed at least four years of secondary school mathematics'.

Engineering continued to be an organisational challenge in terms of course content and duration, although the Institution of Engineers had decided to grant 'graduate membership' to cadets who had passed the diploma-level course as a general rule. The Institution indicated that its membership criteria were, however, set to change and that by 1980 only four-year qualifications would be acceptable for graduate membership. The Council decided to delay any formal decision on the Engineering program until 'the Army's requirements for the training of engineer officers had been definitely established'. It did consider the suggestion that 'candidates for the Engineering course might complete the first year at a university before entering the College, possibly under an extension of the RMC Scholarship Scheme'. This suggestion would allow Engineering undergraduates to complete their degree within three years and ensure the fourth year was entirely devoted to military training. After the meeting an informal proposal was devised that would provide for the creation of a Bachelor of Engineering degree and the delivery of its final year at the Kensington campus. The problem with this, however, was that part of the program had to be delivered in Sydney, when Duntroon's autonomy was sought 'in a relatively short time'. Plainly, the Council subsequently noted, 'the College could not become an autonomous body teaching engineering if it did not offer the final year of the course'. The Professor of Engineering at RMC Kingston in Canada was invited to spend a semester at RMC Duntroon to teach Australian cadets and to give some first-hand insights into Canadian solutions to a problem that remained vexing.

The Council's fifth meeting was delayed as a consequence of public and political controversy flowing from allegations that 'Fourth Class' (i.e. first-year) cadets had been the victims of severe bastardisation. At the centre of what became a national controversy was a newly arrived lecturer and former national serviceman, Gerry Walsh.[5] Concerned by what he had heard from junior cadets who had been bullied and even beaten by their seniors, Walsh wrote privately to the Commandant (Major General Colin Fraser) in late August 1969 with information about the conduct of certain 'First Class' (i.e. final-year) cadets. Although he had heard stories of junior cadets being tormented, he had not taught a Fourth Class course until 1969 and was concerned that informal military 'training', which he considered nothing more than cowardly 'bastardisation', was adversely affecting the academic performance of the whole cadet body. Walsh told General Fraser:

> On several occasions during the last week of Term, I was ashamed of my association with this College for the first time [...] No self-respecting university teacher can tolerate such interference with his work [...] We cannot have serious academic courses of study and 'bastardisation' as it is practised, since one is the negation of the other.

As Faculty Dean, Martin called a meeting of both the academic and military staff on 5 September to address the issues of concern to Walsh and a growing number of his colleagues. The meeting resolved to establish a Board of Inquiry that would include one Faculty member, Mr Alec Hill, Senior Lecturer in History. Hill had served in the British Army during the Second World War.[6] A Faculty meeting held three days later (8 September) passed a resolution condemning bastardisation and lamenting the effect of such conduct on the College's academic standing.

Although the establishment of the inquiry was not public knowledge, a copy of Walsh's letter was leaked to the journalist Maxwell Newton some time before 12 September, when he published a story headed 'Duntroon – tradition of torture'. Newton, who knew how to create a scandal and sustain a controversy, quoted the entire contents of Walsh's letter and appeared to have an 'insider' account of the 8 September Faculty meeting. Walsh became the focus of personal anger and institutional hostility, mainly from the military staff. He was physically threatened by at least one cadet and faced the prospect of unemployment as the worsening scandal had the potential to become an issue at the Federal election to be held in late October 1969. One Duntroon graduate, Geoffrey Solomon, thought the intent 'was not inflammatory but the contents highly combustible'. Although the University was under pressure to take some action against Walsh for bringing the College into disrepute, mistakenly assuming he had provided a copy of his letter to Newton, he was strongly supported by his departmental head, Len Turner. (Walsh consistently denied he had leaked the letter and was believed by his UNSW colleagues.)

As Darren Moore observes in his very thorough history of RMC, this incident seemed to vindicate opponents of the UNSW affiliation 'who did not believe that a liberal university education was possible in a military environment'. This was certainly the reaction at the Kensington campus. Willis recalled:

> just as things were getting nice and the opposition was dying down and [...] of course, then all of the people said: 'well, there you are' and it all started up again. The die-hard opposition said that we should never have had anything to do with [RMC], disregarding the fact that, though it was not so bad, our own Colleges on the campus were having some rather difficult times which some of us knew about, but it didn't hit the headlines.[7]

The UNSW Students Union Council met with the UNSW Staff Association on 17 September 1969. They devised a joint petition that both organisations hoped might end the affiliation if substantial reform was not promised and delivered by the Army. The petition called for RMC cadets to reside on the Kensington campus for at least one year and for a chapter of the Student Union to be established at Duntroon. The aim was not so much to reform RMC but to see the affiliation terminated. The petition attracted over 1000 signatures in the space of one day and was presented to the Vice-Chancellor on 18 September. A week later, representatives of the Staff Association met with Professor Myers and urged him to terminate the agreement if RMC rejected the demands for reform set out in the petition. Myers had no intention of moving in this direction. In fact, he had been heartened by the seriousness the Army had itself attached to the allegations.

The Minister for the Army, Phillip Lynch, had also acted decisively throughout the unfolding controversy. (Lynch had succeeded Fraser as Army Minister in February 1968 after Senator John Gorton became Prime Minister when Harold Holt drowned in December 1967.) Lynch received the Board of Inquiry's report on 23 September and tabled the document in Federal Parliament two days later. He readily agreed that the practices Walsh had described were 'humiliating, stupid or simply a waste of time' and reported that disciplinary action would be taken against a number of uniformed personnel. Because the scandal was an indictment of the corporate culture at Duntroon, Lynch appointed a Committee of Inquiry to examine the training of new-entry cadets. The committee was headed by Justice Fox (an active member of the Interim Council) and included Professor Len Turner.

The Federal Opposition sought to embarrass the Government over the scandal. The Deputy Opposition Leader and future Defence Minister, Lance Barnard, told Federal Parliament that because RMC had an 'association with UNSW, which has an excellent tradition of academic freedom, it becomes even more important that bastardisation is stamped out'.[8] The Labor Member for Yarra and wartime Army educator, Dr Jim Cairns, asked Lynch in Parliament whether this episode was the product of the 1967 cadet entry believing, wrongly as it turned out, that they would graduate with a degree. Cairns claimed that their behaviour steadily deteriorated when they learned that their academic studies would not be recognised by UNSW.[9] Cairns contended that this connection of cause and consequence 'is commonly believed' at the College. Lynch acknowledged that some cadets had lost interest in their studies after learning they would not lead to a UNSW degree and that some had sought only the minimum result required to graduate from RMC. He was adamant, however, that nothing had been brought to his attention that suggested or confirmed Cairns' assertions about the reasons for the bastardisation.

The disappointment among cadets that their studies would not lead to a degree started with the 1965 entry. They had been told by Traill Sutherland (who was not party to the UNSW affiliation discussions, partly because he considered the affiliation an unnecessary and unwelcome intrusion to the academic program at Duntroon) that they would be the first RMC class to graduate from UNSW. It was not until the arrival of the 1968 entry that the 'promise' was fulfilled, leading to some residual resentment from earlier intakes. Whether that resentment was the 'tipping point' for serious misconduct was not canvassed in the inquiry. The Faculty certainly did not believe it was the main cause.

The Committee of Inquiry provided the new Minister for the Army, Andrew Peacock, with an interim report before Christmas 1969 with the expectation that its recommendations might be implemented in time for the arrival of the 1970 entry. (Peacock had succeeded Lynch as Army Minister after the 25 October 1969 general election.) Other than its endorsement of the Commandant's actions, the Interim Council made no official comment on a letter that the Commandant had received from 'a member of the faculty' (that is, the letter from Gerry Walsh) or the conduct that was the subject of the inquiry. A special meeting of the Interim Council was held on 23 January 1970, just before Major General Fraser was posted to command the Australian taskforce in South Vietnam. While the minutes were carefully worded to disclose very little of substance, the Army plainly felt the need to assure the University that it had taken firm steps to deal with the allegations. The Interim Council was assisted by the presence of Justice Fox, who was able to give a first-hand account of the Committee's work which, he noted, was continuing.

The final draft of the 'Fox Report', based on 1300 pages of evidence transcript, was presented to the Minister on 24 April 1970 and tabled in Federal Parliament seven weeks later. There is no record of any comment from the University members of the Interim Council on the controversy or whether there was any sense in which the University's community standing had been affected by press reporting of the incidents. The Council did, however, discuss an earlier letter written to the Army Minister (Lynch) by the UNSW Staff Association expressing its members' concern about 'Fourth Class Training'. The outgoing Commandant advised the University's representatives that he had issued new directives on the conduct of cadet discipline and had developed policies that would assist all cadets in achieving the best academic results of which they were capable. Notably, Fraser's successor, Major General Sandy Pearson (who had been part of the Committee of Inquiry) later revealed: 'I was never once informed or directed or advised to implement any of the recommendations of the report, never once. I think the politicians, having approved the report, sort of forgot about it'. But change did occur. Study time was protected and military activity was circumscribed. The positive side, Professor Al Willis reflected, was that it gave the academics a greater role in shaping the culture at RMC. That may have been true but it was a role that some members of the military staff resented, believing that academics did not understand the needs of the Army and that they had exceeded their brief. Tragically, given his avowed desire to avoid a public scandal, Walsh was made to pay a high price personally for precipitating the controversy. In addition to being shunned by cadets who ought to have been chastened by the adverse attention their actions had brought to the College, Walsh was accused of being a 'card-carrying Marxist' – which he wasn't – and denounced for being anti-military – which was untrue.

When it met for its next scheduled meeting on 26 February 1970, the Council noted with some concern that the 1971 cadet intake would be limited in size because accommodation at the College had reached capacity. While pleasing progress had been achieved in recent building works, the lack of accommodation was now the principal impediment to expanding the student body and advancing the cause of autonomy. The shortage of space affected the University's desire to expand its academic offerings to postgraduates as well. The Council asked the Faculty to produce a memorandum that included a detailed justification for offering postgraduate studies at RMC, an explanation of whether postgraduate work would be better funded by the Australian Research Council than the Army, and nominating areas of priority for postgraduate work in 'fields of defence interest'. A comprehensive memo was prepared and circulated on 3 July 1970. The 'case for developing postgraduate studies' claimed it would be difficult for UNSW 'to retain existing staff and to attract new staff unless postgraduate studies are developed as soon as possible'. Without postgraduates 'there is the marked absence of seminar work', which the memo's authors felt was crucial to 'keeping senior undergraduates in touch with developments on the frontiers of contemporary knowledge'.

The memo pointed to the award of postgraduate qualifications through RMC Kingston's Graduate Studies and Research Division, while the Royal Military College of Science (RMCS) at Shrivenham offered masters degrees and doctorates. Indeed, the staff at RMCS had concluded that only when undergraduate studies were combined with postgraduate education could the College share 'a common aim and climate with the universities […] In a University, which aims to be a true center of learning, teaching on the one hand, and research and advanced studies by the teachers on the other, are complementary and are deemed to be part of the essential climate of the place'. The absence of postgraduate teaching at the three Service Colleges in the United States was a source of criticism rather than a basis for commendation. The memo noted John Masland and Laurence Radway's 1957 study *Soldiers and Scholars* and quoted their observation that 'the training approach to officer education shows insufficient understanding and respect for truly intellectual enterprise' and their conclusion that 'American military schools [...] are not pushing out the frontiers of knowledge in their professional field'.[10] Then followed a plea:

> The Department of the Army, after long and careful deliberation, chose to provide university education for its cadets by seeking cooperation with the University of NSW [and] [...] presumably wishes to ensure there could be no doubt about the value of the degree obtained by cadets [...] Had the Royal Military College not obtained degrees backed by the name of an established university, its degrees would have met difficulty in securing recognition. Should a new institution at Duntroon fail to have postgraduate studies, there would be a considerable danger that its bachelor degrees would lose the recognition that they will gain in the early 1970s [...] because it is part of the philosophy of Australian education that effective undergraduate education at full university standard gains much if staff are also engaged in the supervision of postgraduate studies. This matter was debated at length when the ANU incorporated Canberra University College. It was decided that the College, which became the School of General Studies, must have the right to engage in postgraduate work if

its bachelor degrees were to secure recognition. The right was guaranteed by an Act of Parliament.

The memo also warned the Army that a refusal to offer postgraduate study might prompt the departure of academics who believed in good faith that postgraduate teaching and supervision would follow the consolidation of the undergraduate program. The memo's 'modest proposal' was to aim for 25–30 full-time postgraduate students by 1974, with the majority being civilians, including Defence public servants. The memo conceded that uniformed postgraduates would initially be few in number. The (Australian) Chief Defence Scientist (Howard Arthur Wills) mentioned 'his belief that fruitful collaboration in research could develop between the science and engineering departments at RMC and the laboratories connected with the Departments of Defence and Supply'. He believed that sharing facilities was a possibility and that Australian Research Council grants involving RMC and the Department were an attractive proposition. The way to generate a research culture was to offer scholarships, some funded by the Army, that would attract candidates mainly from the civilian population. But the research conducted by Faculty members should extend beyond the immediate interests of Defence. The memo defended this recommendation by citing the 1958 RMCS Report: 'their advanced studies should illuminate their teaching work, and there can be no general rules about this'. The Council noted 'a strong body of opinion that the College would not attract good staff unless the opportunity was provided for them to take part in postgraduate work, and here was agreement that, at least initially, it would be necessary to rely on other than young soldiers to take part in the postgraduate program to ensure continuity'. Minds had been changed and attitudes had been altered.

By 10 September 1970, the seventh meeting of the Interim

Council was told that the accommodation shortage at Duntroon was a continuing problem that was severely impeding the College's overall development. The Council noted that 'a decision on the establishment of a tri-Service academy' should not determine the 'properly-planned development of the College' because the 'College would undoubtedly form the basis of a new academy'. The absence of cadets at the Kensington campus had not made much of a difference to the accommodation problem. In June 1970 there was 'one cadet at Kensington and shortly another 24 would be attending there for a month'.[11] Professor Willis reported that the four-week visits 'had proved useful and informative and revealed that the physical association [...] was very real, valuable and necessary'. The short-term visits would continue, although the arrangements were costly and complex. Professor Paul George observed that 'the closed-shop attitude at Duntroon had gone to a great extent and the need for a period on Kensington had less validity'. He thought the experience of the difficulties encountered in the 1970 exercise was worthwhile because it provided a convincing argument to discontinue the arrangement. The Interim Council heard it was an 'enjoyable experience' which would continue on a trial basis into 1971.

The eighth meeting of the Interim Council, held on 25 February 1971, resolved that it would 'begin considering autonomy for the college' and appointed a sub-committee to explore possibilities, noting that 'it should bear in mind the possibility of [RMC] becoming a tri-Service academy'. The Council was advised that autonomy would require an 'Act of Incorporation' to establish RMC as a 'separate autonomous degree-granting body'. Existing legislation that had been used recently to create new universities would be modified for RMC's purposes. The main issues were the nature of the entity, finance and property, governing body, staffing arrangements, transitional provisions and statutes. Lines of responsibility were not straightforward. Who would control the curriculum and assessment and what was the basis of their claim to competence in doing so? The initial proposal was for a military academy and an academic college, with neither transgressing into the areas of the other's responsibility. The sub-committee suggested that 'the Academy structure replaces the University of NSW and the Interim Council in their present relationships with RMC'. The head of the College would be styled the 'Rector'. But who would exercise ultimate authority? The sub-committee was divided over 'whether the Commandant be designated CEO of the Academy and take precedence over the Rector, or whether they should be equal in status', with the Rector reporting directly to the governing body.

The Council's Executive Committee met on 10 June 1971 and noted that 'a degree of dualism appeared to be inevitable: the governing body in a university had direct and undivided authority over all staff, but in a service academy it would not exercise full authority in respect of service personnel'. But it also noted that a 'consequence of placing the "Rector" below the Commandant' was that the incumbent's status would be less than that of a vice-chancellor, and the Academy might fail to gain representation on the Australian Vice-Chancellors' Committee (AVCC). In its final report, the sub-committee recommended the Rector be under the Commandant but report directly to the Governing Body. Ahead of the next step, an 'Autonomy Committee' would be established consisting of senior officers from the Army and RMC, the Vice-Chancellor of UNSW, a member of the ANU, 'one of the members of the Interim Council from a university other than UNSW' and two members of the Interim Council.

The ninth meeting of the Interim Council on 23 September 1971 noted that UNSW was continuing to assist the College towards achieving autonomy within ten years and was 'planning towards this end'. The target date for autonomy was 1975, noting that the 'legislative processes could take two years'. A draft act of parliament was to be prepared by March 1973 and would be modelled on the legislation that created the ANU and Macquarie University. Six separate committees were established to achieve autonomy. UNSW was working conscientiously with the Army to achieve autonomy. There is no evidence of reticence or resistance from any UNSW employee. Even the RMC Staff Association, headed by Associate Professor Sid Hodges, entered into the exercise with a desire to co-operate, the only concern being staff entitlements and ensuring that no staff member was disadvantaged by the transition to autonomy. But would the new institution be entitled to call itself a university and would that word appear in its official title?

If RMC's academic enterprise was not referred to as a university, the Interim Council feared, 'it was possible that it might be regarded as similar to a [College of Advanced Education] CAE. The name could have some influence on the recruiting of staff and in attracting postgraduate students'. The Staff Association 'felt that "academy" implied something less than a university, as in the United States'. But Professor Charles Moorhouse of Melbourne University feared using the term 'university' might 'lead to a misunderstanding of the nature of the Academy. The Academy would not be another of the Australian universities of standard pattern, but a special and unique institution'. He did not object to 'academy' and preferred it to any title that implied a 'conventional university'. Jo Gullet, a former Liberal parliamentarian, thought the 'Royal Military Academy' might include 'Duntroon University'. Myers 'thought it would be preferable to reverse the terms in such a title: 'Duntroon University with the affiliated Royal Military Academy'. John Robertson, then senior lecturer in history, contended that Duntroon 'as an educational institution will be bracketed, not with universities but with the CAEs' if it were branded solely an academy.[12] He argued: 'If the faculty does not become a university there would be a breach of faith with me and with others in my position; and this would represent a reversal of the present policy of giving cadets a university education'. There was no consensus on the name of the new institution and the matter was left open.

Curiously, little work was done on fashioning a vision statement for Service officer education. The nearest to such a statement was a paper on 'The Requirement for Higher Education in the Army' produced by one of the autonomy sub-committees. It claimed that 'the armed services require at least a percentage of their officers to be highly educated – certainly not less highly educated than the civilians with whom their activities are intertwined'. It also argued that the:

> Separation of military training from academic education by, e.g., sending recently commissioned officers to civilian universities, would both lengthen the period of preparation for a career in the Army and weaken the value of the initial period of military training. It would also be almost certain to increase the rates of early resignation.

Thus, training and education had to be concurrent. Notably, there was no mention of the other two Services and their arrangements.

At the Interim Council meeting on 4 May 1972, Brian Beddie offered his thoughts on the general direction of the transitional arrangements. Beddie's role in what followed was crucial, as he had formed many insightful perspectives on matters ranging from principles to policies to practicalities. Deeply respected by his colleagues, Beddie was an unlikely participant in these negotiations. He was the foundation secretary of the Philosophy Club formed at Sydney University on 13 June 1939. Personally influenced by the controversial philosopher John Anderson, Beddie joined the Australian Army as a gunner in 1942 but seemed temperamentally ill-suited to military life.[13] He was selected to become a diplomatic cadet in 1944 but decided instead on an academic career after being appointed to the Canberra University College in 1948. 'Fleeing a politically fractious department at ANU', according to his colleague Dr Hugh Smith, Beddie became the foundation Professor in Government at RMC during 1970 and defended his department 'against bureaucratic pressures from government and university administrators'.[14] He was well connected with the Commonwealth Public Service and knew a number of its rising stars, including WB 'Bill' Pritchett, who would later become Secretary of the Department of Defence. It was said of Beddie that 'like his mentor, John Anderson, he was moved neither by ideology or fashion – only by the spirit of inquiry [...] As a man, he was an uncommon mixture of modesty and stubbornness'.[15]

As the University and the Army moved towards autonomy, Willis credited Beddie with observing 'a self-contradictory proposition, an autonomous self-governing university under the military, plus transfer of conditions without any deterioration of the conditions'. Willis saw the proposed arrangements as a dichotomy that could not be readily resolved: education had to be unfettered but military training had to be subject to ministerial control. Turning to administrative systems, Beddie suspected that the Autonomy Committee's proposals were 'influenced by the cases of Newcastle and Wollongong. The University of NSW, in fostering those two universities, had built up not only the academic staff but also the necessary administrative infrastructures. This was not the case at Duntroon, where (except for the College Registrar and his Deputy) the whole of the [administrative] servicing was provided by the Department of Army'. He was concerned that the administrative arrangements were less than ideal. Myers envisaged that the

University would:

> create the essential administrative structure for the Academy [...] once the Act was passed, it was likely that the University [of NSW] would negotiate a supplementary agreement with the Minister to enable the appointment of administrative staff in the interim period.

Myers thought that UNSW would actually run the Academy for the first six months after autonomy was achieved to ensure a smooth transition. All that was needed now was an Act of Parliament.

The push for autonomy would be unexpectedly halted by a prospect Fraser had first mentioned in 1967: the amalgamation of the three Service colleges. It was one thing to negotiate an affiliation agreement with a civilian university; it was quite another to work against the strong loyalties inculcated by the Service colleges. They were institutions charged with maintaining each Service's professional ethos and ensuring their corporate identity was not watered down by the intrusion of 'jointery'. While the common enemy may have been Soviet-backed communist infiltration of democratic states, the Services regarded each other as adversaries engaged in a protracted struggle for recruits, resources and responsibilities. There was strong opposition to any proposal or project that diminished the ability of one Service to resist the pretensions of another. The Navy and the Air Force shared a dislike for the Army's corporate culture. As the Army was the largest Service in terms of personnel, the Navy and Air Force feared that its physical size would lead to its organisational dominance. A tri-Service academy would see the Army exert such dominance to the detriment of their interests. The idea of creating a tri-Service academy was always about much more than the effective and efficient delivery of higher education.

4

The University as a unifying agent in Defence, 1967–1975

The review of Australian universities conducted in 1957 by Sir Keith Murray, Chairman of the United Kingdom University Grants Committee, was paralleled by a review of the Defence 'group' of Departments. Lieutenant General Sir Leslie Morshead, a distinguished former citizen-soldier and businessman, chaired an inquiry that recommended amalgamating the Departments of Supply and Defence Production and absorbing the Departments of Navy, Army and Air into a unified Department of Defence. The first recommendation was accepted but the second, which was far more complicated and controversial, was not. A clearly frustrated Prime Minister Menzies issued a confidential memorandum to the Minister of Defence on 11 December 1958 in which he addressed 'some uncertainty about the strength and the extent of the authority of the Department of Defence':

> The Minister and Department of Defence have overall responsibility for the defence policy of the country [...]
> I expect the Department of Defence to be more than a coordinator. I expect it to make clear recommendations in the defence field and when these have been accepted I expect

it to act strongly to see the policies are carried out [...] In brief, I do not want to be told that there is some doubt about the authority of the Department of Defence in any matter pertaining to Defence.[1]

The Minister's resolve was tested the following year. A committee chaired by Major General Ronald Wade recommended the integration of all Service officer training at a tri-Service academy located at Duntroon, with each Service preserving its own college for post-graduation specialist training. The Services agreed in principle but the devil was in the detail. Australian planners were also precluded from taking a lead from the United Kingdom, as the education and training of British officers was in a state of flux. In fact, the British officer education system was in turmoil. Nothing practical eventuated from Wade's recommendations until the Minister for Defence, Senator Shane Paltridge, had all parties agree in 1965 that a tri-Service academy was to be the ultimate goal. Five years had elapsed since the principle was first agreed in 1960. Such was the strength of inter-Service rivalry and bureaucratic inertia.

Although they were outwardly committed to collaborative effort and joint operations, the Navy, the Army and the Air Force, supported by their own separate government departments, tended to function in a manner that militated against co-operative efficiency. The Department of Supply existed alongside the Department of Defence, which was set 'above' the Service departments but was powerless to compel decisions or to command resources. Sectional interests and partisanship were clearly hampering the co-ordination of Australia's national defence resources.[2] Rivalry between the Services for strategic pre-eminence and funding supremacy became more apparent after April 1965, when Australia committed combat forces to the continuing conflict in South Vietnam.

The University as a unifying agent in Defence, 1967–1975

General Sir John Wilton, the Chairman of the Chiefs of Staff Committee after May 1966, argued that the Services needed a unified command structure so that they could conduct complex operations with clear lines of command and control. To that end, in September 1967 Wilton revived the Morshead Review proposal that the single Service departments (together with their ministers) be absorbed into a unified Department of Defence. Wilton was acting with the strong support of the Minister for the Army, Malcolm Fraser. But as neither the Minister for Defence, Allen Fairhall, nor the Secretary of Defence, the long-serving Sir Edwin Hicks, were known for creative thinking or courageous leadership, and given Australia's escalating involvement in South Vietnam, the prevailing mood was against substantial organisational change. Although Hicks' successor, Sir Henry Bland (who was appointed in January 1968) supported Wilton's reform agenda, they were unable to pursue a thoroughgoing program of administrative reform until Prime Minister John Gorton appointed the 39-year-old Malcolm Fraser to the Defence portfolio in November 1969. Wilton, Bland and Fraser worked to create a Joint Services Staff College (1969), the Joint Intelligence Organisation (1970) and the Australian Joint Warfare Establishment. They also commissioned the design of a joint services badge, pre-empting the recognition of the Australian Defence Force (ADF) as a formal entity in February 1976.

With the appointment of the energetic and reformist Sir Arthur Tange as Secretary of the Department of Defence in March 1970, the amalgamation of the three Service Departments had a champion whose close personal relationship with Fraser saw the launch of a revolutionary program of change. The release in 1973 of *Australian Defence: Report of the Reorganisation of the Defence Group of Departments*, known as the 'Tange Report', provided a blueprint for major change. On 1 December 1973, with the complete support of Prime Minister Gough Whitlam and the Minister for Defence, Lance Barnard, the three Service Departments and the Department of Supply were amalgamated into a unified Department of Defence.

The general move towards 'jointery' precipitated progress towards the formal creation of a tri-Service academy in 1970, although the forces arrayed against it were entrenched and extensive. Within the 'Defence Planning' section of the Department of Defence, the known proponents of 'jointery' were Malcolm Fraser, Sir Arthur Tange, Admiral Sir Victor Smith (the Chief of Defence Force Staff) and Mr Graham Wright, a former naval captain turned civil servant. Indeed, progress on the tri-Service Academy proposal was seen as a barometer of the mood for change among the Services. That UNSW had entered into an agreement with the Navy and the Army meant that negotiations for a tri-Service arrangement were a little less complex. Two of the Services already had a formal relationship with a university. Assuming that this relationship was satisfactory, there was a hope that the Air Force would be persuaded to join the UNSW arrangements. A tri-Service academy would bring together cadet officers at the beginning of their careers and lead to friendships that would work against the single-Service rivalries that were hampering Australia's ability to mount effective joint operations.

The challenge was to reconcile several objectives. The Army had entered into a relationship with UNSW with the specific aim of Duntroon becoming an autonomous degree-granting institution within ten years. Defence's pursuit of the proposal to establish a tri-Service academy was a complication that the Army essentially ignored, for three reasons. First, many within the Army doubted that such an academy would ever be built. Second, if Duntroon became a degree-granting institution, the education of officer cadets for the other two Services would probably be subsumed into

Duntroon. And third, the possibility of educating Navy and Air Force cadets alongside those of the Army might threaten the coherence of military indoctrination and dilute the Army's officer ethos. These fears were partly assuaged by the fact that Wilton, as Chair of the Chiefs of Staff Committee and an Army officer, would hardly act against the Army's best interests.

The first practical steps towards a tri-Service academy were taken in 1967. As the agreement with UNSW was signed, the Holt Government announced the formation of the Tertiary Education (Services' Cadet Colleges) Committee and appointed Professor Sir Leslie Martin as its chair. The original Committee members were Martin, General Sir John Wilton (Chairman, Chiefs of Staff Committee) and Professor Sir Hugh Ennor (Secretary of the Department of Education and Science). They were joined during 1968 by Gordon Blakers, Deputy Secretary of the Department of Defence; Professor George Russell, head of the ANU English Department; and Professor Percy Partridge, Professor of Social Philosophy at the ANU. The Committee members were all located in Canberra, making it easier for them to meet regularly. Each was highly regarded and considered well qualified for the task. The original ministerial decision was to have an 'armed forces academy which will operate with separate wings at Duntroon and Point Cook, but with its headquarters at Duntroon'. The committee focused on the Army and the Air Force but their terms of reference mentioned 'the future of the Royal Australian Naval College, Jervis Bay'. By the end of 1967, the committee was convinced of 'the advantages of establishing the academy at one location only, rather than with separate wings at Duntroon and Point Cook'. The Minister then agreed to amended terms of reference, with the committee concentrating on 'the establishment of an armed forces academy which will operate at Duntroon'. Its deliberations were guided by the Howard–English

Report on the tertiary component of service education in the United Kingdom, which was released in 1966.[3]

This review looked like an effort to achieve reform from within. From the time of his appointment to the Australian Universities Commission in August 1959, Martin had co-ordinated the rapid expansion of the tertiary education sector, marked by the establishment of five new universities. Two years later he headed the Committee on the Future Development of Tertiary Education in Australia that led to the formalised 'binary divide' between the universities as research and teaching institutions, and other tertiary institutions which were designated colleges of advanced education. At the same time he negotiated the establishment of the RAAF Academy as an affiliated college of the University of Melbourne. His appointment to Duntroon at the end of 1966, essentially a post-retirement appointment, was timely and influential. Martin was highly regarded by politicians and public servants, who acknowledged his personal integrity and appreciated his professional insight. A biographer noted that 'his reputation went far towards winning academic acceptance for the college and smoothing relations with its military authorities. He took to lecturing again, overseeing the appointment of staff and encouraging them to develop research programs'.[4] But Martin was about to embark on one of the most complex negotiations of his career.

The Martin Committee released an initial interim report in November 1967 which flagged the direction of its thinking and the likely shape of its principal recommendation. Although the RANC association and the RMC affiliation with UNSW had not even begun, the committee contended that:

> In their anxiety to preserve academic standards the
> Universities have sometimes encouraged standards which,

The University as a unifying agent in Defence, 1967–1975

for Service requirements, are unrealistic. Coupled with this there has been an under-estimation by the Universities of the educational content of the military studies which must be undertaken by cadets. In our opinion the RAAF Academy provides an illustration of the difficulties which arise from a close association between a university and a Service college. A baccalaureate course in Science [...] with its emphasis on physics, is not wholly suited to the needs of the RAAF.

The Air Force was already persuading itself of the need for a tri-Service academy because its relationship with Melbourne University was becoming increasingly problematic. But the Air Force's concern was not dissimilar to the Navy's – the dominance of the Army and the potential diminution of the RAAF's distinct ethos. The Minister for Air, Peter Howson, noted in his private diary on 22 August 1967:

> had a meeting with Al Murdoch [Chief of Air Staff], McFarlane [Secretary of the Department of Air] and McLachlan [Air Member for Supply and Equipment] on the problems confronting the future of the Academy at Point Cook and the proposal that Defence are examining for a Joint Services Academy. The main point that we have to decide is whether we make do with the present set-up – that is, buildings at Duntroon and buildings at Point Cook – or whether instead we should suggest the ideal solution which is to have one college in one place [...] if the ideal solution is not favoured, then to come back to the proposals for making the best of the present situation, which is that we should handle Science and basic Engineering at Point Cook, leaving the Arts students to be taught at Duntroon. The main difficulty

that I foresee here is the problem of control of two separate institutions.⁵

Just over two years later, the review was finally completed. On 27 January 1970, the 'Report by the Tertiary Education (Services' Cadet Colleges) Committee' was presented to Malcolm Fraser, who had become the Minister for Defence on 12 November 1969.⁶ The conclusions were surprising and the recommendations were unexpected. In no more than a few paragraphs that were long on assertions and short on arguments, the Committee concluded that young officers were best trained in a tri-Service academy setting. Service-specific requirements would be met by specialist courses of study that reflected the particular needs of the Navy, the Army and the Air Force. The key – one might say determining – issue for the committee was the unsustainability of the three Service colleges and their future as independent bodies. At the time of reporting, there were only 87 full-time officer cadets receiving accredited tertiary education across the three Services. This made it difficult to justify the existing arrangements with an academic staff-to-student ratio of 4:8 to 1 across the three Service colleges, whereas the Australia-wide ratio was between 8:1 and 11:1 lecturers to students. The committee recommended establishing a single academy (rather than a multi-campus arrangement) to provide a 'balanced and liberal education' at a tertiary level in the social sciences, humanities, physical sciences and engineering. The committee felt that Duntroon was most suitable in terms of infrastructure and its co-location with a number of stakeholders.

But why create a separate institution given the costs associated with establishing a new facility? The affiliation model with UNSW had yielded good results in 1968 and 1969 but there were 'educational disadvantages' in not being able to offer courses 'wholly suited to Service needs'. The committee also claimed that university representatives 'would be among the first to admit that the education and training of young officers presents special problems, which call for a unique educational approach'. While the committee accepted that 'there would be disadvantages in separating them from their peers in the formative years of young manhood and educating them in a Service academy, which must be to some extent a closed establishment [...] nevertheless, no university now provides the kinds of courses [...] that are best suited to the education of officer cadets'. Furthermore, a tri-Service academy that taught History, for instance, could justify concentrating on Military History, whereas a civilian university would be obliged to offer a range of options across the discipline. The final consideration was cultural: 'the Services regard it as important that cadets should remain in a Service environment so that they may be properly motivated to a Service career'. Although the views of the Services were not disclosed, the Committee concluded that it was unjustifiable:

> for any one college to contemplate educational autonomy as a means of securing freedom to design courses of its own. We think that this problem will be resolved most effectively by the establishment of one tri-Service academy that has power conferred on it by legislation to grant its own awards for courses of study designed specifically to educate young officers for a career in the profession of arms.

Although the Academy would be free to design its own courses, the committee 'envisaged that close links would be established and maintained between the academy and outside educational institutions. We believe this would do much to avoid the danger of the academy's becoming a closed establishment'. Notwithstanding concerns about isolation, the committee concluded that a tri-Service academy that was not linked to any university would have the freedom to shape its own curriculum and design its own courses. Such an institution needed, however, to offer bachelors programs that reflected the elements of a 'balanced and liberal' education that were organised to allow professional training to take place between semester breaks and during a consolidation year on completion of degree studies. Professional training would remain the responsibility of each Service and take place at Jervis Bay, Duntroon and Point Cook.

Fraser was pleased with the report, although he had some reservations on the 'balance of emphasis between the military and academic sides of the proposed Academy'.[7] His ministerial colleagues were, however, unconvinced. Senator Tom Drake-Brockman, Minister for Air and a decorated member of the wartime RAAF, was uneasy about academic standards and whether other Australian universities would respect the Academy degree. Jim Killen, Minister for the Navy and a wartime RAAF Flight Sergeant, was concerned about the preservation of single Service identities and whether the proposed course content would suit the Navy's needs. He was far from persuaded by the arguments being presented for the Academy.

Despite these anxieties, Fraser thought he had gained the backing of the Services, which were keen to see a reduction in the overall cost of cadet education. If his colleagues wavered in party room meetings, Fraser thought he could cajole them into endorsing his proposal when a decision was needed. He took the proposal to Cabinet on 29 May 1970, fully expecting to gain approval to proceed, if only because the tri-Service Academy was a cost-effective solution to an expensive problem. The initial capital cost was estimated at $1.3 million, with another $5 million needed when cadet numbers reached 700. Recurrent costs would be $3.3 million a year against the present cost of $4.3 million to run the existing separate colleges. But the submission did not receive the reception Fraser expected. Fraser's first biographer (Philip Ayres) reports:

> The Service Ministers were co-opted into the Cabinet for this item, though Fraser points out that at least two of them were persuaded by higher authority not to turn up [...] although Fraser had the backing of his Service Ministers he decided that it would be prudent to withdraw the proposal so that any decision might be deferred to a more propitious time.[8]

While he notes Fraser's attribution of the proposal's failure to his deteriorating personal relationship with Prime Minister Gorton, Ayres suggests the existence of more compelling factors, including 'distrust of over-emphasis on academic as against leadership training' within the Services, who wanted to manage their own people – separately. Howson was against a 'tri-Service college to replace Duntroon and Point Cook' because 'some boys wanted to get a degree and others were content with a diploma' but thought a 'Joint Staff College for the mid-career education of more senior officers' made sense. He went on:

> Point Cook had an affiliation with Melbourne University which provided a very much higher standard of degree than anything that could be obtained at Duntroon. I believed that this should not be sacrificed until we were going to obtain something better, and nothing in the present proposals provided that.[9]

Fraser was angry that his proposal was failing to gain Cabinet support and was embittered against Gorton. Although Fraser had rallied behind Gorton's bid to succeed the late Harold Holt as Prime Minister in early 1968, his support for Gorton had quickly ebbed away to be replaced with deep disappointment well before May 1970. The two men became and remained enemies. As Prime Minister, Gorton was not attentive to the benefits of Cabinet solidarity and found it difficult not to interfere in Defence-related matters. Gorton had served as a pilot in the RAAF during the Second World War and been injured in air combat operations over Singapore. His first ministerial portfolio, and the one he held the longest, was the Department of Navy. In fact, Gorton is widely considered to have been the best Minister for the Navy and the architect of its revival in the early 1960s, when the Australian fleet could have been reduced to little more than a coastguard. He was creative and energetic although interested more in practical outcomes than policy development. Refusing to be bound by conventions, Gorton was dedicated to the Navy's best interests and was loyal to its senior leadership. By way of comparison, Fraser was too young to have served in the Second World War (he was born in 1930) but had a long-standing interest in Defence and foreign policy issues. Fraser's ministerial style was not unlike that of Gorton. The difference was that while Gorton was inquisitive and direct, Fraser could be intrusive and dictatorial. Fraser frequently exceeded his authority as minister by involving himself in operational matters that were in the professional domain of the Service chiefs, including the conduct of the war in South Vietnam.

When I interviewed Gorton in 1991 for a study of the 1964 HMAS *Voyager* tragedy, the former Prime Minister restated his complete opposition to the whole notion of a tri-Service academy. It seemed his position was based less on the merits of the institution and more on his abiding disdain for Fraser, the individual most responsible (other than Gorton himself) for his subsequent downfall as Prime Minister in March 1971. Gorton was aware that the Navy and the Air Force were still unconvinced of the need for a tri-Service academy. He also thought there was too much emphasis within the Services on tertiary studies at the expense of vocational training. But there was one objection to the Cabinet submission that Fraser had not expected. There was a widely held view (which proved to be accurate) that the proposal had not been properly costed, and his colleagues wanted more detail. This was not unreasonable.

Cabinet did not consider the matter again until 14 October 1970. Gorton, John McEwan (the Deputy Prime Minister) and William McMahon (the Treasurer) led the opposition. Again the discussion was deferred 'until the matter can be further elaborated and the Chairman of the Chiefs of Staff Committee and the Chiefs of Staff can be available for consultation if necessary'.[10] Reports at that time indicated that Fraser was annoyed by the summary dismissal of the proposal but denied the rebuff was a motive for a later move against Gorton in the party room. Fraser believed that Gorton had pressured the Services to oppose the scheme.[11] Sir Leslie Martin said his report 'failed to survive the prejudices of the past'.[12] This was an unfairly critical view of those who thought that too many questions remained about the kind of institution Fraser proposed to establish. Service loyalties were strong and some commentators saw inter-Service rivalry as a form of healthy competition that helped to hone arguments for and against equipment acquisitions and strategic decision making. Lacking intuition at times, Gorton was unaware of the extent of Fraser's anger at the deferral of his proposal. The relationship between the two men was further inflamed by Gorton's continuing interference in the affairs of the Defence portfolio, principally his personal contact with the Chief of the General Staff, Lieutenant General Sir Thomas Daly. Without Fraser's knowledge, Gorton had tried to influence the Army's commitment to civic action in South Vietnam and the possible call-out of Papua New Guinean troops to suppress local unrest.

Fraser's relations with Daly had been difficult since the time he was Minister for the Army. They remained so when Fraser was Minister for Defence. Fraser did not support Daly becoming Chairman of the Chiefs of Staff Committee, and the position went to the Chief of Naval Staff, Vice Admiral Sir Victor Smith, who was appointed in November 1970. While Fraser's strong work ethic was universally admired, his personal style did not endear him to senior uniformed officers, who resented his insistence, his interference and his impatience. Fraser had a particular view of ministerial authority and its expression in relation to the Services. He dismissed criticisms of his style as the inevitable reaction of those who were trying to resist his leadership because they mistakenly believed the Services were beyond the reach of political direction. It was Fraser's perceived lack of loyalty that prompted Gorton to assure Daly that he would personally respond to any unfair ministerial attack on the Army. The media began to run stories that Fraser had lost the confidence of the Army's senior leadership. They certainly had concerns about his judgment and his willingness to support them publicly and politically, if necessary.

Personal disappointment over the failure of the tri-Service Academy proposal had certainly propelled Fraser towards a public falling out with Gorton. It was one of several matters that had led Fraser to conclude he could no longer work with the Prime Minister. For his part, Gorton felt that Fraser consistently misunderstood his role as Minister for Defence and unilaterally imposed his will on people and procedures. Following Fraser's resignation from Cabinet on 8 March 1971 and Gorton's replacement by William McMahon a week later as Prime Minister, Fraser's successor as Defence Minister did his best to bury the now-divisive tri-Service

The University as a unifying agent in Defence, 1967–1975

Academy proposal. The new Defence Minister was none other than the Academy's chief opponent – John Gorton. When Fraser eventually returned to Cabinet in August 1971, he served as Minister for Education and Science while David Fairbairn, who was awarded the Distinguished Flying Cross (DFC) during wartime service with the RAAF and was a former Minister for Air, replaced Gorton in the Defence portfolio. Fairbairn shared the Air Force's general lack of interest in establishing a tri-Service academy and made no effort to present revised costings to Cabinet.

With the politicians placing the Academy at arm's length, its chief advocate was the Secretary of the Department of Defence. Sir Arthur Tange personally esteemed the value of a tertiary education, and had completed his own degree part-time after business hours at the University of Western Australia while working full-time in the Perth branch of the Bank of New South Wales. Tange's biographer, Peter Edwards, reported that Tange was:

> excited by the study of logic, a training in clear and precise thinking. Forty years later, when espousing the cause of a broad and liberal tertiary education for young officers in the armed services, Tange was to startle the service chiefs by recommending that philosophy be part of the curriculum [at the tri-Service Academy].[13]

Tange believed that philosophy 'gets you asking questions'. While noting the need for Science and Engineering, Tange tried to steer the Service chiefs 'towards making room for their cadets to have their minds opened by the humanities'.[14] He felt that many within Defence confused training with education, 'between learning how to do things (often technologically complex) and reasoning about objectives and consequences and the fundamentals of society'.[15]

Edwards notes that after Fraser's cabinet proposal was rejected, Tange was 'working hard behind the scenes to ensure it did not disappear altogether'. In addition to believing it was important to socialise young officers from the three Services together, Tange spoke constantly of the 'need for more officers to have a "broad and liberal" tertiary education'. He was a keen advocate of all disciplines, not just Engineering and the Physical Sciences, believing that officers should be taught 'not so much what to think, but how to think, particularly how to think about broader strategic issues rather than more technical military matters [...] [with] a strong grounding in the history and politics of the region to which the Australian Defence Force was most likely to be deployed'.[16] Edwards explained that while 'some elements of the military grumbled that he was trying to exclude those in uniform from positions of influence, Tange was explicitly seeking to prepare officers to have a greater role in the preparation of policy advice on strategic and higher defence issues'.[17] Tange wanted Australian officers to be 'adept at handling political and strategic concepts as well as military challenges', like the American generals 'who had impressed him in Vietnam'.

With Tange's help, consideration of the tri-Service Academy proposal was quietly revived. The RMC Interim Council produced its own 'Special Report' on 7 July 1972 and claimed its concept for an academy was:

> readily adaptable to tri-Service conditions, if the Government should decide in that direction. The academy will bring together original and unique features in the spheres of both service colleges and universities. It is designed to preserve the heritage of the Royal Military College and consolidate the benefits of the joint experience of recent years during which the College and the University of NSW have been affiliated.

The University as a unifying agent in Defence, 1967–1975

Notably, civilian undergraduates would be excluded from the academy because of their potentially detrimental effect on discipline and the fear they would disrupt the College ethos. The University accepted this condition without comment, although some academics believed a small number of civilians would help rather than hinder the learning environment and felt the potential disruption (which was not canvassed in detail) was overstated.[18]

A copy of the revised proposal was given to the Minister for Defence, David Fairbairn, on 27 September 1972 but no action was taken. The Coalition Government was headed for electoral defeat and would lose the 2 December election. Without a Cabinet minister being personally committed to the project, the tri-Service Academy proposal had no chance of success. There was little point in UNSW appealing to the Department of Education for its support because the Department resented Commonwealth funding for tertiary education delivered at both RANC and RMC going directly to UNSW from the Defence budget.

The incoming Whitlam Labor Government was not committed to any particular course of action on Service education and had not gone to the electorate with a policy platform that included creating a tri-Service academy. With the new administration busy terminating national service and ending Australia's involvement in South Vietnam, it was not until 7 February 1973 that a copy of the RMC Interim Council report on the tri-Service Academy was sent to the Defence Minister, Lance Barnard, whose personal attitude to the proposal was unknown. Professor Myers explained to Barnard that:

> [the] commitment with which UNSW has undertaken in assisting in the education of service officers has been prompted by its desire to serve the Australian community in this very important area. If perchance you feel it might be able to assist

in the extension of this objective towards bringing into being of an educational establishment for services other than the Army, I hope you will not hesitate to give us the opportunity.

Myers reiterated the support of UNSW for a tri-Service academy 'if the government considered this appropriate' and reminded Barnard that 'the time remaining before the agreement [with UNSW] terminates (1977) is not much more than would be necessary to carry out all the actions needed to establish an autonomous institution'. Plainly, service education was not a priority for the new minister, who did not reply or even acknowledge Myers' letter.

Barnard eventually visited Duntroon and the Faculty of Military Studies on 24 May 1973. He was made aware of the University's anxiety about the future and the lack of a clearly articulated way ahead. The Minister received a letter from the Dean, Professor John Burns, on 5 June 1973 pointing out there had still been no follow-up consultation between the University and the Department of Defence about either the Academy or the UNSW agreement. Burns emphasised the importance of the 'tri-service proposals' conforming with the 'principles of academic autonomy' set out in the existing agreement, and assumed that the Labor Government would support the general thrust of Fraser's original proposal, given its public commitment to Defence unification and the benefits of 'jointery'. The Faculty were plainly becoming restless. Partly out of frustration with inactivity on the part of the Department of Defence, the academics argued that the Act to establish the autonomous institution Duntroon would become as part of the tri-Service Academy would best be administered by the Minister for Education, who exercised 'responsibility for other Commonwealth educational institutions, and that the funds for education and research purposes should be supplied through the Department of Education'.[19]

The University as a unifying agent in Defence, 1967–1975

This was the last thing the Defence Department wanted to hear. The matter was complicated enough without a second government department becoming involved.

There was another problem: the University's continuing role in the movement towards autonomy was unclear. The University assumed it should be an advocate for autonomy as the 1967 agreement was predicated on Duntroon becoming a degree-conferring institution in its own right. But there was a slight sense of ambivalence creeping into the University's dealings with Defence. As UNSW would not be the educational 'provider' after 1977, was it entitled to insist on a particular course of action? If it was not seen to support either autonomy for Duntroon or the creation of a tri-Service academy, the University could have been accused of either indolence or self-interest. To show no interest in autonomy for the proposed academy could be interpreted as a lack of regard for Canberra-based faculty and might imply that Kensington had abandoned them. Inactivity could also have been interpreted as an attempt to prolong the 1977 affiliation and to secure a permanent arrangement by default. Too much interest in autonomy might have been seen as an attempt by the University to walk away from its responsibilities to the Navy and the Army, or as failing to show a genuine long-term commitment to officer education.

In the absence of any clarity from the Department about its preferred option, Myers assumed the role of advocate for the academic staff and guardian of the original vision of an autonomous degree-granting institution. He believed UNSW had a responsibility to the staff who had been recruited to the Faculty of Military Studies since 1967 to see that an autonomous degree-granting body was established. Myers insisted that UNSW's original commitment was for a specified period, based on the Government's pledge to create an autonomous institution. Myers had not expended the

University's resources nor invested its time for anything less than autonomy. Assuming that autonomy would be achieved by late 1976, a memo from the UNSW Registrar (Godfrey Macauley) to Myers dated 18 May 1973 advised that the University should be prepared to award its degrees to students who enrolled in 1973 or earlier but that those who enrolled later would receive degrees from the institution in which they were enrolled at the time of graduation. The transition period was to be brief as the University did not want to be in the awkward position of awarding degrees for study undertaken mostly within another institution.

With uncertainty continuing, the Commandant of RMC, Major General Bob Hay, wrote to Myers on 12 June 1973 explaining that Barnard wanted to make a statement of the Government's 'intent' in the near future about a tri-Service academy, although Barnard 'is not aware of the total cost'. The Minister was waiting for the report of a Special Defence Department Committee and his Cabinet colleagues' endorsement of the proposal before making any public utterance. During the hiatus, Professor Richard Bearman (Head of Chemistry) wrote to Myers about autonomy, expressing private doubts about its wisdom. He thought 'the development of Duntroon to University status, even if in a tri-Services content, can only be justified economically on the grounds of importance to the national welfare'. Bearman attributed the difficulty of 'attracting outstanding academics' to the small number of staff and students, and Defence's disinterest in having 'an excessive number of officers with postgraduate degrees' and the small number of civilian postgraduates. Contrary to some opinion, he insisted, 'prejudice against us by the academic community does exist'. Because of these 'severe difficulties', he thought it was 'essential to compensate for our disadvantages through continued good equipment support, professional assistance and favourable staff/student ratios'. He argued that generous concessions needed to be made if he and his staff were to:

> maintain university standards in a unique institution. I do not believe we should have the best of both worlds. On the contrary, to give us university status, as we have been promised, will require special care to avoid the worst of both: we must steer clear of the Scylla of economic ruin via funding through the AUC [the Australian Universities Commission] and of the Charybdis of academic ruin via complete control by the Defence Department.

Commenting on preparations for autonomy, Bearman claimed that:

> a vocal segment of the academic staff believe Duntroon should be completely absorbed into the Kensington fabric prior to the end of the Agreement. Although the achievement of autonomy will tax your negotiating skill to the limits, I have the firmest opinion that we should not now adopt Kensington procedures. They are more suited to a NSW institution than to a Commonwealth one located in the Australian Capital Territory and in no way do they reflect our special needs.

As the renewed push for autonomy gained momentum, academic freedom at Duntroon was tested for the first time in early June 1973. Professor Brian Beddie floated the idea of a seminar to be held on self-government in Papua New Guinea following its independence from Australia. It was to be a joint New Guinea Council, ANU and RMC initiative, with the Army being approached for assistance with some minor travel and staging costs. After starting work on arrangements for the seminar, Beddie was advised by a Departmental official that the subject of the seminar was politically too sensitive to be held at Duntroon. In a personal call from the Defence Minister's private secretary, Beddie had been asked whether he was willing to suppress the fact that his department at RMC 'was a joint sponsor of the seminar'. Beddie responded with an emphatic 'no'. In a later telephone call, Beddie agreed to keep the initial approach from the Department of Defence confidential and make no public reference to it. Beddie was also determined to resist any requirement that he seek Departmental approval for future seminars, later complaining to Burns that: 'I do not wish to hold large public and highly publicised conferences at Duntroon [...] Although I made clear that the seminar I proposed to hold was to be "small" and "closed", the authorities in the Department of Defence constantly referred to it as if it were to be a public meeting'.

Beddie, who was not content with the proposed seminar arrangements, asked Burns to refer the matter formally to the Minister for the Army. While Beddie was in Papua New Guinea on an unrelated matter, a high-level meeting was held involving senior officers at Duntroon, the University and the Department of the Army. A compromise suggestion was then put to Beddie: could the event happen at RMC but a 'special briefing' be given to any members of the press who expressed an interest in reporting the proceedings? They would be advised that the seminar would be an academic event and would not involve discussion of Army operations or specific details of Government policy. Barnard separately advised Myers that because preparations for the gathering were quite well advanced: 'I agree to the seminar going ahead [...] I have asked the Department of the Army, at an appropriate time, to give a discrete briefing to the press, emphasising the academic nature of the seminar'.[20] Notably, Barnard told Myers he was aware an Army officer might give a presentation but made no comment about whether he condoned it. Beddie was willing to compromise but feared the wider implications of a 'special briefing' on his work in the Faculty of Military Studies. This episode highlighted some of the complications arising from the structure and composition of the Duntroon Faculty, which included a number of serving Army officers such as the Commandant and the Director of Military Art. Although the role of uniformed officers was to provide information about cadets and to answer questions about their training, their participation in Faculty meetings meant they were involved in Faculty decisions that were determined by a vote. As the majority of the Faculty were civilian academics, it was possible for the military members to be embarrassed by a majority resolution of their civilian colleagues. Thankfully, the need for any decision making by vote did not arise in this instance.

While UNSW funds were being arranged to replace those the Army would not be offering, Beddie promptly cancelled the seminar and said he would confer with his colleagues about the reasons. Beddie was distressed that one of his Masters students, Major Paul Mench, who was to be a key contributor to the seminar because of his recent service in Papua New Guinea, might not be permitted to participate. With Mench's contribution uncertain, Beddie felt he had no choice but to cancel the event. Plainly angered by the exchange, Beddie asked the Commandant to advise him 'more precisely on the restrictions that may be applied by the Department of the Army to the holding of seminars at Duntroon'. In subsequent discussions, Beddie was advised that he was free to conduct seminars but that Mench's role was circumscribed because he was a military officer and that the Army was applying the rules relating to 'any serviceman taking part in public discussions on controversial or other important issues involving Government policy'. As Faculty Dean,

Burns was sympathetic to Beddie's frustrations. There was some genuine concern about the fate of postgraduate work at Duntroon if the proceedings of seminars and the outcomes of research could not be published or even presented in public forums. He was also worried that Duntroon students might come under close supervision as a consequence of their work on controversial subjects. Beddie accepted the need for regulations and restrictions but thought they should be 'interpreted liberally, especially in relation to postgraduate students, to prevent the development of an intellectually closed atmosphere in which Duntroon graduates might fail to develop certain of the essential qualities of other university graduates'. He also deplored 'the apparent failure by the Department of the Army, even after some years of experience with the Agreement with the University, to appreciate the need to treat the Faculty of Military Studies as an academic rather than a public service organisation'.

After meeting privately, Beddie's colleagues wanted clarity on their academic freedom within Duntroon and their entitlement to discuss and publish on matters of academic interest. Furthermore, these discussions were relevant to the culture of the proposed autonomous body and the kinds of conversation it would promote. The RMC Commandant was quietly supportive of Beddie and appreciated the principles he was seeking to defend. Censoring students, or making them fearful of censorship, was Beddie's abiding concern. While Beddie was absent overseas, Dr Hugh Smith arranged a seminar focusing on 'Problems of the Australian Military Involvement in an Independent PNG'. It was promoted internally but not externally. Mench presented a paper that he had previously submitted for official clearance. It was slightly modified at the request of the Department.[21]

By this time, Mench had sought access to official documents for a thesis entitled 'The Role of the Defence Forces in PNG'. Mench

The University as a unifying agent in Defence, 1967–1975

had served for two years with the Pacific Islands Regiment in Port Moresby and maintained close friendships with a number of PNG officers, including the future Commander of the PNGDF, Ted Diro. Access was refused even though Mench believed his request would receive favourable treatment after a similar approach he had made was approved for a related but minor work. Mench was dissatisfied that access was denied on the grounds that he was bound by the Military Board's instruction on public comment. He was fully aware of his responsibilities and did not need to be reminded. He had also agreed to submit his completed thesis to the Army for comment and clearance before any public release. In response, the Army stated that the provisions of the 30-year rule governing official records applied to Mench because his research could not be considered official. His request was, therefore, denied. The University felt that such restrictions would deter rather than encourage inquiry, would hamper the overall work of the Faculty and constitute a lost opportunity for the Army with respect to what the Faculty could provide in terms of thoughtful analysis and creative scholarship. By this time (March 1974) Beddie wrote to the Army to implore a change of heart given its likely interest in deploying the post-independence PNG Defence Force. Noting that it had recently funded two Defence fellowships at the Strategic and Defence Studies Centre at ANU, he could not understand why the Army was being so obstructionist with Major Mench. Beddie had good reason to lament:

> I know of civilian students at ANU who, though not subject to official controls, have had relatively free access to official and classified documents and I examined one PhD thesis in which the student was permitted to supply a confidential appendix citing official sources. I would also point out that the public servants who have been granted leave to pursue

postgraduate studies at ANU have [...] been allowed during their leave to consult the files in their departments.

Beddie mentioned that Mench had already prepared an article based on his research and was waiting for Defence to clear its contents. After ten weeks, no decision had been made. Sir Arthur Tange then wrote to Beddie and pointed out that Mench had served in the Papua New Guinea Defence Force Headquarters and 'would be known to have so served'. While there were reasons to support publication, Tange conceded, 'I see that strong arguments have been advanced to suggest that the views expressed may well be associated with official Australian views or even military plans in a manner injurious to Australian Government relations with the Indonesian Government or the PNG Government'. Tange told Beddie that although the outcome had not yet been determined, his staff would be 'exploring further' whether they could help Mench by 'suggesting adequate emendations'.

This problem was becoming more serious. At Kensington, Professor Al Willis wrote to Myers summarising the situation. His aim was to avoid a showdown. He told Myers that, in his view, 'a) a confrontation on the issue would be counterproductive and b) the issue is less real than it can be made to sound'. But there were delicate issues involving the provision of documents and whether seminars that included uniformed officers needed to be 'closed' and made confidential. John Burns felt a great deal could be learned from the matter and suggested that:

> a statement should be prepared of principles which must be observed when the tri-Service academy is set up. Such a statement would, in my view, be for internal use only, a kind of checklist to guide those who are involved in the

negotiations to set up the new academy and to ensure that at least the known difficulties are avoided.

At the outset, Beddie would have been satisfied with a single authoritative statement from Defence, noting that what happened within the Faculty was academic in nature and in no way purported to reflect Government thinking or official policy. This episode had been poorly handled by the Department and naturally left some academics anxious that freedom was more honoured in principle than practice. Many academics now believed that autonomy was mandatory if the integrity of their research and the standing of their publications were to be preserved.

For different reasons, Myers was becoming frantic. By early 1974 Barnard and his staff still seemed to be unmoved by the looming end of the 1967 affiliation agreement and appeared unable to fathom the complexities of establishing the tri-Service Academy. It was time to raise the place of the Air Force in the negotiations. The Vice-Chancellor explained to the Minister that there was no formal or comparable agreement between Melbourne University and the RAAF, although the Commonwealth Government gave an undertaking to fund the positions of staff needed to deliver courses at Point Cook leading to the award of Melbourne University degrees. He also pointed out that 'Melbourne University would not wish to "absorb" the academic staff now at Point Cook and that every effort would be made to negotiate such staff into any new tri-Service institution'.[22] Myers then revealed:

> From recent discussions with my colleague Professor
> Derham, Vice-Chancellor of the University of Melbourne,
> I have gained the impression that he feels it must fall to
> UNSW (rather than to the University of Melbourne) to do

> most of the University-level thinking about the future of the tri-Service institution. Naturally he is interested in the outcome of discussions about the academic staff at Point Cook but he is not so concerned about future organisational matters. Knowledge of Professor Derham's views has encouraged me to think carefully about the ways in which we at UNSW may be able to assist you and the Government.

Derham's patience with Defence had essentially evaporated. The Academy proposal seemed destined to fall victim to inactivity. It would not happen by itself but the Department seemed unmoved. Derham was content to see Melbourne University's relationship with the RAAF Academy continue indefinitely. Myers was destined to play a near-lone hand in holding the Department to its pledges on autonomy, at least for the Faculty of Military Studies at Duntroon. He then offered his own suggestion, embodying a significant new element in his thinking. The Department and UNSW might consider:

> negotiating an agreement along the lines of the present agreement between the Minister for the Army and UNSW for the development over a few, say five years, of a tri-Service institution. Under such an agreement the Minister for Defence and UNSW could agree that there would be set up what I shall refer to as a 'Tri-service Interim Council' with agreed membership. This body could be responsible for advising the Government and UNSW on the academic content of any proposed new tri-Services program and on any other matters pertaining to the organisation of the new institution and its material development [...] Perhaps the greatest advantage of the suggestion I am making is that [...]

The University as a unifying agent in Defence, 1967–1975

UNSW could continue as the interim employing authority and could, in particular, be the employer during the planning stages of the 'academic head designate' selected by the Tri-Service Interim Council. We are already committed to solving the transfer problems of all of the academic staff at Duntroon so the task would not be much increased by the small number of additional staff members we have for the time being.

Did Myers privately want UNSW to have a permanent role? The answer was an emphatic 'no', as the affiliation had always been transitional and this remained his public and private position. He had adopted this stance for the sake of the new institution. But he was offering the University's assistance to the new institution until at least 1979, when it 'could become completely independent operating under a new governing body'. But if it didn't, UNSW was well placed to continue with the agreements already in place. He concluded his letter to Barnard by stressing he was writing in a personal capacity and was not committing the University Council to any specific action. He was merely making a suggestion, but added:

> The University sees it as a duty to foster the development of educational facilities which will lead to the adequate preparation of professionals for the Defence services. While my senior and academic administrative colleagues and I in the University have been looking forward to the time when we would be free of the additional load which arises from our commitment at Duntroon, I confidently assure you that we would not wish to leave the venture in an unfinished condition and that we would be prepared to carry the extra burden for a little longer if the Government felt this to be desirable.

Myers later told the RMC Commandant:

> My approach is to find the best way to achieve autonomy for RMC within the tri-Service context and also having regard for the Government's time frame for ADFA. Whether this would involve a fresh agreement or an appropriate modification of the present one (or both) cannot be stated at the moment.[23]

It was now up to the Government to make a decision. A clear statement of intention was well overdue.

5

The Academy commitment, 1974–1976

After some pleading from the University, Federal Cabinet finally gave in-principle approval to the creation of the Australian Defence Force Academy (ADFA) on 19 March 1974.[1] The Academy would begin operating in 1979; the precise timing of its opening would depend on the pace of building works. It was envisaged that the first full year of teaching – meaning all four years of an Engineering degree – would be 1982. Barnard thought this was too protracted – more than eight years from the original decision to its realisation. One thought was to establish the Academy as a three-campus entity until the buildings were completed – accommodation being the determinant of progress. He wanted the three colleges 'brought together' by the beginning of 1978 as part of an 'autonomous degree-granting institution' that would employ all staff employed at Jervis Bay, Duntroon and Point Cook. Although the original press release mentioned Albury–Wodonga as a possible site for the Academy in line with the Whitlam Government's policy of public sector de-centralisation, it was never a serious proposition and a detailed examination of the site's merits was never conducted.

As the UNSW–RAN College agreement would also expire in mid-1977, the Navy would continue to send second- and third-year students to Kensington until the Academy was ready to receive them. To avoid the possibility that the Academy might continue to operate at three separate locations, Barnard proposed drafting legislation and appointing a Development Committee to ensure the Canberra site was built and the unification of academic staff achieved. The Development Committee would include a substantial number of stakeholders – the three Services, the central Defence organisations, the single-Service colleges and the universities – and would be intentionally large. The stakeholders were also aware that the educational gap between the uniformed and civilian staff was widening, with 70 per cent of the Australian Public Service Division 1 and Division 2 officers possessing tertiary qualifications while fewer than 15 per cent of equivalent-level uniformed officers were university graduates.

The Royal Military College (RMC) Interim Council discussed the Minister's announcement of the Cabinet's decision on 21 March 1974. It resolved to 'request that the University and Department of Defence examine the implications of the Minister's statement for the Agreement [with UNSW] and the objective of achieving autonomy for RMC'. The Vice-Chancellor addressed the Faculty on 26 March 1974. Professor Myers explained that the University Council had discussed the statement and resolved:

> that in the light of the announcement by the Australian Government that it intends to set up an Australian Defence Force Academy to begin operation in 1979 at the earliest, the Council authorises the Vice-Chancellor [...] to explore ways in which the intention of the Agreement between the Minister for the Army and the University 'to develop the College into a separate, autonomous, degree-conferring body' can now be achieved and to make appropriate recommendations to Council.

The Academy commitment, 1974–1976

Barnard's announcement included concrete steps and practical actions but it raised a series of questions and provoked a raft of concerns. Why did the Minister not use the word 'university' in relation to the new institution? When he said that new courses would be career-oriented, would this turn academic studies into vocational training? The new institution would be an 'Australian Government instrumentality' but could such an entity operate as a genuine university? What role would existing universities, especially UNSW, play in developing the new institution? And would those presently employed at the single-Service colleges have assured employment in the new institution? The President of the RMC Academic Staff Association, Dr Roger Thompson, informed Myers that the Association:

> 1. deplores the absence of the word 'University' from the proposed title of the new Tri-service institution, which omission is contrary to the unanimous recommendations of the twelfth meeting of the [RMC] Interim Council 2. inclusion of the word 'University' in the title would be an important guarantee that the present university education of officer cadets will both continue and be seen to continue in the future.[2]

Dr Hugh Smith, a lecturer in the Department of Government at Duntroon, tried to shift the conversation towards vocational considerations in a two-part article published in the *Canberra Times*.[3] After noting that military careers were becoming shorter under the influence of more generous superannuation provisions, the 'prospect of a recognised degree is a good guarantee to the potential recruit that he is not painting himself into a corner where his career is concerned'. He also thought the offer of a degree would attract 'officer cadets from a wider social and intellectual base', thereby avoiding drawing only from 'a narrow group of conservative, traditionally minded young men'. To ensure the cadets made a 'mature commitment to their profession', Smith thought they 'need to feel that they are not being denied the education which their peers in universities are receiving. They also need to know they are not being compulsorily shut off from the rest of society'. He stressed how important it was for the new Academy to have university status to preserve teaching standards, protect intellectual freedom and promote lively debate. While the new institution was being referred to as an 'Academy', it needed to function as a university to ensure the Services were not distanced from the host society from which its members were drawn. It was for this reason that Smith shared the general concern that the full title of the tri-Service Academy had to include the word 'university'. He explained that:

> [the] question is not merely one of names but involves a number of substantive issues [...] Given the objectives of high academic standards and of integration in society, there seems great merit in the suggestion that instead of setting up a monolithic academy there should be an independent university within the organisational walls of the larger academy.

A *Canberra Times* editorial was less concerned with the issue of the new institution's name than its location within Australian public life. Its question was:

> whether it is desirable to set up a new tertiary institution reserved exclusively for the people who will constitute the officer corps of the Australian defence forces, or even an

institution that would open its doors also to the few – public servants and specialists – who have a professional interest in the military sciences.[4]

The editorial argued that because military operations were increasingly political in nature, those in uniform needed to be an integral part of the 'wider national society'. Thus, junior officers should not be 'insulated during their training years from civilian moods and questionings, in the rarefied atmosphere of a military institution'. Having concluded that 'Duntroon has ably demonstrated that a military school can function successfully under the wing of a university', the editorial suggested that the Academy should be open to civilian students but doubted whether the ANU was 'the institution to take on the burden'. It is apparent that the editorial writer seems to have misunderstood the Government's intention, which was to create a new institution and not an affiliate of another. But the newspaper warmly welcomed the development of 'the second university which Canberra can expect'.

While Barnard's vision of the Academy was being deciphered, there were immediate practical problems to resolve. RMC was blighted by the persistence of temporary arrangements. Although the facilities at Duntroon were poor and deteriorating, repair and maintenance work had been delayed until a decision was made about the new Academy. The Army had hoped to avoid spending money on buildings that might not be used. The Commonwealth could pay for an upgrade to the Duntroon facilities as part of the Academy project. But this 'wait and see' approach was shortsighted and highly dangerous. The Faculty Dean, Professor Burns, was increasingly concerned that RMC staff would start to look elsewhere for employment because their working conditions would not improve until the late 1970s.

After yet another administrative hiatus, this time caused by the May 1974 double-dissolution election, Myers wrote to Barnard expressing personal delight that Barnard would continue in the Defence portfolio and again offering to assist with establishing the Academy. On this occasion he suggested appointing a UNSW academic to serve as the Development Committee's full-time Executive Officer. Barnard did not reply. Undeterred, Myers continued to work on an extension to the 1967 affiliation agreement to cover the period from its expiry in late 1976 to the date the new Academy was established and operating. The University Council had empowered Myers to negotiate an extension and a period of five years was discussed. Despite all parties agreeing to such an extension and conducting the negotiations in an atmosphere of goodwill, there was continuing unrest among the Faculty about the long-term future. The Cabinet decision and the lack of detail in Barnard's subsequent statements had added to anxieties. The mood among faculty at Duntroon led the Pro Vice-Chancellor, Professor Al Willis, to write Myers a confidential letter on 21 June 1974:

> I am told there is little enthusiasm for ADFA from the RAAF, service or civilian, and the Department is extremely concerned about what may happen at Point Cook [...] the Department fears that academic staff will leave for other jobs, leaving too few to carry on the physics degree. Mainly because of the apprehension about the Point Cook Academy, the Department is looking for some way of bringing forward the development of ADFA. This would involve setting up the new Academy without physical facilities but with the legal authority to accredit the degrees of the other colleges. On this limited basis, the Academy could be in operation by mid-1977, when our own agreement ends. There is some fear, however,

The Academy commitment, 1974–1976

that the sun-and-planet arrangement might be seized on as a permanent solution to tri-Service officer training.

The key date remained mid-1977. Myers was worried that there was still much to be resolved before then. He met with Ferdinand Mahler, First Assistant Secretary of the Department of Defence, on 2 August 1974 to try to hasten a number of practical matters. Myers told Mahler that 'the essential academic elements of the Academy already exist [...] the task is to form those elements into a single institution, in due course to be located at Duntroon'. They agreed there was merit in creating the Academy by legislation (probably enacted in 1977) before the buildings were available, which would not be before 1979. Myers thought the Academy could operate temporarily in the three separate college sites, Jervis Bay, Duntroon and Point Cook, although if this arrangement functioned well he shared the fear that there might be pressure to abandon the new building project in Canberra. He relayed one other complication to Mahler: he did not want to take on the Point Cook staff and did not believe they would want to become 'interim' employees of UNSW. Myers and Mahler met again on 13 August 1974. They canvassed possible names for the new institution. Myers told Mahler:

> You may recall I mentioned 'Duntroon University' or a possible double-barrelled name eg., 'Duntroon University – the Australian Defence Force Academy'. Recent events incline me strongly now to support the use of 'University' – and in any case I am sure there would be very strong objections from the Faculty of Military Studies staff, and perhaps Melbourne University staff too, if it were not used.

Willis visited Duntroon on 4 September 1974 and addressed the faculty. He and the Commandant convened a special meeting of academic staff and were greeted with what Burns described as 'a widely pervasive spirit of pessimism' about the future and their prospects.[5] It was clear the Academy would be a military establishment rather than a university campus. Major General Hay explained that 'officer-like qualities could only be practiced by living within a disciplined military environment where military and personal characteristics were watched and developed'. Faculty and students were then invited to express their opinions.

The first speaker was Major Paul Mench, representing the postgraduate student community within the Faculty of Military Studies. They had met and had passed a very long and detailed resolution. In essence, the postgraduates were concerned that the alternative models for providing officer education had not been adequately assessed. Further, there had been 'inadequate consultation with staff, students and the Australian public on the shape of the important new institution'. They were also concerned that 'the new institution should have genuine academic autonomy with an appropriate relationship to both the Departments of Defence and Education'. Finally, they proposed the appointment of a Commission of Inquiry consisting of academics, Service officers and community leaders to conduct hearings and receive submissions, and to report before the end of 1975. In what constituted a virtual vote of no-confidence in the Department of Defence, the postgraduates recommended that the proposed Commission's report be made public. Mench conceded that the appointment of an Academy Development Committee addressed some of these concerns, but the views of students – undergraduate and postgraduate – had not been canvassed. Furthermore, he retorted, the staff at the Service colleges had not been involved in any of the discussions. The postgraduates thought it was

The Academy commitment, 1974–1976

'potentially disastrous that the new institution should be established by what amounted to an administrative fiat when important issues of academic and military autonomy [...] remained in serious doubt'. Mench also spoke strongly of the need for the new institution to have broad-based community support to ensure it helped to 'enhance the status' of uniformed officers in Australia.

Students and staff believed they had been shut out of the development process, leaving them feeling alienated and suspicious. The bulk of the actual liaison within Defence was conducted by the Chief of Army Personnel (Major General Sandy Pearson, a former RMC Commandant), the Special Deputy Secretary of the Department of Defence (a Mr Livermore) and the committee secretary Len Hume (who did most of the 'hands on' work while reporting to his departmental superior, Ferdinand Mahler). Hume was an important figure. He was the departmental 'continuity man' throughout the Academy's long gestation. Hume joined the Australian Army in 1943 and served with the 2/31st Infantry Battalion at Balikpapan before becoming part of a garrison for Japanese prisoners of war in Rabaul until demobilised.[6] After the war he joined the Department of Army until he transferred in 1965 to the Department of Defence. He was minute secretary for the Martin Committee and provided administrative support to all subsequent committees concerned with the tri-Service Academy until his eventual retirement in 1984. His gradually expanding knowledge of proposals, problems and people was an invaluable resource to academics and uniformed officers, as was his comprehensive personal archive of every official document relating to the project dating from 1967.

Plainly, the membership of the Development Committee and the attitude of those serving in the Interim Council (who would appoint the Vice-Chancellor Designate) were crucial to the shape of the new institution, since the Minister's office was still vague on details. Barnard's concern was more to establish a tri-Service academy and to standardise training than to settle on a particular model for delivering education. In speaking with his Canberra colleagues, Professor Willis explained that there was no 'standard procedure' for establishing a University and no extant blueprint on how it should be organised or structured. Willis tried to reassure the Faculty, however, that he thought it:

> highly unlikely that the new institution would differ greatly from the present organisation at Duntroon where a faculty structure existed, linked with an orthodox university, where academic freedom was recognised, and where courses had been developed to postgraduate level [...] Duntroon was academically sound, it had an established relationship with the military and an agreement to preserve certain standards.

He pledged that the Faculty would be most influential in shaping the internal structures, although their role to date had, in fact, been negligible.

Professor Brian Beddie was not as positive as Willis about these developments, explaining that the 'the establishment of a military university would be a difficult undertaking'. Beddie did not think that the Faculty of Military Studies was actually 'winning recognition from other universities. The Faculty had made good progress but there was a feeling that this progress was now halted'. He contended that 'considerable opinion [existed] in the Faculty which wanted university education in the services backed by an outside university' and claimed that 'fears existed that if the University of New South Wales were not behind the Faculty, academic freedom and quality of performance would not survive'. Dr George Gerrity endorsed Beddie's laments about independence, noting that the new institution would be established within Defence and not the Department of Education. David Daw suggested 'there was more than one acceptable means of developing the new institution but the concept behind ADFA was not one of them' and was deeply concerned about staff conditions of service and guarantees about their preservation in the new institution. Professor Harry Green shared Beddie's general misgivings, explaining that the new institution 'would not be a university nor a College of Advanced Education' and feared that Defence would take too long to make decisions affecting academic programs. Associate Professor Boyd Dempsey declared his support for Beddie's view that the best model was a Faculty supported by an existing university. Dempsey believed the problem they were facing derived from negotiations in 1967:

> when the unfortunate phrase 'autonomous institution' was coined. The Army had reached this situation because earlier arrangements had been rejected by the Government. The college had been hell-bent on gaining autonomy since 1967 but history had shown that it was better served by being part of the University of NSW. An extension of what was already in existence would be better than ADFA.

But John Tardif, Senior Lecturer in Chemistry, took a different view. He thought that the integration of military and academic programs had achieved substantial gains and that separating training from education was a mistake.

Willis believed it would indeed be possible to extend the 1967 agreement but that this would involve negotiations that the University Council would need to approve. Myers had sought this authority and was preparing an extension agreement for consideration. Associate Professor Sid Hodges noted that previous proposals involving the ANU had involved that institution virtually 'taking over' the Royal Military College. He felt this was 'a wholly unacceptable arrangement and would view with concern any proposal which would lead to the College being absorbed into a university in Canberra'. Hodges believed that Duntroon had been 'favourably treated by the University of NSW' and hoped the agreement would continue indefinitely.

Professor Geoff Wilson, the new Faculty chairman, concluded the consultation by drawing attention to the composition of the Academy Development Committee. While he questioned the entitlement of Sydney University to have a representative because it had 'no special commitment to officer training' and was generally hostile towards UNSW, there was merit in Melbourne University being represented because of its association with Point Cook. He stressed that UNSW 'has a clear commitment to the setting up of the new institution and engaged staff committed to service that institution'. He felt that UNSW should be the majority voice from the education sector on the Development Committee. The Commandant then closed the meeting. He suggested that academics were being 'unduly pessimistic' about the new institution and said they would be consulted as part of the planning process while the Vice-Chancellor of UNSW would have direct communication with the Minister and the Department at every stage.

In conveying the essence of this discussion to Myers in a letter dated 6 September 1974, Wilson said the Faculty shared the lament of the postgraduates that not enough attention had been paid to alternative models of delivering officer education.[7] For its part, the Faculty did not share the enthusiasm of the postgraduates for a formal inquiry, believing the Executive Committee of the RMC Interim Council and the Academy Development Committee could examine the matters that were causing concern. He also urged the

The Academy commitment, 1974–1976

Vice-Chancellor to emphasise the value of the Faculty's achievements over the previous seven years in establishing customs and building traditions. Wilson hoped these would not be discarded in the Academy transition. Starting the new institution from a 'clean slate' would 'negate most of the work put into the development of our Faculty'.

Anxious about what he had heard at the meeting and noting disinterest in the many alternatives, Beddie produced his own detailed paper on possible models of educating young officers. He began his paper with a warning:

> The Government's scheme for the tri-Service education of cadets envisages that functions previously performed by three single Service colleges and two universities can co-exist within a single degree-conferring academy. That an arrangement of this kind can work might seem to be confirmed by pointing to overseas institutions (i.e., RMC Kingston) which carry out both military and academic functions. Australian planners will, I believe, be seriously misled should they attempt to take another institution such as RMC Kingston as a model of the way in which military training and instruction can be combined with academic education.

He went on to note that academics at Kingston were public servants responsible to the Canadian Minister for National Defence. Because UNSW staff recruited to the Faculty in 1967 and thereafter were assured employment in the new institution with comparable conditions, it was not possible for the new institution to have the same status as Kingston. In effect, a decision had already been made to preclude that possibility, Beddie concluded.

Beddie also contended that if the Commandant and the

academic head were to have complementary rather than competing authority, they needed to be the heads of separate entities. One could not be subject to the other. The military could not regulate curricula or determine academic staff appointments but neither could the University Council claim any authority to comment on operational training or to influence military discipline – not that it had any interest in doing so. Drawing on his experience of the 'PNG seminar' controversy, Beddie pointed out that because academics were ultimately responsible to the University Council (which was itself responsible to the Commonwealth Department of Education), the Defence Minister could not claim authority over the staff of an entity within which he had no formal standing. If the Minister did claim authority, Beddie pointed out 'the unfortunate consequence which will follow from this arrangement is that the academy will be cut off from its lines of communication with the Minister for Education and so isolated from all other degree conferring bodies in Australia'. While it was possible to pre-empt the likely areas of conflict and controversy, he thought 'it would be pointless to predict in which particular direction it will go most wrong'. Beddie expressed the concerns of many when he observed that the Department of Defence:

> is not one that is qualified to protect academic autonomy and freedom, to meet the special needs of a small university institution or to make reasonably swift decisions on academic matters which, when placed in the context of national security, will always appear to be of peripheral importance.

Beddie's solution was to form two separate institutions: the 'National Defence College (NDC)' and the 'Australian Defence University (ADU)'. He did not 'rule out the possibility of seeking an agreement with a university under which it would set up special faculties at Duntroon, an arrangement analogous to that now in being with UNSW', which he thought remained the best option. Contending that 'such an arrangement would not be acceptable to any civilian university' – a curious notion given that Willis had said an extension of the extant agreement was possible – the ADU needed its own separate campus, albeit co-located with the NDC. He conceded the NDC would need to be 'admitted into the community of universities and obtain their enthusiastic support'. There were two main challenges to be addressed with this proposal. First, would the ADU 'be sufficiently responsive to the needs of the Services'; and second, could a 'body so small and so exposed to military and bureaucratic influences maintain its autonomy and academic freedom?' What Beddie neither asked nor answered was the question of who would pay for these institutions. If Defence were paying, it would want control of the programs it was funding. If the Minister for Education was the source of funding, it was highly probable that the Education Department would take a minimalist view of what it provided, quite apart from having no capacity to assess or to fund specific Defence-related projects.

Beddie's protégé, Hugh Smith, was also troubled about the status of the new institution, contending that whether it was an academy or a university would hinge 'on the importance to be attached to academic excellence'. He explained: 'it is not that the label "university" automatically confers high standards but that it creates favourable conditions for developing and maintaining them'.[8] He included in these conditions recruiting able lecturers and attracting good students, the dynamic interaction of teaching and research, the involvement of professional associations and the application of external validation, an openness to ideas and a willingness to debate. Smith was one of the first academics to question the Academy model, fearing a loss of academic respectability for his faculty colleagues and a diminution of academic standards for the programs on offer.

On 26 November 1974, Myers' patience with the Department of Defence was exhausted: 'I must now express my concern that the Government has not yet indicated what role, if any, it wishes the University to play in the creation of the new academy'. He reported the existence of 'considerable disquiet [...] about the lack of progress in establishing the "autonomous degree-conferring body" which the Government and the University have undertaken to create'. He feared the lack of action would lead to 'transfer of some issues to the public domain' by which he meant Duntroon staff talking to the press about the Government's inability to make a decision and the absence of detail in the decisions it had made. Barnard finally responded on 4 December 1974. He thanked Myers for bringing to his attention 'disquiet among members of the University, particularly those at Duntroon, concerning ADFA'. Barnard said he wanted:

> the university to participate fully in the development of the new academy [...] not only to ensure that the University can discharge its obligations to the staff of the Faculty of Military Studies but because of the wealth of experience the University can bring to bear on what will be a complex planning task.

Barnard was 'personally in favour of establishing the academy as quickly as possible' and said he would bring draft legislation into parliament 'during the budget session in 1975'. One of the practical problems that needed to be resolved first was the fate of the RAAF Academy staff. Barnard met with Myers and Sir David Derham, the Vice-Chancellor of Melbourne University, on 21 March 1975 to consider the matter. They resolved that staff who transferred from Point Cook to the Academy when it opened would receive the same contracts as those offered to UNSW staff already employed at Duntroon.

On 17 April 1975, Barnard delivered the long-awaited ministerial statement on the Australian Defence Force Academy. The details were now to be disclosed. Barnard reiterated that the Academy would 'replace the single-Service cadet colleges' based on the recommendations of the Martin Committee and the identification of certain 'long apparent advantages', of which he highlighted three. The first was better promoting 'the spirit of inter-Service cooperation that is essential in the prosecution of joint tasks'. The second was the 'need for a greater proportion of Service officers to have a tertiary education'. The current arrangements were, he said, inadequate. The third advantage, he explained, was having an Academy 'established as a tertiary institution in its own right [...] [leading] to economies in the use of resources [to] provide a sound basis from which we can plan officer education for the future'. The legislation needed to establish the Academy, which would be located on 'an area adjacent to RMC Duntroon', would be introduced during the 1975 budget session.

Barnard explained that he had conferred with the Vice-Chancellors of Sydney, Melbourne and UNSW, and had discussed various options with the Minister for Education. He announced the formation of the Academy Development Council. This body would 'remain in existence until replaced by a First Council for the Academy'. Barnard approved the Council's terms of reference on 6 May 1975. Its role was to undertake planning for the 'establishment of the Academy as a body corporate and an autonomous degree-granting institution' to the point where its responsibilities could be assumed by the 'First Council for the Academy'. The Council would be assisted by an Academic Planning Committee consisting of several Course Groups, an Academic Facilities Advisory Group, and an Academic Administration Committee that would draft position papers and ensure the Academy would not be an actual drain on Commonwealth revenues for at least five years. The Council would be chaired by Sir Henry Basten, former Vice-Chancellor of Adelaide University (1958–67) and Chair of the Australian Universities Commission (1968–71). Basten's appointment was welcomed by Myers, who described him as a 'most acceptable colleague', although some within the Department of Defence, such as Len Hume, felt that Basten was probably 'past his prime' and incapable of giving firm direction when it was needed. The other Council members included General Sir Francis Hassett (Chief of Defence Force Staff), Sir Arthur Tange, Professors Myers and Wilson (as Chairman of the Faculty of Military Studies) representing UNSW, Professor Victor Hopper (Professor of Physics since 1961 and Dean of University Studies, RAAF Academy) and representatives from the University of Sydney and the three Services.[9] Recently retired Rear Admiral and former RANC commanding officer Bill Dovers was appointed Chief Project Officer pending the appointment of a Council Executive Officer. Dovers had been the Director of Joint Staff within Defence before concluding his naval career. Although there had been tension between Dovers and Tange, the latter recognised the former's abilities and experience. Dovers was chosen not because he was identified with the Navy (which remained uneasy about the project) but because he was approachable, could converse with people of all backgrounds and understood how both the uniformed and civilian sides of Defence functioned. The Council's Secretary was the highly experienced Len Hume. Activity began immediately. There were site visits to two newly formed universities, Flinders and Griffith, for planning purposes, and there

The Academy commitment, 1974–1976

were consultations with the Australian Universities Commission.

Organised opposition to the Academy was inevitable.[10] The Australian Vice-Chancellors' Committee (AVCC) met on 17 June 1975 and expressed concern that if the tri-Service Academy was granted university status then comparable institutions, like Colleges of Advanced Education (CAE), might also seek such status. The AVCC indicated it would need to admit the Academy to the Association of Commonwealth Universities and the Academy's Vice-Chancellor to membership of the Committee. Such admittance could not be assumed. This comment prompted the Federal Education Minister, Kim Beazley (Senior), to enter the discussion. He thought the Academy's vocational focus meant 'its academic program should be related to college of advanced education work [...] more than university-type work' and that CAE representatives should be appointed to the First Council rather than Vice-Chancellors. Neither UNSW nor Defence was interested in establishing a CAE. The aim was always to create a degree-granting institution. This was the first and last time the Department of Education was consulted or allowed to contribute.

At the 20 November 1975 meeting of the Development Council, attention turned to the Academy's name. There were numerous suggestions including the 'Australian Defence Force University'.[11] Myers shared some private thoughts on possible names with Len Hume on 11 December 1975. They included Phillip (the first Governor of New South Wales), Canberra, Menzies, Molonglo and Hutton – 'the first general officer commanding in Australia'. Other names that had been suggested to him were 'George Pearce University' (after the first Minister for Defence) and 'Kokoda University'.[12] But the need for any name was unclear after the Whitlam Labor Government was swept from office at the Federal election two days later. Malcolm Fraser would become Prime Minister and lead the new Coalition Government. At a scheduled Chiefs of Staff Committee meeting, two of the Service chiefs sought to have the decision to establish the Academy re-examined. In a letter to the *Canberra Times* published on 12 August 1978, the then Chairman, General Sir Frank Hassett, disclosed that he had refused to allow any discussion on the matter as the Government had made its decision 'after much deliberation' and the task of the Services was to implement it 'in the best possible manner'.[13]

Although Malcolm Fraser and the Coalition parties had been elected on a platform of fiscal restraint, the incoming Prime Minister had been the foremost advocate of the tri-Service Academy project in 1970. The Commonwealth was facing an enormous deficit and every area of Government activity, including Defence, which was enjoying some respite at the end of the Vietnam War, was being examined for potential savings. Fraser might have been keen to see the Academy established in principle; it was another matter finding the money to pay for it. Fraser had chosen the Member for Moreton, Jim Killen, as his Defence Minister. Killen had not supported the tri-Service Academy proposal when he was Minister for the Navy in 1970 and had 'not been available' for the Cabinet meeting that led to Fraser's original submission being withdrawn. Observers asked: was Killen still opposed to the Academy?

As the Government considered its options, there were suggestions that Australia should examine the approach the British forces were taking towards officer education. After all, they had provided the models upon which Jervis Bay, Duntroon and Point Cook had originally been founded. But the British were in the midst of their own self-searching. At the Royal Military Academy (RMA) Sandhurst, the new entry officer course had been reduced to two years in 1955. There were six terms, of which two were military training and four were broadly academic, and included languages, social studies, mathematics and war studies. The program did not attract any external recognition and did not award degrees. An inquiry into officer education by Professor (later Sir) Michael Howard and Dr (later Sir) Cyril English conducted in 1966 recommended a new structure, fresh content and a different approach to education. A new approach was necessary because of heightened competition for bright young men from the universities and technical colleges after 1963. The military colleges needed to offer better opportunities and greater rewards for study completed. They proposed that the Sandhurst course be reduced to one year devoted entirely to military training and then, after a regimental posting of two or three years, all officers (excluding graduates) would complete one year of academic work at a Joint Services Defence Academy located at Shrivenham. This would equate to the first year of an undergraduate degree. Those who did well and wanted to continue their studies would complete another two years and obtain a degree. But the Services would not agree to the reduction of their own colleges or to a common curriculum. There was thinly veiled doubt that officers really needed degrees. There was also the cost of a new academy. It was deemed too high. The Government ignored the recommendations, leading Professor Howard to call the report a 'complete waste of time'.[14]

In 1968, Sandhurst sought to offer an externally accredited diploma to those who completed their academic studies to a set standard. The diploma would be the equivalent of the first year of an undergraduate degree. But negotiations with four provincial universities failed to secure any agreement. Howard and English were right: the number of young men wanting to become army officers declined to a critically low level by the end of the 1960s and there was greater reliance on short-service commission officers who were, in time, offered regular commissions. The question was asked: why bother with education if minimally trained officers met the needs of the service? The British Army had a crisis in the mid-1970s. It did not appear to know what it wanted from academic education and it was unable to compete with the universities for talented youth.

The Royal Navy was similarly inconsistent in defining its educational requirements. After abandoning secondary school education in 1960, British midshipmen received one year of professional training at Dartmouth, with the first two terms being devoted to general and academic subjects and the third term to professional training at sea in the Dartmouth Training Squadron. Midshipmen spent the second year of training in a seagoing fleet unit before returning to Dartmouth and promotion to Acting Sub-Lieutenant. Seaman and Supply officers would continue with academic courses with a professional focus, while the Engineer officers would proceed to the Royal Naval Engineering College at Manadon for degree studies. During the mid-1960s, a small group of third-year midshipmen were sent to civilian universities rather than back to Dartmouth; the Royal Navy started a direct-entry graduate and short-service supplementary list scheme; and a university cadetship scheme was also trialled and proved to be very successful. There was a commitment to university education within a flexible framework of delivery. There was no agreement, however, on which approach best served the needs of the Navy or was the most efficient in terms of cost or efficacy when it came to academic achievement. A contributor to the *Naval Review* alleged that the Navy:

> does not want graduates with specific degrees, though some degree subjects have obvious importance or relevance – but the Navy needs the sort of people who aspire to degrees. The Navy must therefore enable future officers to satisfy their

enhanced ambitions and to obtain a degree before or as part of naval training. This, or the Navy will not get the officers it must have.[15]

The same kind of turbulence accompanied Royal Air Force (RAF) cadet education. In the late 1950s the first year of the three-year RAF College Cranwell course was devoted to basic academic studies (mathematics, physics, English, history, aerodynamics and navigation) that were designed to bring high school students to the same academic level after twelve months. The academic bridging course would be followed by branch-specific training in the second and third years. In 1959, sensing competition from civilian universities, Cranwell attempted to raise its program to degree standard. The RAF also initiated a university cadetship scheme allowing undergraduates to join the RAF while continuing with their civilian academic studies. The College developed an association with London University, the Royal Aeronautical Society and various professional institutions. But the practical difficulties of managing these affiliations and requirements led the Minister for Defence Administration, Gerry Reynolds, to announce at the end of 1968 that the RAF would thereafter 'recruit its direct entry permanent officers from the universities'.[16] Because Cranwell could not compete with the universities, emphasis shifted to attracting graduates who had attained the education Cranwell could not deliver, while the cadet scholarships reflected a kind of 'partnership' between the RAF and the universities.

The only common theme across the British Services was the desire to recruit as many graduates as possible and to encourage serving officers without degrees to pursue tertiary education in a civilian institution. Dr Cathy Downes, an educational and organisational culture researcher with experience in New Zealand and

Australia, concludes her survey of the British experience between 1945 and 1975 by noting that financial stringencies and pressing practical problems had produced a great deal of 'ad hocery' and 'perpetual tinkering':

> There is as much danger to institutional health and the development of optimal training and education systems from such tinkering and the resultant instability and inability to validate and evaluate changes, as there is from stagnation, inertia and conservatism.[17]

She felt that the British Services in the 1970s took an 'econocentric' approach that emphasised the cheapest option, focusing on what *could* be taught within a stringent budget against what *should* be taught. Education was scrutinised independently and often in isolation by economists, organisational theorists and operational planners, without taking a holistic approach to what was wanted and how it might be delivered. She thought there was little regard for the longer-term costs against short-term savings. The general attitude was one of striving for the least possible amount of training and education and an indifference to the later career performance of the individual and their ability to mature as a person to become a sound and solid leader. The claim that what cannot be quantified – 'such as the intellectual skills of analysis, judgement and flexibility [...] [and] the intellectual qualities of curiosity, initiative and open-mindedness'[18] – cannot be important (or defended) led to reduced investment in intellectual preparedness for uniformed service. Spending time in simulated operational environments was the apparent priority. The 'real learning environment' was the fleet, regiment or squadron rather than the lecture, library and tutorial. The attitude was, according to John Sweetman, 'that graduate officers resemble cream on the recruiting cake: welcome but not essential'.[19]

It was time for the Australian Government to make up its mind on a number of philosophical and practical questions. Was it going to do officer education well or just adequately? Would it develop the best solution for Australian requirements or settle on a compromise that suited the needs of no-one? And would it have the political will to ensure its resolutions became reality notwithstanding the state of the economy? Although the nation had entered a protracted period of peace, notwithstanding the possibility of Cold War conflict being played out in surrogate nations, could expenditure on the nation's uniformed leaders wait until the strategic situation deteriorated or was this period actually the best time to make a firm stand and to insist on a new beginning?

Some within UNSW wanted to dispense with Duntroon and focus on pressing needs at the Kensington campus. The Wollongong campus first established in 1951 had become the University of Wollongong in 1975 and, with good management, the Canberra campus would become the tri-Service Academy in the near future.[20] UNSW would then be better placed to compete more fully in the Sydney-area educational marketplace that now included the rising Macquarie University. What was Prime Minister Fraser's thinking on the controversial matter that had completed his personal and political alienation from John Gorton? The Academy Development Council was desperate to know.

6

The joint educational enterprise, 1970–1980

By 1970, the Army had secured a sound platform for educating its officers in the Faculty of Military Studies at Duntroon. As the Army was the largest and most manpower-intensive Service, it could spread the cost of this initiative across a much larger student base than either the Navy or the Air Force, which struggled with the financial viability of their educational institutions. Jervis Bay and Point Cook were small and expensive but the Navy and Air Force insisted on retaining separate education and training systems worrying that Army culture would dominate, as well as fearing the demise of the particular ethos needed to conduct operations at sea and in the air, and the loss of specific subject matter expertise.

The delivery of first-year UNSW undergraduate courses at Jervis Bay began in February 1968. Preparations were rushed after the agreement was signed in July of the previous year. Lecturers at Jervis Bay were introduced to UNSW's processes and, where expertise did not exist at the Naval College, faculty from the Wollongong campus were made available to teach and supervise. At the same time, a small number of midshipmen undertaking Arts degrees were sent to the shore establishment HMAS *Watson* at South Head on Sydney Harbour, from where they commuted to the Kensington campus for classes. They would return to Jervis Bay between semesters for professional naval training. An RAN

Liaison Officer was later posted to Sydney and given an office on campus. As Liaison Officers, Commanders Jim McKeegan and later Ted Shimmin were active in undergraduate teaching and were given time to complete their doctoral studies as well. They were the principal point of contact between the 'University midshipmen' and the Naval College.

The mood at Jervis Bay quickly changed. Although half of the new intake each year were Junior Entry cadet-midshipmen aged between 15 and 16 years who would complete the New South Wales Higher School Certificate before being eligible for tertiary studies, and half of the Senior Entry midshipmen would elect to study the post-matriculation non-degree course offered at Jervis Bay, the association of the RAN College with UNSW had an immediate and profound effect on the outlook of staff and students at HMAS *Creswell*. Completing a UNSW degree was the new 'gold standard', while the experience of studying on campus was viewed as an attractive prospect, particularly for Junior Entry cadet-midshipmen who would endure Naval College routines for at least two years. External scrutiny and accreditation of the College's programs ended any sense that HMAS *Creswell* was a closed community or that naval education was a matter for the Navy alone. The University opened the College to a discipline and a culture it had never known.

Three years after the association agreement began, the Naval College held a major symposium to review recent progress and to canvass future directions for educating naval officers. The gathering was an initiative of Professor Al Willis on the suggestion of a UNSW academic based at the Wollongong campus. Dr Paul Van der Werf had spent six months at the United States Naval Academy at Annapolis and believed that much could be learned from comparing Australian arrangements for educating young officers with those of the nation's main operating partner. As the RAN was looking more to the United States Navy and less to the Royal Navy for technology and tactics, the gathering could not have been more timely. The delegates included staff from the Naval College and the University, and a selection of officers from various naval specialisations who were serving in the Fleet.

Captain George Histed RAN, the Director of the Naval Education Service, set the tone for the symposium, explaining that the Navy wanted midshipmen 'to study at a university and to share in undergraduate activities', believing that the 'benefits of campus life are hard to state in concrete terms but are nevertheless substantial'. But the Navy also believed there was no sense in pitching a new-entry naval cadet into the freedom of undergraduate living 'with no naval indoctrination, no naval knowledge and lots of pay in his pockets'. Hence the desire to teach the first year of the University program at Jervis Bay 'during which cadets could be inculcated with some naval tradition and their officer-like qualities assessed'. Histed praised UNSW as a 'progressive University' and believed the initiative was proceeding well. Of the fifteen cadets who were enrolled in first-year degree studies in 1968, fourteen had passed their exams. Those who enrolled in the Arts program, and the number was limited to eight each year, were sent to the Kensington campus for their first year of study because suitable humanities tuition could not be offered at Jervis Bay. Science and Engineering students completed the first year of their degrees at Jervis Bay before proceeding to the Kensington campus. In his closing remarks, Histed wondered whether the Navy was:

> being pushed into more and more academic training, and less and less time for professional Service training, to lure the young lads who have academic baits cast in their direction by so many other professional bodies. And because of this

emphasis on recruiting by inducement, are we receiving the types we require – the lads with motivation towards a naval career?[1]

He did not seem to have an answer but felt that the Service needed 'fighting sea officers' who were prepared in 'matters technical, tactical and strategic'. Whether the balance between education and training had been achieved, he could not say.

Paul Van der Werf was clearly impressed by what he had seen and heard at Annapolis. After outlining the organisational structure of the Academy and the way its graduates' aptitudes and abilities were validated within the Fleet, he stressed that undergraduate students at Annapolis were provided with 'an education which is flexible and imaginative enough to give them a start on their way. Long-range academic objectives are not the implantation of scientific knowledge but rather habits of thought and intellectual curiosity'.[2] He also noted that three-quarters of Annapolis graduates would proceed to a higher degree at some point in their careers. He counselled against obliging Australian midshipmen to undertake specific courses and programs of study that were too restrictive and did not allow them room to develop critical faculties that were not necessarily tied to a particular discipline. His closing message was not to educate junior officers too narrowly.

The next speaker was the Director of Studies at the Naval College, Dr Ewart Dykes. He had been an Instructor Captain in the Royal Navy and had previously served with the RAN as Director of the Naval Education Service from 1953 to 1958. Dykes had been Dean of the Royal Naval Engineering College at Manadon when he accepted the appointment at Jervis Bay. He inferred that the system the RAN had abandoned was better than the one it had embraced, suggesting that sending engineering students to Britain made a great deal of sense. He claimed the British experience gave them insights into the corporate knowledge that Manadon had acquired over decades and access to quality facilities. He noted that the RAN had decided to head in another direction in 1968 and to entrust its engineers to UNSW. He did not comment on whether he thought this was a good decision or a poor one but implied that young officers needed to be motivated for life in their service and there was only one place that could happen – in a naval environment. He concluded: 'I consider that the absence of a training ship and sea experience at the beginning of our courses is unfortunate'.[3]

The Chief of Staff to the Fleet Commander, Commodore Neil McDonald, tried to summarise the Fleet's expectations of naval officers. In addition to being technically competent in their chosen specialisation, they needed an 'adequate knowledge of the Australian political scene, diplomatic organisation, international law and public relations [...] the ability to express themselves well, both verbally and on paper [...] [and] the ability to undertake reflective thought'.[4] Did those being sent to the Fleet fulfil those expectations? McDonald consulted with other Fleet staff and commanding officers of seagoing units and concluded that the current crop of junior officers were 'generally satisfactory'. He reported, however, that the 'standard of English expression' was low; they have 'little knowledge of military, maritime, international or criminal law'; competence in basic book-keeping and accountancy was poor; there was a 'lack of knowledge of the history and cultures of the nearby countries'; few officers were even acquainted with modern management theory and practice; and most junior officers lacked a second language, a distinct disadvantage; none were proficient in Malay, Indonesian or Japanese.

McDonald noted that many seaman officers felt frustrated that they did not have a degree but he observed that 'the role of the seagoing officer is essentially a practical one, and it has not yet been demonstrated in the Fleet that higher learning necessarily injects those qualities of leadership and fighting ability essential in the professional military man'.[5] But he conceded that it was 'highly questionable' for ship's captains without degrees to be leading men who had degrees. His reasoning was unclear. Did he really think that command at sea now needed a university degree for its discharge or that the best-educated man (women did not go to sea at that time) in a ship needed to be the commanding officer? Would captains without degrees be embarrassed or perhaps shown up by officers who were university graduates? He might merely have meant that gaining a degree ought to be the new standard for all officers aspiring to command. If he meant to affirm the importance of higher education he might simply have said a degree was valuable in the professional development of an officer. His comments were liable to misinterpretation.

In the case of engineering degrees, he noted that they neither helped nor hindered an officer's career prospects. More problematic, he thought, was the existence of two schools of thought among engineer officers in the Fleet: some thought educational standards were too low and others thought they were too high. One senior engineer told McDonald that the equipment then in naval service did not really require a professional engineer with degree training. For the greatest part, he explained, technicians kept ships at sea and not engineers. McDonald closed with the question: 'what proportion of our officers in general should be trained to degree level?'[6] His answer was surprising: 'frankly I'm not sure, and I wonder whether my uncertainty is not shared to some extent by other officers'. But he was willing to venture the opinion that for those who were destined for senior rank and 'playing an important part in the machinery of government and public life, education to a degree standard is essential'.[7]

The symposium was highly revealing. Outwardly, the Navy was content with the UNSW association. There was no suggestion that collaborating with the University was not a good thing or that the education being provided was lacking in quality. The issue was whether the education midshipmen were receiving was relevant to their employment and beyond what was really needed to lead men at sea and ashore. While none of the delegates went as far as saying education was a luxury the Navy could not afford or that its provision through UNSW was wasteful, there were clearly doubters who quietly wondered whether the new system was flawed. Some evidently preferred a closer relationship with the Royal Navy; others hinted at delaying tertiary education until a midshipman had shown a commitment to the Navy. Degree studies would become a mid-career reward for loyalty. Curiously, the value of education was simply assumed. Its purposes were not canvassed and its relationship to training was not explored. There was no attention to where the University could contribute to naval capability through its research programs, although British universities had been closely involved in naval work since the Second World War.

It was difficult to criticise the University for its performance or to critique the Navy's association agreement with the University when there was a lack of consensus about what educating officers was meant to achieve or how educated officers would make a difference to the Navy. The principal outcome of the 1970 symposium was closer consideration of how much academic education for junior officers was sufficient to meet the RAN's immediate needs. Mid-career education was a subject left untouched, as were the possibilities associated with postgraduate study. The Navy did not have any targets for the percentage of officers with tertiary qualifications that it wanted (or needed), nor did it develop a mechanism to determine whether the money invested in education was well spent. The objective seemed to be producing as many graduates as possible, with a limitation on the number studying humanities and every encouragement given to those interested in engineering. Considering the University's capacity to enrich the Navy's corporate culture and the capacity of its leaders to influence policy was a matter left for another day. That the Navy had secured an agreement with the University and that its young officers were being educated was sufficient for the moment.

While the Minister and the Naval Board were generally content with the arrangement (and did not propose an alternative), there were a series of unforeseen problems which may, or may not, have been attributable to the nature of the compromises the Board had been willing to make to secure the original agreement. The principal compromise was to send midshipmen to the Kensington campus for extended periods, something the Army was very reluctant to do. Five years after the first midshipmen arrived on campus, the Navy was faced with a serious challenge: a growing number of midshipmen were either failing to complete their degrees or resigning before completing their studies. The attrition rate was now a cause of substantial concern and threatened the entire arrangement. The Navy asked the University to help it understand why this was happening and how it might improve the retention rate. In October 1973, Professor Al Willis wrote to all midshipmen and recent graduates asking them to complete a 'Naval Student Questionnaire'.[8] A covering letter explained the reason for the survey:

> The question has arisen: 'Are we doing the right thing the right way?' and clearly the experience of those people who have been directly involved at the 'receiving end' is most critical in this examination [...] you are free to make criticisms of any part of your training without fear that your

opinions will be held against you [...] the position is simply this: we have an unusual but potentially very worthwhile scheme which we believe will be in the long-term interests of the Royal Australian Navy [...] we need your cooperation to make it a better scheme.

Some of the midshipmen thought the questionnaire was provoked entirely by the high academic failure rate, something canvassed in the survey. Others thought the large number of optional resignations at the end of 1972 had prompted the survey. This was the view of a mature-aged hydrography student (a former sailor who had been promoted to commissioned rank) who claimed that the attrition problem was caused by 'too many people being given a chance to study at university without enough checking as to their intellect and temperament'. This officer suggested offering some of the places to sailors or other mature-age officers who would make more of the opportunity than recent school leavers. He wanted to see people with proven ability and a demonstrated commitment to the Navy selected over those needed to 'fill the Navy's quota'. Some 80 completed survey forms were submitted.

The survey was actually driven by disquiet within both the University and the Navy. The University wanted to know why midshipmen were not doing well; the Navy wanted to know why they were leaving. Both wanted to know why so few of those who resigned went on to complete their degrees. They were UNSW undergraduates as well as naval officers, and the fact that they did not complete their studies was considered a reflection on UNSW. The poor academic performance of naval students and their desire to leave the Navy did not make sense. They were well-supported and paid to study. In many respects they were the envy of their undergraduate peers – other than being required to have short hair.

The joint educational enterprise, 1970–1980

The University wanted to understand the reasons and to initiate remedial action; the Navy wanted to make the scheme more cost-effective. The student responses are worthy of close examination because they are a snapshot of attitudes in a time of enormous change. As this was the first and only survey of its kind before 1986, it illustrates how junior officers assessed the attempts of senior officers (most of whom did not have any tertiary qualifications) to prepare them for uniformed service. The survey also addressed whether it was preferable to educate junior officers in a Service college or in a civilian university – this would loom large in the late 1970s and mid-1990s when very strong claims were made for both approaches.

The questionnaire began by asking about the personal and professional importance of gaining a degree; the value and relevance of the degree to subsequent employment; whether more naval service before degree studies would have been beneficial; the role of degree studies in personal and professional motivation; whether the student thought they were capable of undertaking an Honours degree or postgraduate study; the role of degree studies at RANC as preparation for life at the Kensington campus; the nature and the strength of the distractions from study; the perceived attitude of civilians towards naval students studying on campus; whether the naval students believed themselves to be more privileged than their civilian counterparts; the benefits of living at HMAS *Watson* compared to residing at a campus college; and where the naval student would prefer to live if given a choice.

Most responses focused on practical matters. The living arrangements were far from conducive to completing a degree and maintaining a commitment to the Navy. One respondent said of *Watson* that 'the routine of being a university student by day and a naval officer by night was a big façade'. A Senior Entry midshipman said: 'you can't be a university student and an ordinary naval officer at the same time unless you've been one or the other for some time; and last year we have been neither – so there was confusion and discord'. On balance, it appeared that individual self-discipline was the key. Some found study at *Watson* easier; others preferred to be at the University. For some, there was a debilitating contrast between the Naval College, where they were compelled to study, and the campus, where they could do as they wished. Some naval students recognised that they lacked motivation and had not developed much self-discipline. Several Junior Entry students who completed their Higher School Certificate at the Naval College compared their experience at *Watson* and on campus and felt that the latter was better because it had fewer mandated distractions. Another respondent, an ex-HMAS *Leeuwin* recruit who had completed his matriculation at the Junior Recruit Training Establishment in Fremantle, thought that 'naval students are spoon-fed at RANC during the easier first year and have to make the normal first year adjustments in the more difficult and demanding second and third years'.

Some of the respondents were adamant that being on campus was the key difference:

> All midshipmen should go straight to a UNSW [residential] college. The reasons for this became apparent from my first year at New College. The attitude to work becomes much more serious. The drive to learn and obtain as high results as possible is instilled in the students. Talking with tutors here, all doing postgraduate work, and meeting and getting to know fellow civilian students has shown or introduced me to new fields of research or interest.

The same student suggested 'more cross talk between the University syllabus and our [naval] training period at Christmas' to connect academic work with naval training and to enthuse students about their future naval careers. He lamented that the Navy has 'proferred no incentives for students to obtain their degrees. In fact, we have been told our degree will be almost useless for our positions in the Navy'. This student reported that discussions with senior naval officers had led him to believe that the prevailing 'attitude towards the degree stream is a means whereby people – us – can obtain letters after their name to be used as a bargaining weight in Navy Office, Canberra'. He was led to conclude that 'I cannot see myself working in a Navy where "intellectual stimulus" worked in a negative direction or was simply non-existent'. His final words were blunt:

> unless the Navy can tell me in more specific terms my role or career lines in the Navy, I will resign, because I can see I will be of greater use outside of the Navy. This attitude is nearly universal throughout naval students on campus and it is only a matter of time before more and more naval students will resign.

Another complained: 'degree students, particularly Science and Arts, should not be constantly told specifically by senior officers that their degree is useless and that their time would be better employed at sea'.

There was a widely held view that the first year of undergraduate study (which most completed at the Naval College) was too easy and that over-confidence loomed large in second year. One respondent said the main problem was not the course but the aptitude of naval students: 'the quality is deficient not only in academic terms but, sadly, also in personal terms. Many naval personnel simply do not have the character, determination, call it what you will, to actually pass this degree'.

WIDENING MINDS

The absence of any chance of doing postgraduate work was also the subject of comment. This was frustrating for those who had done well as undergraduates and were being encouraged to enrol in an Honours program by lecturers who perceived their potential. Another irritant was the Navy's personnel system: midshipmen who elected to complete the non-degree 'Creswell Course' and who remained at Jervis Bay were promoted to Acting Sub-Lieutenant while their degree colleagues remained midshipmen. Not only were they promoted more quickly, they were financially better off as well. One respondent suggested making promotion to the rank of Commander contingent on having a degree. He suggested that the degree could be gained by whatever means but graduating from a university was each officer's responsibility if they aspired to senior rank. This would also deal with those whose progress at Kensington was too slow. If a midshipman had not completed his degree in the allocated time because he had failed a subject or was indolent, he would be obliged to finish his studies while serving in the fleet or in a shore establishment. He could seek leave without pay to do so.

Some naval students suggested the solution to better results and higher retention was inculcating greater self-discipline. The Navy was apparently making the university experience too comfortable. A proportion of students, and they were a mixture of Junior and Senior Entry, believed that new-entry midshipmen should 'spend six months at the Naval College and six months at sea getting to know about the Navy [...] [after which] they should then choose their specialisation'. One respondent suggested 'three months at the Naval College and then [...] the Fleet for a year to do some sea time'. Several respondents thought that Jervis Bay ought to be about training naval officers and the university about 'training minds'. Plainly, the Navy had not clearly communicated to them the approach it was taking to their education and training.

The joint educational enterprise, 1970–1980

Most of the naval students attributed poor academic performance to the proliferation of distractions in and around the Kensington campus, principally women and 'the pub'. One complained about the lack of campus parking for their motor vehicles, another noting that 'the number of thefts from cars parked in High Street and Barker Street is phenomenal'. Many said the accommodation at HMAS *Watson* and other naval establishments was far superior to what the campus colleges provided but this was offset by the greater 'personal freedom and expression of individuality afforded to naval students at the colleges' and the general hostility of naval personnel of all ranks at *Watson* to the University contingent. One student spoke of being treated there as a 'leper'. Others felt that *Watson* was better than the campus colleges because it was more isolated and there were fewer distractions. To others, *Watson* was too far from the campus for convenient access to the library and lecturers. Some simply enjoyed the respite from uniformed life. A naval student resident on campus confessed: 'RANC poisoned my attitude to studies, not through the academic staff so much as the naval training side and the lack of freedom'. The same respondent spoke of the Naval College as 'narrow-minded' and dictatorial in not allowing those who changed their mind about a naval career (he joined when aged 15) to leave when they had finally come to an adult decision. Some thought that their colleagues deliberately failed their courses so they would be dismissed from the Navy rather than resigning and having to pay out their return of service obligation.

One naval student was adamant that at RANC and *Watson* healthy competition among peers was lacking; on campus most of the naval students found they were studying with people more able than themselves – and this was a positive incentive to work harder. Another remarked that 'there exists an unhealthy attitude among the naval students as many people are aware, that even by failing university they suffer no hardships but return to the Navy and go straight to the fleet, at the same time there is no properly functioning incentive system'. Time spent in the fleet or at shore establishments between leave periods was also deemed to be a disincentive because the time was not well planned and the ships and establishments receiving midshipmen did not know what to do with them, especially in the case of one midshipman who was sent to a ship in dry dock and was told to amuse himself.

The inability of senior officers to explain the vocational purposes of tertiary education loomed large as a reason for apathy. Many of the midshipmen felt they had been banished to another place for a season – without knowing what the Navy wanted or needed from their studies. It was as though they needed to be educated but no-one had made a compelling case for why. The absence of discussion about individual subject choice was a case in point. They had not previously been consulted about the subjects they ought to study, the major or minor sequences within their degree programs that they might consider, and they appreciated the chance to express some of their frustrations with lecturers on campus. One respondent thought the 'Navy should have a professional outlook towards degree course officers. The impression we obtain is that the Navy considers [degree studies] a privilege it bestows on certain personnel' but that the possession of a degree did not lead to better pay or necessarily quicker promotion. There was a sense in which neither the naval students nor the Navy really knew what to make of a university degree. One student lamented the attitude of a senior officer, who said that naval students should be 'eternally grateful' for the opportunity they had been given. The engineering students, all claiming their degrees were the most demanding of the three undergraduate programs, pointed out that unlike their Seaman and Supply Branch peers, they needed a degree for their profession

The joint educational enterprise, 1970–1980

and would certainly be better paid outside the Navy once qualified. Thus, they challenged the alleged 'privileged status' they possessed. Others claimed that engineering studies were not directly relevant to what they believed they would be doing as naval engineers and questioned whether they should have been sent directly to the Royal Naval Engineering College at Manadon or to a trade training establishment. One said: 'we are being over-trained for what the job demands'. Another remarked:

> quite candidly, except for Engineers, I think from the Navy's point of view sending future naval officers to a do a Science or Arts degree is a complete waste of time and money. The pass rate is low and because of disillusionment the Navy loses too many potentially good naval officers.

Changing attitudes to the status of naval service were also apparent among the naval students. One respondent commented that 'when some of the senior officers realise that the RAN now is merely a job and not a penance because you signed on the dotted line, the better it may be'. It was clear that most naval students did not think that senior officers understood their challenges or problems, and fewer seemed prepared to address them. After all, compared to what the senior officers endured as junior officers during the Second World War or the Korean War, those being trained in the 1970s were privileged and pampered.

The naval students had their say and few were complimentary. But little was done to change the environment in which they studied. The Navy appeared to accept that some midshipmen would be lured away from the Service during their time on campus while others would need to be thoroughly re-enculturated into the naval ethos after they graduated and were sent to the Fleet. The scheduling of professional courses during academic breaks helped to reconnect midshipmen with the Service and to revive their desire to serve at sea, but it also broke whatever momentum they might have achieved in their academic work. Although many senior officers feared the consequences of relocating midshipman education away from Jervis Bay, which provided opportunities for short visits to Fleet units and the occasional 'sea day', just as many had come to the view that sending junior officers to a university campus was disrupting their preparation for naval service. While there were complications associated with educating young naval officers alongside their military counterparts, at this formative stage in their lives it was probably better that they interacted with other uniformed young men than with civilians who would question if not ridicule their commitment to serve. By the end of the 1970s and with the attrition rate still high, the Navy's senior leadership recognised the need to overhaul its approach to educating young officers.

The first step was phasing out secondary education at the Naval College and abolishing the Junior Entry. At the Passing Out Parade on 5 July 1979, the Commanding Officer of the Naval College, Captain John Snow, announced the end of an era. As a cadet-midshipman on parade that day, I recall my shock and surprise at hearing his words:

> The Junior Entry was long recognised as the mainstay of College activity. With the growth of Senior Entry and Short Service intakes, in today's changing social and economic climate this source no longer justifies the overheads involved, consequently 1979 will be the last occasion of a Junior Entry selection.[9]

The final Junior Entry would be admitted in January 1980. My class had found its motto. We were now 'Second to None'.

The second step was to review the UNSW association. The prevailing view held that it was expensive and inefficient. While many midshipmen responded well to the Kensington experience and made the most of the academic opportunities, the Navy wanted a new model that was both financially viable and vocationally appropriate. There were practical challenges associated with educating naval officers alongside their Army and Air Force counterparts but they were deemed to be of a lesser order than the problems of continuing with the Kensington arrangement. Circumstances, not of its own choosing, had obliged the Navy to think more positively and creatively about a tri-Service academy. The same was happening with the RAAF.

Educating Air Force cadets

The conduct of officer education was also becoming an acute problem for the Air Force.[10] The Martin Committee contended in 1969 that the academic burdens imposed on RAAF Academy cadets by the University of Melbourne were 'unrealistic'. A single degree course was being offered – the Bachelor of Science – and there were only two majors – mathematics or physics. The Committee explained: 'In our opinion the RAAF Academy provides an illustration of the difficulties which arise from a close association between a university and a Service college. A baccalaureate course in Science [...] with its emphasis on physics, is not wholly suited to the needs of the RAAF'. There was a widely held belief in the Air Force that the Martin Report 'signalled the demise of the RAAF Academy'.[11]

Three years later, former RAAF Squadron Leader and

long-serving Academy Warden, Mr Walter Hardy,[12] produced a confidential report entitled 'Decline in Academic Standards, RAAF Academy', which noted that cadets could not see the value of their degree beyond it being a graduation requirement. Nor did the Air Force seem very interested in its content. The '51 per cent syndrome' (in which candidates would do the bare minimum to pass) had taken hold even among cadets capable of achieving a high distinction average in an Honours program. It seemed in the mid-1970s that the academic staff believed it was inevitable that officer education would be concentrated in Canberra and the RAAF Academy closing, with the effect that they were seeking new positions. Even within Air Force Headquarters there was some diffidence about the RAAF Academy and whether sending cadets to a university campus was not a better option if they were to mature socially and intellectually. While Point Cook did not have the bastardisation problems that drew national attention to Duntroon in 1969, RAAF Academy staff were certainly concerned about the conduct of some cadets and the proliferation of group behaviour that was deemed inconsistent with Air Force values.

There were other reasons for the Academy's eclipse. Some aeronautical engineers were being educated at Sydney University after completing their first year at Point Cook. The RAAF connection with Sydney University dated from 1947, the year the RAAF College was established. Sydney had the country's only professorial chair in aeronautical engineering. Within the next 25 years, over 40 RAAF aeronautical engineers had graduated from Sydney University and many had done very well in their studies, with two Rhodes Scholars among the group. By way of contrast, cadets who had joined the Air Force to fly were frustrated that flying hours were strictly rationed.

With the likely advent of a tri-Service academy and the possible closure of the RAAF Academy, Walter Hardy resigned in January 1976, and was closely followed by his Deputy and short-term successor, Bill Gravell. By the end of that year, remaining staff at the Academy realised they had no more than five years of service before the Academy would close, with the last cadet intake to complete the Melbourne University Bachelor of Science degree and Graduate Diploma of Military Aviation (the fourth-year qualification that covered most of the professional development elements in the educational program) arriving in January 1978. Those who entered the Academy in 1979 would undertake a three-year degree course with the RAAF, hoping that the subjects on offer would be more vocationally appealing. Also under consideration was shifting the emphasis from pure science to applied science, given that Melbourne University had previously offered such a program. If the University was not interested in this innovation, the RAAF would seek an accreditation arrangement with another institution. Yet another possibility was awarding a non-externally recognised RAAF Academy degree. It seemed that anything was preferable to the existing Melbourne University Science degree program. By 1979, the RAAF and the University finally accepted a three-year course that integrated academic and aviation subjects. But there were, however, some mixed messages about education being sent across the Air Force, such as inconsistencies in the treatment of degree- and diploma-qualified officers, with the training system seeming to favour the latter in terms of promotion and postings.

For its part, Melbourne University became increasingly concerned about the status of tenured staff who were either unwilling to relocate to Canberra or were unwanted at the tri-Service Academy. Indeed, some of the Point Cook staff believed their qualifications had not been fairly or consistently evaluated by those who determined whether they could transfer to the tri-Service Academy should they wish to do so. With unrest and resentment among the civilian academics, the opening date for the tri-Service Academy being delayed and the facilities in need of substantial upgrading, Point Cook was an unhappy place by the late 1970s. The substantial reputation it had earned for innovative research that had led to significant national and international grant funding was ebbing away.

Consistent with corporate attitudes in the Navy, the Air Force did not promote higher degree study among its younger officers. The prevailing view was that the education and training pipeline was already too long. There were also longstanding cultural conflicts. The Academy's historian, Air Vice-Marshal Roy Frost, concluded that:

> The problems of amalgamating the spirit of individual inquiry and intellectual integrity with military discipline and unqualified regard for authority were in evidence throughout the life of the Academy. The conflicts between the objective of education which seeks to stimulate discussion and independent thoughts, on the one hand, and training on the other, which looks to vocational relevance and group orientation, were never satisfactorily resolved from an Air Force point of view.[13]

Frost felt that the Air Force had not realised the significance of its decision to affiliate with Melbourne University, nor did it track the impact of the affiliation on the mood at Point Cook. It was not just another military base, it was also an academic establishment. But given that many non-graduates had achieved senior rank in the Air Force, Frost thought there were 'reasonable grounds for querying the wisdom of establishing and maintaining' either the RAAF College or the RAAF Academy. In sum, the Air Force might have met its needs just as well without the cost of an educational institution to provide a university degree. But perhaps, he mused, many of those with degrees who left the Air Force might have done so before the vocational value of their education was apparent. By 1980, the Air Force realised its troubled affiliation with Melbourne University needed to end and it embraced the possibilities that would flow from establishing a tri-Service academy.

In contrast to the Navy and the Army, the Air Force had little previous contact with UNSW. The RAAF New South Wales University Squadron was formed on 16 October 1950.[14] The Squadron provided service training for students on the campuses of the University of Sydney, the University of NSW and Macquarie University, preparing them for appointment to commissioned rank in the Citizens Air Force (CAF). The establishment strength was three Permanent Air Force officers, two warrant officers and three airmen, with up to 120 cadets in various stages of training. There were 28 days of allocated training each year, of which 14 days were served with a RAAF unit. In August 1973, Air Force Headquarters directed the disbanding of all university squadrons. All permanent staff were posted, all training programs ceased and all cadets were discharged by October 1973. Although the RAAF had no institutional preference for or against UNSW, it could only presume from the attitudes of naval and military officers that it was suited to advancing the interests of the Air Force.

7

From autonomy to uncertainty, 1975–1978

The election of the Fraser Government on 13 December 1975 led many within both the University and Defence to presume that progress on establishing the tri-Service Academy would continue and, because Fraser had been its progenitor, that it would proceed with added urgency. Although establishing the Academy was not a policy the Coalition took to the election, Fraser had not spoken against the project or hinted that its development would be halted by the austerity drive his government implemented to deal with the enormous financial deficit left by the outgoing Whitlam Labor Government. The project was, however, effectively halted while the Government determined its funding priorities. The Development Council was not idle while it waited for the Government's advice. On 9 February 1976, Len Hume (secretary of both the Development Council and of its Executive Committee) wrote to Myers explaining that the Development Council meeting scheduled for later in the month was to be deferred because 'it had not yet been possible to obtain an indication of the Government's attitude to the project' although the target date to open the Academy officially remained 1980.

Rear Admiral Bill Dovers continued as Chief Project Officer. His duty statement effectively subsumed that of the Executive Officer, a post the University assumed would be filled by someone

From autonomy to uncertainty, 1975–1978

from within the academic community, employed by UNSW, with the equivalent status of a Deputy Vice-Chancellor. Dovers, who knew many naval officers staunchly opposed a tri-Service academy, visited the Dean of the Faculty of Military Studies, Professor John Burns, at Duntroon and told him he was 'one hundred per cent in favour of the concept of a single Defence Academy'. Burns believed his actions had already demonstrated his personal commitment.[1] In late February 1976, Burns told his colleagues in the Faculty of Military Studies that the appointment of the Executive Officer had been deferred. In a letter to Myers dated 1 March 1976, Burns explained that the reaction:

> was very hostile, the attitude being that the appointment
> had been seen by all as the major safeguard against undue
> Defence influence in the development of the Academy
> and its abandonment without any alternative arrangement
> unthinkable [...] it is unfortunate that this move comes at a
> time when the handling by the Department of Defence of
> various government economies is causing intense irritation in
> the Faculty [...] I think this would intensify the indignation
> about the Executive Officer and there is likely to be a wave of
> allegations of Defence hostility and University indifference,
> all of which would be doubly unfortunate when ADFA
> planning seemed to be going so well [...] I expect a reaction
> – a revulsion even – against the idea of having a University
> institution 'planned by an Admiral' [...] The trend of events
> seems rather to have been that the [Development] Council
> has been undertaking much work which should be done by
> the First Council with the assistance of the 'Vice-Chancellor'
> elect.[2]

Myers had known for some time that the appointment was being deferred and that it might never be made. He explained to Burns:

> Towards the end of 1975 there were a number of developments which ultimately led to the present Minister for Defence, Mr Killen, writing to the University with the suggestion that it may not be the wisest course to persevere with the notion of making an appointment of an Executive Officer and that it would perhaps be better to wait until the project had advanced with sufficient definition and sufficient assurances of parliamentary support to attract a person of academic excellence and high administrative ability on a continuing basis. Having regard to all of the circumstances the University agreed with this approach.[3]

Myers went on to say that the existing Faculty representatives were doing a good job conveying the concerns of their colleagues and these arrangements ought to continue. He told Burns: 'I am sorry John but you will just have to wear it. Off you go and keep them quiet'. Burns later recalled that the Faculty 'had pretty clear ideas' about who they wanted as Executive Officer although a name was never discussed publicly. In a private note to the RMC Commandant, Myers remarked: 'I think I have convinced Professors Burns and Wilson that the step to defer was a good one'. But Myers did not allay concerns within the Faculty. Indeed, he became the focus for their hostility, with the following resolution sent to the Professorial Board:

> That Faculty express serious concern over the decision not to appoint an Executive Officer in the light of the role which the Minister, the University and members of the Faculty expected

the Executive Officer to play in the crucial early development of the tri-Service institution; and express dismay at the failure of the University to consult with representatives of the staff of this Faculty in the making of this decision.[4]

The UNSW Professorial Board noted their Canberra colleagues' concern and referred a resolution to the Vice-Chancellor on 4 May 1976. Myers wrote to the Chair of the Faculty of Military Studies, Professor Harry Green, again asserting that the existing arrangements were working well and reiterating his opinion that Geoff Wilson was exercising 'an outstanding role' especially on the Academic Planning Committee. On his copy of the agenda for the Development Council meeting on 29 June 1976, Myers briefly recorded:

> Wilson: Fully understands rationale of decision. Decision became known by accident and caused disquiet. Tange: Decision not made because belief in no need for senior academic but because realities are realities [...] senior academic step into dark.

Dovers, aware of the disquiet, highlighted the important contribution of the 'Heads of Department' at Duntroon and inferred that their role met the need for academic input and that the academic discipline groups were doing thorough work as well. In sum, the extant arrangements were satisfactory. Sensing finality, the Faculty did not raise the matter again, partly because Dovers was so effective in the role. Despite their initial misgivings, he was warmly praised by the academics who appreciated his genuine sensitivity to their needs. In June 1976, Basten confirmed that an Executive Officer would not be appointed, advising the Development Council that UNSW understood and concurred with the decision although it had advertised the position and received applications. Nonetheless, the Faculty was still concerned about the general direction of the project. The academics seized the initiative and formed their own 'Committee on Drafting a New Affiliation Agreement'. This committee submitted a report to the Development Council.

The report noted the assumption that the 1967 affiliation agreement between the University and the Department of the Army was now the responsibility of the Department of Defence, which had subsumed the single-Service departments. It also noted that the extension agreement, which the University rather than Defence had drafted and pursued, did not have an end date as the Government could not confirm a target opening date for the new institution. This lack of clarity was inevitably affecting people and programs at Duntroon and had created 'doubts in the minds of staff of the Faculty about the *bona fides* [of Defence] and its ultimate willingness to proceed to the creation of the autonomous degree-conferring institution'. The report claimed that abolishing the Department of the Army and the Military Board had adversely affected the 'natural evolution' of the Faculty of Military Studies into an autonomous institution because the new entity needed to be a tri-Service institution. It claimed that the needs and conventions of the Navy and the Air Force were different to those of the Army and needed to be accommodated. In any event, the Faculty's administrative machinery was weak, its procedures were under-developed and equipment acquisition was still unsatisfactory. The existing arrangements at Duntroon could not be expanded to include the Navy and the Air Force because they did not adequately serve the needs of the Army. The report contended that:

> the Department of Defence has generally been insensitive to the detailed provisions of the Agreement and has chosen to ignore its obligations whenever it has suited it to do so [...] in clear contradiction to the Agreement, the Department of Defence has insisted that in those areas of the Agreement that it directly administers, it has the right to treat the staff of the University of NSW in precisely the same way as its own public servants.

The Library was said to be adequate but the faculty buildings were seriously inadequate:

> Considering the time and effort that have been wasted on fruitless planning and, for so many years, Faculty has been obliged to make do with temporary and makeshift facilities in the expectation that new construction would soon commence, it is perhaps not surprising that there is now deep cynicism about the competence of the Department of Defence to physically plan any new academic institution.

The report claimed that autonomy would take at least five years and urged UNSW to take the lead with its 'assent to the new Agreement only on the understanding that it will not be used as a recipe for more of the same. There is a limit to the time for which a promise of a better future can retain its allure'. In the absence of a specific academy agreement, the Faculty drafted a continuation agreement that would run until 31 December 1982 or the establishment of the Academy, whichever was sooner.[5] Notably, it related only to the Duntroon Faculty and did not include the academics employed by the Navy and the Air Force. The Development Council received but did not respond to the report. It had other priorities.

The Development Council meeting at the end of June 1976 was the first to grapple with practical planning issues, most importantly the number of students the Academy would need to accommodate and the economies of scale that could be achieved by concentrating officer education at one venue. By the mid-1980s, they estimated that the total student population would be around 1200, with some external and part-time Service students. The estimates were: officer cadets 951, foreign students 57, serving officer undergraduates 151, serving officer postgraduates 41, civilian postgraduates 40 and part-time enrolments 130, giving a total population of 1370 students for planning purposes.[6] This figure also presupposed a continuing shortage of engineering students. The number did, however, fluctuate. The Acting Head of Civil Engineering, Dr John Sneddon, explained to the Army that 'there were a lot of matriculants with the necessary tertiary entrance scores to enable them to complete an Engineering degree. Getting them to join RMC was a matter of sharpening the recruiting campaign. The following year we had a very much larger student intake, and the group of students recruited were the best group I had ever observed'.[7] The initial planning made no allowance for female students but noted that 'such a provision should not have a significant impact on design'. There were no civilian undergraduates:

> It has been assumed that the institution, which will exist
> for a Defence purpose, should not encroach into the field of
> education of the general community. Moreover this could lead
> to certain problems in administration and the maintenance of
> the correct military environment.[8]

The 'size of the academic staff considered appropriate' for the Academy was considered by the Chairman of the Universities

Commission, following a request from Dovers, who had received a proposed staff establishment from UNSW. Professor Burns was advised that the Commission's suggested staffing profile of 150 academic staff, 130 general staff, 40 library staff and 125 'other' staff (totalling 445) reduced the teaching and research staff by some 53 persons. Here was another matter to be addressed in more detail.

In terms of curriculum, the Academy would offer a traditional Arts degree with strategic studies to be covered in postgraduate programs, most probably a Master of Arts coursework degree. The Navy and Air Force wanted management subjects but the Army expressed no view. For their part, the Arts academics did not want to teach management, although Beddie thought that management economics could be offered in the Department of Economics and organisational theory in the Department of Politics.[9] A traditional Science degree would be offered although it would not include behavioural science (which might be offered once the Academy was operational). There was broad agreement that 'a psychology department could not be justified but that the possibility of including an appropriate general studies subject should be the subject of further investigation'. Marine atmospherics and oceanography were to be electives. The desire to create a Commerce Department was greatest in the Navy, although there was broader interest in commerce subjects being offered in postgraduate studies. Similarly, foreign language studies were desirable 'but in terms of staff and running costs this may not be practicable'. It was resolved that a language department would not initially be included. The possibility of students completing an Honours program was problematic, with disagreement on whether it would be an additional year of study (as it had been at Duntroon) or a separate program that could be undertaken some time after a cadet graduated. The Services' requirements were unknown at this point, probably because they did not know themselves. This would become a familiar theme.

The Government's intentions were still unclear by the middle of the year. On 16 July 1976, Sir Henry Basten wrote to all members of the Development Council, explaining that its meetings had been delayed because:

> it would be prudent not to give an impression, by continuing to hold meetings of the full Council, that continuation of the project was being taken for granted [...] The situation now is that the Minister for Defence has decided that a submission seeking the Government's approval to proceed with the project should be placed before the Government in the next few weeks.[10]

In fact, it took another two months for Cabinet to make a decision and another fortnight for Minister Killen to table a statement in Parliament.[11] The original submission dated 22 September 1976 was simply to 'obtain Cabinet agreement to establishment by legislation of the Academy *as an autonomous university* on a site adjacent to RMC'.[12] The formal decision on 12 October 1976 omitted the italicised text for reasons that are not clear.

Addressing the House of Representatives on 20 October 1976, Killen said the project would proceed in principle and the estimated cost was $45 million. The Coalition had also gone further than the previous Labor Government in announcing that the Academy would have full university status. The Academy would open in 1982 and, Killen said, the Service Chiefs were keen to have more officers with degrees as a matter of priority. The local Federal Parliamentarian, Ken Fry, welcomed the announcement. He lauded the Faculty of Military Studies as a 'unique experiment in Service education' before remarking that:

From autonomy to uncertainty, 1975–1978

> It would also be of interest to Canberra people to know whether the new Academy is to be a closed shop available only to educate Service personnel or whether in appropriate circumstances when places are available other people in the community may be accepted for enrolment. I refer particularly to the proposed engineering degree course which is not available in any other institution in Canberra.[13]

He went on to contend that UNSW 'has done more to advance the cause of officer education than any other Australian university'. The Shadow Defence Minister, Bill Hayden, expressed reservations about the prospect of the Academy being a university and predicted opposition to the concept, especially from the Vice-Chancellors' Committee. The Labor Party was concerned that the Government's approach to Defence planning appeared to lack consistency and coherence (not that Labor had done any better in the Whitlam years). How did the Academy fit into the Government's vision for Defence outlined in the 1976 White Paper? The only mention was a single paragraph in the Defence Personnel section, which noted the Government's decision in principle to establish the Academy which would 'enable the advantages of scale to be exploited'. The document continued: 'there will be advantages in the range of degree courses which could be obtained; in the calibre of the staff that can be attracted; in the recognition that can be gained from degrees conferred; and in the economic use of resources'.[14] Critics in the ex-Service community who preferred to retain the single-Service colleges noted the flimsy grounds upon which establishing the Academy was based and questioned whether the benefits would be even greater if young officers were educated at a civilian university. Why replicate on a small scale what existed in a more advanced and diverse form on an existing campus? These views did not, of course, address the twin matters of poor results and high attrition or the substantial cost of refurbishing the single-Service colleges.

The announcement gave the Duntroon faculty limited reassurance. Burns told Myers on 22 October 1976: 'while it is disappointing that we only have a decision in principle the statement seems in general terms to be satisfactory. Nevertheless it is unfortunate that its wording is in places less than felicitous and in some respects seriously deficient'. Burns had the absence of a clear statement on autonomy most in mind. (When the staff and the Development Council expressed concern that there was no mention of 'autonomy' in the published decision, Killen went back to his colleagues on 10 November 1976 asking that Cabinet 'vary wording of Decision 1636 by making it plain that the Academy was to be established as an autonomous university'.)[15] But Burns remained concerned about the proposal for a central administrative unit that would handle financial and business management. Burns was adamant that academic and military administration needed to be separate but accepted that an 'Administrative Coordination Group' could identify common functions and provide shared solutions.

Dovers had produced his own paper entitled 'ADFA Proposed Administrative Structure'. Burns wrote to Myers on 3 September 1976 offering his views on Dover's proposals. He noted that the academic head would be the 'Principal'. The Business Manager would co-ordinate administration and control finance, leading Burns to fear that in some matters the Business Manager's authority transcended, and was superior to, that of both the Commandant and the Vice-Chancellor. He was concerned that the paper was written from a uniformed perspective and gave the military staff greater if not ultimate authority, effectively rendering the academic staff subordinate or subservient. It was a military establishment with a university element rather than a university campus with a military presence. The phrase a 'university within a military academy' was the source of most anxiety. At Kensington, Professor Al Willis separately noted: 'I am concerned that the idea persists that the military curriculum is sacrosanct – imposed and controlled in a dictatorial fashion. (A good way to cover up some dreadful shortcomings!)'.[16] Wilson wanted it made clear that the Business Manager was a co-ordinator and not a controller.

At the meeting of the Development Council's Executive Committee on 6 September 1976, Dovers accepted that his proposals were likely to attract disagreement and was not surprised when they did. He stressed that 'we were striking out in a way no other country had attempted to do. The proposed arrangements now before the Committee would be a purely Australian concept, which are both challenging and appropriate'. Myers noted that 'no attempt had been made overseas to bring the Defence and military interests into direct consultation with universities, as is being done in this case'. This marked out the Australian initiative as unique and something to which many other nations had aspired.

The status of academic staff at Jervis Bay and Point Cook and whether they were suitable to appoint to the new institution began to consume an enormous amount of time. It was a complex question, and each individual's circumstances were different. Some held academic appointments; others were Public Service appointments. Did they have the necessary qualifications and the right experience? Answers were not straightforward. Some were teaching at the single-Service colleges but were deemed unsuitable to appoint to the Academy because standards had changed and some were no longer qualified. An Accreditation Committee was established to deal with the complexities and the negotiations that would need to occur. The problem was less pressing at Point Cook, where more recent appointments had been non-tenured because the Academy's future was uncertain. The future of these staff members was, however, a long-running saga that caused great angst to the affected individuals, who eventually combined to submit jointly signed letters of protest at the lack of clarity about their future.

The actual shape of the Academy began to emerge with the release of the 'Functional Brief' in 1977. This highly detailed document explained that the new institution:

> will be a university and a military academy. In its university functions it will operate in its own right as an autonomous university providing undergraduate and postgraduate degrees and encouraging and providing facilities for research and scholarship [...] the primary function of the Academy will be to provide a balanced and liberal university education for officer cadets of the ADF [...] it is [also] intended that there will be Service and civilian students studying for higher degrees. No provision is being made for the admission of civilian undergraduate students at the outset but provision will be made in the design for the admission of female students at a later date [...] The environment which will exist in the academic areas of the Academy will be that of a university. The pursuit and preservation of knowledge and the encouragement of teaching and research will be emphasised.[17]

There were to be three faculties and 12 departments. The Arts Faculty would include four departments: Economics (later renamed Economics and Management), Language and Literature (later re-named English), Politics, and History. The Engineering Faculty would contain three departments: Civil, Electrical and Mechanical Engineering. The Science Faculty would incorporate the departments of Physics, Chemistry, Mathematics, Geography, Computer Science and Operations Research. The possibility of another department in Arts, that of International Relations, was being considered. Notably, the RMC Department of Government would become the Department of Politics at the request of its staff. Apart from long-standing confusion with Commonwealth agencies, the new name for the department highlighted the intention to focus on the broader field of politics, including international relations and diplomacy. By then, the Army was satisfied that a Department of Politics did not mean the Army could or would be politicised.

In terms of site planning, there was to be a designated academic zone to emphasise and protect the university environment. The first design for the entire site was produced by a group of government architects who planned for a high-rise building that juxtaposed the Mechanical Engineering workshop with the Department of English. After they presented an initial plan that placed all of the academic departments in a single building, which made no allowance for the special needs of the technical disciplines, and put the academic tower in the flight path of the east–west runway at Canberra airport, the Academy Project Office engaged a commercial firm. This aspect of the project had not begun well but was soon directed to a more constructive trajectory: distributed buildings, uniform roof colours and off-white bricks, with each complex to be designed by different private architectural firms, chosen on the basis of competitive design outlines.

Another critical issue was whether the Academy would have a separate officers' mess and academic staff club. The original plan was to have three uniform messes: officers, senior non-commissioned officers, and other ranks, in one building with a shared kitchen. Academics would be able to join whichever mess was comparable with their university standing. This was unacceptable to the

University staff, who refused to countenance such divisions in their community. Dovers insisted: there would be one staff facility and it would be shared, despite the preference of some military members of the Development Council for an officers' mess that academics could join if they wished. The decisive factor was the very positive experience at RMC. Much informal consultation and co-ordination between Army officers and UNSW staff happened at shared morning teas, in addition to the firm friendships that were formed in Duntroon House.

Echoing a previous dispute about membership of the RMC Officers' Mess, the position of General Staff who had been employed within the Faculty of Military Studies at the Academy was the source of deep anxiety as well. A memo from Ms J Shapcott to Burns on 10 October 1977 relayed the extent of feeling:

> The General Staff Committee is strongly opposed to the stratification of civilian members at ADFA. We reject the proposition that stratification of civilian staff is inevitable in a military academy and argue that stratification of civilian staff in the area of staff facilities can only lead to disharmony and dissatisfaction between academic and general staff.[18]

Defence was left in no doubt as to the solidarity of the University's employees.

The only other vexing administrative issue was staff superannuation. Negotiations were complex, with Burns complaining privately to Myers that Department of Finance staff 'could not have cared less about the issue'. Many staff employed by the University were able to retain Commonwealth Superannuation Scheme (CSS) membership but Defence was unresponsive to requests from staff subsequently promoted by the University to have their promotion recognised in their CSS benefits. It took a formal approach to the RMC Interim Council to have the Army fulfil previous pledges. It looked like sound practical progress was being made as matters of principle intruded again.

At its meeting on 23 May 1977, the Development Council noted that the Chairman of the Universities Commission had questioned the principle of autonomy and expressed reservations that may have been raised at Cabinet. Myers told the meeting that 'if the autonomy of the proposed institution was to be circumscribed as a result of a Cabinet discussion, then Cabinet would be deciding to destroy the whole concept'. He also asserted the importance of postgraduate teaching and research, pointing to the ANU. Initially, it did not have any undergraduate teaching at all. It was absurd, he argued, to assert that the ANU was indeed a university and then deny the title to the Academy. The Council asked the Parliamentary Counsel to 'find words which ensure, without legal equivocation, that the institution is established as a university, while at the same time creating something which is more than a university in the usual sense, i.e., a military academy'.[19]

Formal Cabinet consideration of the Academy submission began on 31 May 1977. Cabinet was asked for 'firm approval' in place of 'approval in principle', priority in the drafting and presentation of legislation, approval for the expenditure of funds for professional services, and referral to the Parliamentary Works Committee. An interdepartmental committee was formed and told to report in one month on the status of the new institution, the timing of the Academy's establishment and more detailed information on when funding would be required. The Council received a report on 1 July and made its decision on 12 July. Hume reported to the Development Council on 15 July 1977 that Cabinet would be asked to approve the establishment of the Academy as an 'autonomous university' with the legislation to go before parliament in the next budget session. The proposed opening date remained 1982. The Cabinet also wanted more information on the 'provision for postgraduate training and research facilities, having regard for under-utilised capacities in existing institutions'.[20] Cabinet formally approved the establishment of the Academy on 20 July 1977. The Government's decision that it would have the autonomous status of a university created immediate controversy within the academic community. While Myers was naturally pleased, his response to a letter from Killen (dated 21 July) advising him of the Cabinet decision revealed some lingering concern:

> I hope that reference in your letter to discussions with the Minister for Education about 'planning for the provision of studies and research for higher degrees' does not indicate that there will be an opportunity for opponents of the proposed academy to so de-nature it as to make it unlike the normal conception of a university in the minds of people who are experienced in assessing 'universities'.[21]

Myers closed his letter by reminding Killen that 'adequate provision for postgraduate studies for higher degrees and for research is crucial to the success and acceptance of the new university'. Burns had been close to the Government's discussions and tried to reassure Myers by explaining that:

> Cabinet did reach a rider to the effect that a constraint be considered to prevent ADFA from competing in postgraduate activities with similar groups at other universities. This is one of those letters written 'just in case'. How formal the rider was

> I'm not sure but it was obviously mainly to 'protect' ANU! Len Hume informed me of this in strictest confidence and asked me for some assistance in preparing a case against this […] It does seem to me (and to Len) that it shouldn't be very difficult to quietly sort this out.[22]

Wilson had already produced a detailed paper on the need for postgraduate study and research at the Academy.[23] He began by noting that much of the specialised research undertaken at Jervis Bay, Duntroon and Point Cook was not being considered at the ANU, and that preventing the Academy from having a lively postgraduate program and an advanced students' community would directly and severely harm the 'success and reputation' of the new institution. It would have a bearing on teaching, staff professional development and recruiting. As the Faculty of Military Studies had existed for nearly ten years and was conducting valuable research, creating the Academy continued work that was already underway. This work would enhance academic life in Canberra, particularly the physical sciences, rather than impede it. Wilson did not believe that the Academy would compete with the ANU which, he noted, was not equipped nor demonstrably interested in the areas in which research at the Academy was likely to proceed.

The first meeting of the Development Council's Executive Committee after the Cabinet decision considered the matter of postgraduate research. In his report to the Committee, Dovers referred to the Cabinet decision:

> that there should be further examination of the requirement for postgraduate training and research facilities, having regard to under-utilised facilities existing at other institutions. There was a specific Cabinet attitude to research and there

appeared to be some implication that there should not be any postgraduate students.[24]

Later in the meeting, a statement in the functional brief that appeared to 'conflict with the reported Cabinet decision' provoked concern. Tange intervened and explained that 'there had been no direction from Cabinet in relation to higher degrees'. Tange went on to state that:

> there is to be an examination of the extent to which there may be excess capacity in facilities for higher study and this should be read as an injunction to avoid duplication at ADFA of facilities readily available elsewhere, in the interests of sound resource management and perhaps of education as a whole.

Notably, the newly appointed Chief of Defence Force Staff, General Arthur MacDonald, and Tange both objected to the phrase 'university level education' in the functional brief and suggested it be replaced with 'university education' as this, they believed, was the Cabinet's intention. This change was more than a cosmetic matter because it touched on an important point of principle. The cadets would receive a university education that would be recognised as such across the country.

Unlike North America, where institutions were at liberty to call themselves universities, the Australian Government reserved the authority to bestow the title 'university' on an educational institution – although it sought, and usually followed, the advice of the AVCC on deciding whether an institution should be deemed a university. Myers told the Chief of Naval Staff, Vice Admiral (later Sir) Anthony Synnot, that the new institution had to have 'University' in the title and was equally vehement that it not have a qualifier.

From autonomy to uncertainty, 1975–1978

> In the early days, [UNSW] was rejected by many – especially the very conservative ones – because of the qualifying phrase in its name [NSW University of Technology]. It has lived through this period and has thrived but I would not want to see such a fate visited on an institution of such national importance as ADFA is [...] the name 'Defence Force University' does not appeal to me. Frankly I think it would be a great mistake. I am increasingly of the view that the special nature of the institution can be emphasised, and the virtue of having the word university used can be best achieved, by the combination '[...] University – Australian Defence Force Academy' [...] I have been especially taken with the suggestion which has been made that the University in fact should be 'Hume University'.

Myers was responding to a memo which suggested that 'the name of an explorer would avoid undesirable connotations [...] to the contrary it would suggest qualities of enterprise, leadership and endurance. The name of an explorer who conducted expeditions in the Canberra area, to Jervis Bay and to Port Phillip would be eminently suitable'. Hamilton Hume seemed to fit the bill. The same memo counselled against using the name of a politician to avoid charges of partisanship, a former serving officer to avoid alienating two of the three services, or a place or person connected with one of the single-Service colleges. It was for this reason that 'Duntroon University' had already been rejected. The proposed name was always a heated issue and led to acrimony between Myers and General MacDonald. MacDonald did not want the word 'university' in the title. A sub-committee was formed on the motion of Myers and MacDonald to consider the matter. It later settled on the name 'Casey University' to honour Baron (Richard Gardiner) Casey, the former soldier, diplomat, Liberal Party politician and Governor-General, and their difference was resolved. Casey's public standing evidently transcended party politics.

Burns later told Myers that Tange had sealed the issue at the 9 August 1977 meeting by remarking that 'there had been a certain coyness in the past about this matter but now there had been a "resounding decision" by Cabinet that ADFA will be a university', therefore all references should be to a 'university education'. This decision was greeted with relief by Myers, who had earlier discussed the Academy's standing with Professor Sir Bruce Williams, the Vice-Chancellor of Sydney University and President of the Committee of Inquiry into Education and Training. Williams had initially believed that the Academy should be designated a College of Advanced Education (CAE). This was unacceptable to both the University and Defence, which decided together to avoid close contact with the AVCC until the Act was passed by Parliament and the AVCC had no choice but to live with the reality of the new institution.[25] Although Myers had just become the AVCC's new chair, at its meeting on 20 September 1977 the AVCC noted that Cabinet's decision to establish the Defence Force Academy as an autonomous university 'was made without advice from the AVCC and [resolved] that the AVCC would welcome an early opportunity to consider and advise upon the detailed proposals'. The prediction from Bill Hayden, by now the Opposition Leader, of strong opposition from the vice-chancellors was fulfilled. Myers was disappointed by this resolution and wrote privately to the Prime Minister stating: 'I am of the strong opinion that the Government has made the correct decision'. He went on:

> In my view the Defence Forces of Australia must have within their ranks at least their fair share of highly intelligent and

well educated and trained people. I am convinced that this can
only be properly achieved by pursuing vigorously the course
on which your Government is embarked.

Professor Anthony Low, the ANU Vice-Chancellor, was critical of the proposal. It would see, among other things, another university 'set up shop' in the national capital. Low's criticisms were not new. He contended that a university ethos could not co-exist within a military environment. Plainly, he had either not familiarised himself with the success of the Faculty of Military Studies or preferred the objection his university had first raised in 1959. For his part, Tange:

> suspected that the vice-chancellors were no less concerned by
> the loss of student numbers, and therefore funding, implied
> in the establishment of a new university, and especially
> that ANU would regret the creation in Canberra of an
> engineering faculty for which the ANU had itself been
> denied funding.

The opposition of vice-chancellors continued into the new year, when draft legislation for ADFA was prepared. The problem for Myers was to secure the Government's authority to show the Bill to the AVCC while knowing the text was complicated and the wording was still provisional. The Bill was an amalgam of several existing pieces of legislation and 'bolt on' additions to satisfy every individual and institution that had been asked to comment on the draft text. For UNSW, the bill created superannuation problems, principally the proposal that UNSW staff would be obliged to transfer from the University's employment to the Academy's, with possible loss of rights and value.[26] These concerns related to creating a new entity, that of Casey University, and the need for faculty members to join a Commonwealth superannuation scheme. Myers wanted transferring staff to be on 'permanent secondment' to the new institution so they could continue contributing to the University's scheme. But these were problems of process. Prior questions of principle were still being debated.

After some lobbying, Myers was eventually authorised to show AVCC members the text of the draft Bill on 1 March 1978. Several of the vice-chancellors were decidedly unhappy with the Government's whole approach, such as Professor Ray Martin, the Vice-Chancellor of Monash University. He told Myers that:

> the Minister has erred by not inviting the AVCC to express formally its considered views on the proposed ADFA. The establishment of a new and autonomous university is a matter of the greatest importance to the AVCC; its Vice-Chancellor becomes our 20th member. In this matter the proper role and standing of the AVCC has been debased.[27]

Professor Low of the ANU, the university most affected by the proposal, could readily see why Myers would endorse the Bill. He seemed less anxious about its consequences:

> I can readily appreciate that the proposed legislation will create an Academy of the kind the Services have been seeking, and the general outline looks to me very sensible and appropriate. It will be a tri-service Academy – of which the advantages are patent and manifold.[28]

But in a tone reminiscent of the mood within the ANU 15 years earlier, he questioned Casey University's capacity to manage its own affairs and to resist intrusions from other arms of government. He identified the reference to the new university being autonomous and independent as his main concern alongside what became familiar laments about the size and composition of its Council. Low privately confided to Wilson his judgment that it would be best to continue the relationship with UNSW, and that with the Vice-Chancellor located more than 100 miles (161 kilometres) away, the Academy would have sufficient operational autonomy to meet the Services' needs. He was not alone in holding this view.

The principal objections to the Bill were about preserving academic standards, and the absence of student organisations in the new university. Notably, 'balanced and liberal' appeared in the draft Bill alongside the word 'suitable', which seemed to some vice-chancellors to be a dangerous limitation. There was also disquiet about the respective roles of the Commandant and the Vice-Chancellor in the context of governance with a Council that was deemed far too big. The Vice-Chancellor of Griffith University and decorated wartime Fleet Air Arm pilot, Professor John Willett, exclaimed in his commentary on the draft legislation: 'The Lord help all who serve in her! The Council is far too large, far too heavy in *ex officios*, far too liberal in the provision of capacities to appoint temporary or permanent alternatives [...] I can see it as an unworkable monster'.[29]

Professor John Scott, the Vice-Chancellor of La Trobe University, asked whether there ought to be reference to academic standards in the draft Bill.[30] La Trobe's act specified standards comparable to Melbourne University. Myers said there would be resistance among the Duntroon Faculty to Casey University being based on the ANU, given its proximity. Professor George Davies, Vice-Chancellor of the University of Queensland, sent a telex to Myers after the Senate of his university resolved without dissent that Casey University was being established at an 'inopportune time' given the reduction of funding to Australian universities, that the Government had already announced that no new tertiary institutions would be established in the triennium, and the opinion that the existing arrangements met the needs of the services.[31] It was apparent to Myers that the vice-chancellors were divided. Some could easily live with Casey University; others were implacably opposed.

Two university councils passed formal resolutions on Casey University. They were Monash and Adelaide. The Vice-Chancellor of Monash, Professor Ray Martin, wrote to Killen on 9 May 1978 on behalf of the Monash University Council asserting that use of the term 'university' for a military academy was 'inappropriate and undesirable' and objecting to the proposed constitution of the governing body of Casey University which 'was heavily weighed in favour of the special interests of the armed services and not of the broadly representative character normally provided in universities'. Killen responded by saying the student body would be drawn from a diverse section of the community and that academic staff would have the same intellectual freedom as their colleagues elsewhere.[32]

Francis West, Professor of History and Government at Deakin University, prevented his university's Council from passing a formal resolution because he considered it 'political, rather than professional or scholarly and, therefore, inappropriate'.[33] In his judgment, Casey University was a cost-effective way forward while the draft resolution had misunderstood the nature of the university as a social and cultural institution. West had been a visiting lecturer at the Joint Services Staff College and the Army Command and Staff College. He explained in an ABC Radio National commentary entitled 'Notes on the News' that:

> whatever the mythology about universities, in practice the professional faculties (especially theological and medical schools) are segregated. Still, those who object on this ground may say that a 'true' university at least allows the possibility of exposure to 'community' points of view which a Defence Academy does not. But universities do not reflect community points of view. By their selection processes in fact they contain an unrepresentative section of the community. In any case, university training is not about 'points of view': it is about disciplined study by professional standards. The exposure to community points of view comes from outside the university: in the press, on the radio and television.[34]

West felt the proposed resolution was politically motivated rather than professionally justified. The Institution of Engineers Australia was also concerned about Killen's statement, although it was not aware of most of the details, a fact that Professor Green ascertained from meeting with some of the Institution's executive officers.

The full committee of the AVCC discussed the Casey University proposal at length in June 1978. The committee was opposed to the proposal by a margin of 10–9 or 9–8 according to a report in the *National Times* headed 'Dons split over plan for Defence Academy'. The AVCC asked the Government on 26 June 1978 to 'institute an inquiry at its earliest convenience' on officer training and tertiary education in Australia. Myers was obliged as the AVCC's chair to convey the resolution to the Minister for Defence. Only Professor Low of ANU and Professor Donald Stranks of Adelaide went public with their views. Low was opposed to the governance and Stranks to the academic environment. They did not support the Academy being funded by Defence. They thought its funding should come from the Department of Education because the

Academy would stand outside the control of the Tertiary Education Commission (TEC) which controlled funding to universities and CAEs.[35] They also noted that the Commonwealth Department was opposed to the Academy, as was the Education Minister, Senator John Carrick, who had served in the Australian Army during the Second World War and been taken prisoner by the Japanese on the island of Timor. Carrick did not oppose the evolving educational needs of the Services but thought the Casey University proposal was misguided. Tange later revealed that:

> Low came to see me several times to warn me of the likelihood that his colleagues would be sparing in their cooperation. There were doubts about matters of scholarly independence in a military based institution, and so forth. I entertained the thought that other motives might be at work. Was there room for a new rival in Canberra? Would the ANU continue to be denied an engineering faculty while the new institution would have a well-endowed one? While unspoken, these motives might be there.[36]

The Head of TEC, Professor Peter Karmel, claimed the AVCC's real reservation was 'in having a university in the full sense operating in a military environment. It seems to be a contradiction in terms'. The Chairman of TEC's Universities Council of the Tertiary Education Commission (Professor 'Noel' Dunbar, former Dean of Science and Deputy Vice-Chancellor of the ANU) had earlier written to the Minister for Education expressing its opinion on Casey University:

> The Council acknowledged that the military training role of the academy is not a matter on which it is competent to

express a view. It is, however, deeply concerned at the inclusion of the word 'University' in the proposed title of the academy. This term confers an academic flavour uniquely associated with a university. In the Council's view this is inconsistent with the principal role of the academy which is military training. The Council believes the intrinsic nature of the academy to be incompatible with the proper concept of a university because of the academy's single-purpose nature, its restriction of entry to a narrow section of the community, and the potential for conflict between the military and academic roles.[37]

Professor Roger Russell, the Vice-Chancellor of Flinders University, told Myers on 26 June 1978 that he disagreed with the 'Low–Stranks' opposition to the Academy:

> I do not believe that the AVCC has ever had any carefully thought out background position presented to it on the subject [of Casey University]. I do not like to become involved in actions which are not soundly based and I believe that, in this instance, there is a tendency to think with the adrenal cortex rather than the cerebral cortex.

Myers later expressed surprise that an opposition 'movement' gathered momentum in the AVCC to disassociate the entire higher education sector from the needs of Defence:

> I'm not really a military man, but I think having high quality people running your military services is a must for a nation. You might hope to never need them but if you do need them you want the best. And we had a lot of criticism for

having anything to do with Duntroon [...] It was a double contribution because not only did UNSW have courage to do what it did and create a tri-services college but the ANU didn't! And if ever there was a case for the Australian National University to have done something, then that was it.

When asked why he thought the ANU didn't pursue a place for itself within the Academy, Myers replied: 'thou shalt never do anything for the first time'. He claims to have ultimately changed opinions among his AVCC colleagues with the crucial consideration that the Defence student load would not be borne by the universities from extant funding but from new funding. This decision 'overcame any fears inside the university, and I think some fears in the Universities Commission circuit and also generally among vice-chancellors. They then recognised that the given sized cake was not going to be cut up into yet another piece to provide for the [Academy]'.

The relationship between the University and Defence had moved from focusing on affiliation towards autonomy but the goal of establishing a stand-alone degree-granting institution was still some way from being achieved as the 1967 agreement expired. It seemed that opponents of a 'Defence university' had become better organised and more strident. The Casey University concept had provoked scepticism from sections of the uniformed community, it had attracted critics from the ex-Service community, and there were a raft of detractors from the academic community. Its public advocates were few. The most energetic public supporters remained Professor Rupert Myers and Sir Arthur Tange. Its most influential political backers were Malcolm Fraser and Jim Killen, the latter in support of his Prime Minister. The scope and substance of the project was unprecedented. The remit of the three Service Colleges would change completely, the relationship between UNSW and

Defence would end, the cause of 'jointery' would take a quantum leap forward with a tri-Service curriculum and common uniformed training, an educational entity without precedent would be created and a very large building project undertaken. The project consisted of five major elements. The possibility of compromise had not been contemplated because each element of the project was closely linked. But given the forces arrayed against the proposal, compromise seemed inevitable.

8

Parliamentary works and political will, 1978–1981

As some of the vice-chancellors continued to oppose it, the tri-Service academy project was referred to the statutory all-party Parliamentary Standing Committee on Public Works (PWC) in March 1978.¹ As the project involved expending more than $2 million of public money, formal PWC consideration was required as a matter of administrative procedure. Mel Bungey (the Liberal Member for Canning) chaired the Committee and its members included Keith Johnson (Vice-Chairman), three senators and four members of the House of Representatives.² Exceeding its usual brief to examine proposed expenditure and ensure the Commonwealth would get value for money, the Committee took the unusual step of considering the intrinsic merits of the project. Over the ensuing twelve months, the Committee received evidence for and against the proposal before hearing from General Sir Arthur MacDonald and Sir Arthur Tange, who reiterated their strong support for the Academy. Notably, MacDonald was critical of the single-Service mindset, which was resolutely against change despite the benefits of collaboration, while Tange returned to his familiar theme of widening minds and the need for young service officers to be thinking creatively.

Tange had emerged as the most vocal advocate of a close university connection. He continued to believe the Services needed more officers with university qualifications and chided them for the recurring confusion between training and education, which he defined as the difference between 'learning how to do things (often technologically complex) and reasoning about objectives and consequences and the fundamentals of society'.[3] He noted 'scepticism about the advantages of liberal education at a higher level' and an 'unwillingness to disturb existing arrangements with some universities in certain disciplines'. He agreed that training and education had to be delivered concurrently, preparing cadets for leadership was vital and instilling self-discipline was crucial but found his suggestion that a course in philosophy would help to get junior officers 'asking questions' was greeted with stunned incredulity. While he did not want to dictate the contents of the curriculum, he believed the subjects to be studied:

> were matters of policy [...] and not to be left to the academics' views of their responsibility to this unique cadre of students, however much they might (and did) talk about freedom from interference on academic matters. I was in yet another minefield of professional egos. On a learning visit of my own to the Faculty of Military Studies at Duntroon, I learned from some academics that I belonged to a breed called 'bureaucrats'.[4]

The Committee's hearings were extended over five days in late April–early May 1978 and resumed on 8–9 August. It was clear that the Service Chiefs had not persuaded the Committee members of the need for tri-Service officer training. The Committee was also aware of the depth of feeling against the project among the vice-chancellors. Myers appeared during the first round of public hearings on 11 May 1978. He carefully explained that Defence's affiliation with UNSW had to end and that Casey University would be accepted on the basis of the quality of its staff, noting that UNSW had done a great deal to attract talented faculty since 1967. With great conviction he argued that:

> It is essential that a high proportion of the cadre of officers be drawn from the pool of most intelligent and able people available in Australia. It is my firm belief that these objectives can only be met in the way at present foreshadowed by the current Government decision to establish Casey University – ADFA [...] I can say with confidence that the general concept of establishing the new university in the manner proposed has the full support of the governing body of my University.

In subsequent questioning Myers stressed that the Faculty of Military Studies had done well but needed to become a university to realise its full potential and to meet the needs of the nation. Myers highlighted the need for 'separation' from UNSW as a function of the extant agreement and its necessity in ensuring the new establishment stood on it own feet: 'I believe the recognisable independence and autonomy of the new institution is important to its stature, and I believe it will establish a stature by virtue of its separateness and its objectives'. Murray Sainsbury (the Liberal Member for Eden-Monaro) asked Myers whether UNSW could extend its relationship with RMC along similar lines. He replied:

> That is a very difficult question. I am not in a position to answer on behalf of the only authority that matters – that is, the governing body of my university [...] the original agreement envisaged its conclusion sooner than now seems probable and because succeeding governments had committed themselves to the same kind of development [...] We are not

committed to perpetual association with that development. I believe it would not be right for the university to do it. I would advise my governing body against it. I would do that especially because I think it is not right for the new institution. I believe Australia needs an identifiable institution and I believe my governing body would support that view.

When asked by Bert James (Labor Member for Hunter) about the 'generous financial handout' that UNSW was receiving from Defence, Myers was indignant and replied:

the University is expected to provide a program; the University does not charge a 100 per cent profit for handling fees. It does not charge anything; it does it for the love of it. The Government merely meets the cost of doing what the Government and the University have agreed to do. I do not think this is generous. The people at Duntroon would not say it was generous and I certainly would not.

Keith Johnson (the Labor Member for Burke) argued that the existing arrangements were more than satisfactory and that the proposed Academy was simply about 'the prestige of having the 19th or 20th university in Australia'. Myers argued there was no alternative to a new university. Johnson countered by contending that the idea for a separate university was never properly considered and that 'slowly people convinced themselves that there was a need for a separate university'. Myers told the Committee that an alternative to Casey University had not been put to UNSW and that it was the only option on the table. He was pressed on whether it was possible to extend the existing agreement. Myers reiterated that he was likely to recommend declining to the University Council. He also said that transferring the Engineering Department at Duntroon to Canberra CAE would be a breach of the extant agreement. Johnson went on, noting what UNSW had already achieved at Duntroon:

> This place already has a reputation; it is the reputation of your university. Then why, in your view, must we disband that which exists? What advantage do we gain out of disbanding what we have and establishing something in its place that has to grow up?

Myers thought bringing the three Service colleges together would produce an institution greater than the individual entities. 'The standing of a university is not in its name but it is in the sum of the qualities of the staff at the time. It is the people who teach and do research in universities who are the standing.' Myers believed that the Faculty of Military Studies was doing a good job but it needed to evolve to the next stage and become a stand-alone university to ensure it had the right internal dynamics. He told Sam Calder, a Northern Territory Country Liberal parliamentarian and former wartime RAAF fighter pilot, that he had no doubt that Casey University would attract the faculty it needed to establish the reputation it deserved. He also said that the agreement was transitional and it was unlikely that the University would have accepted the role if it had been more than 'passing'.

When asked about the disadvantages of the extant arrangements, Myers remarked:

> The disadvantages are the very heavy additional burden on the executive. I, and some of my senior officers and some of the most senior academics in the university, have devoted a great deal of what almost would have to be called personal

time to this development, a development which we saw as being in the national interest [...] What are the advantages? The advantages are just a feeling of satisfaction on the part of the corporation that it is responding to the Government and what it sees as a national need. It is a service to the community; there is no joy in it except in the joy of seeing a successful outcome. There is no profit in it; there is a huge amount of expenditure in unpaid time. The Government does not pay me for the time I spend on this and I would not expect it to; the University is making its contribution to society.

Myers said there was no other option of which he was aware, 'just an obligation' to educate Service officers adequately. He felt the planned expenditure on the proposed Academy was crucial to the national interest. When asked what he made of representations from other universities, such as that expressed by the Council of the University of Tasmania that Casey would 'not be a university in the traditional sense' and that it would 'consume too much public money', he rejected the arguments. The nation needed to invest in officer education if it was to attract high-quality candidates. In terms of staff attitudes at Duntroon, he was unequivocal:

All of the academic staff recruited into the Faculty of Military Studies are committed to doing one of two things: accept an offer of the governing body of the new institution to join it or cease their employment. There is no option to return to the University.

Myers later said his support for the Academy was:

against the advice of friends and colleagues, both inside and outside the university, who couldn't see why I should regard it as vital. I think they misjudged the importance of giving those who will defend us a proper education [...] it took a long time to get the idea across to some of my colleagues but in the end I succeeded. Some people in the government, too, wanted to make it more or less a technical college but, in my view, this completely distorted the needs.[5]

In addition to Myers' evidence, Professors Beddie, Duggins, Wilson (on returning from study leave in Germany) and Dr Hugh Smith all appeared before the Committee. Wilson had recently visited the Royal Military College of Science at Shrivenham and was also familiar with the role played by West German universities in educating that country's young officers. He believed that it was preferable for academics to be members of an autonomous university than public servants and that it was better for education and training to occur in parallel rather than cadets being first educated and then trained. Dr Robert O'Neill, Duntroon graduate, Vietnam veteran, Rhodes Scholar, former academic staff member at RMC and Director of the Strategic and Defence Studies Centre at the ANU, spoke to the PWC and was congratulated by Myers on the 'lucidity of your representations'. Faculty members had also contributed to drafting the Bill establishing the Academy introduced into the Parliament by the Minister for Defence on 12 April 1978.

The most eloquent uniformed proponent of the Academy during the PWC hearings was the Assistant Chief of Defence Force Staff, Air Vice-Marshal Sam Jordan. His remarks deserve to be quoted at length:

What I must also say to some critics is that cadets at the new Academy, like cadets at the present Colleges, will not be restricted in the range or depth of their enquiries. They will be actively encouraged to be analytical, skeptical, critical and inquiring of both their academic and their military work. These are qualities that we encourage now, and that will become even more important in our career officers of the future.

But there is a further quality that we must impart to members of what is, and must be, a disciplined profession. That quality is that of judgment – including judgment as to the circumstances in which, the times at which and the forms of which scepticism, criticism and inquiry are appropriate, and should be expressed or made.

That essential quality of judgment which we strive to impart in no way fetters the critical intellectual faculties we encourage in our officer cadets. Indeed, it presents them with a further intellectual and moral challenge. It is simply an acknowledgment of what is necessary and appropriate for men who are both members of a disciplined, demanding and dangerous profession and at the same time, members of a liberal democracy. To those who persist in seeing some serious conflict here, I can only say that they have little knowledge of the realities of Service life.

Jordan would later replace Dovers as the Academy Chief Project Officer.

He might also have added that expenditure on the single-Service colleges was minimal after 1970. Establishing the 'impending' tri-Service academy had meant that essential repairs and maintenance were deferred. This meant that by the late 1970s the facilities at Jervis Bay, Duntroon and Point Cook were below standard and that the cost of converting them for use in the combined educational establishment, the so-called 'divided campus' option, would be considerable. Defence argued that building a single new facility would be cheaper than refurbishing three unsuitable sites.

As the PWC's deliberations continued with no end in sight, Fraser needed to intervene to prevent the project being lost in endless reviews and disagreements between parliamentarians and departmental officials who plainly had different views. Between the April–May and August 1978 public hearings, there was press reporting of financial pressures that might delay the establishment of the Academy and of the PWC's obviously negative attitude to the project. The *Canberra Times* complained about the 'enormous capital cost (at present estimated at $60.6 million) when seemingly more urgent defence needs are not being met'.[6] The *Canberra Times* editorialist also objected on a point of principle, despite being badly misinformed about the range of courses that would be on offer:

> The proposed institution should never have been called a university, for the simple reason that it will not be one. No university can be oriented to a single profession and have in its curriculum an inbuilt bias, for instance, towards the military arts.

A flawed appreciation of the planning continued in the ensuing paragraphs, which took the Government to task for not settling the question of whether the function of the Academy was to 'train military officers or to educate them to degree status'. The answer was, of course, to do both concurrently. This had been settled some years earlier. And while the *Canberra Times* was uninterested in the academic standing of the degree produced by the new institution, the Commonwealth and UNSW were intensely concerned. Neither wanted to create a second-class institution that would produce second-class degrees for what would become a second-rate profession as a consequence. The editorial called on the Government 'not to proceed hastily', claiming it needed to explain the 'precise functions and the character of the proposed institution', despite the amount of material that had been made available since 1974 and had remained largely unchanged.

Burns informed Myers on 1 June 1978 that a reliable source had told Professor Wolfgang Kasper, the Professor of Economics in the Faculty of Military Studies, that the Senate might not pass the Casey University–ADFA Bill because of concerns about Defence cost over-runs in other areas.[7] Casey University was a project that non-technical specialists could understand and, in this instance, find reason to oppose. Burns also explained that:

> John Stone of Treasury has told Kasper that it is unlikely that there will not be adequate provision of funds for ADFA, over the next five years at any rate. This is perhaps not surprising but he added that when ADFA featured in budget discussions, the PM did not defend the project at all strongly. The suggestion is that word of this attitude will get down to the backbench and may influence the Senate vote. Speculation then follows (I think this is Kasper's opinion rather than Stone's) that the Government might see a defeat in the Senate as a graceful way of bowing out of the project.[8]

Kasper's behind-the-scenes contacts with senior bureaucrats and Federal politicians were significant and influential. He had become foundation Professor of Economics in 1977 after moving from a Senior

Visiting Fellowship at ANU to a new chair in the Faculty of Military Studies. Kasper maintained a close professional relationship with many Senators and Members of Parliament, including a number of the 'Economic Dries', who were part of the informal 'Crossroads' group,[9] so named because its constituents supported the recommendations of the 1980 report *Australia at the Crossroads* (of which Kasper was the lead author) advocating trade liberalisation and other micro-economic reforms.[10] Kasper was able to explain to his diverse contacts in Canberra that the UNSW affiliation with RMC was working well and had the support of his colleagues. He also argued that a continuing association between the University and Defence to develop and support a tri-Service academy would be mutually beneficial and serve the nation well.

The focus on cost projections continued to dominate press coverage. The *Canberra Times* reported on 11 June 1978 that:

> The Parliamentary Standing Committee on Public Works has challenged the validity of Department of Defence accounting in the estimates for the development of the $52 million defence force academy in Canberra. It has sought the assistance of Treasury in providing an independent assessor. An official of the Auditor General's Department is expected to give an assessment within three weeks.

In response to this report, Burns advised Myers the next day that 'Kasper had been told that a Cabinet directive required a 20 per cent reduction in all government projects and Casey would be affected. Not sure how directly but it was considered possible that Kasper would be asked for unofficial advice'. The Member for the ACT electorate of Fraser, Ken Fry, asked the Minister on 13 September 1978 whether 'it is true that many appointments, including at least two chairs, cannot be filled because of a breakdown in the original agreement and that courses for 1979 may be abandoned?'[11] Neither was true but frustration was growing. Burns asked Myers to speak with the Faculty. They were becoming as anxious about the future as they had been in 1974:

> There is a difference however as it seems that an adverse decision now would really mean that the tri-service academy would never be established. What then would happen to the faculty? I regret that this letter seems to be about the least helpful that I have ever written to you. I can only hope that this is in itself a sufficient reason for sending it.

The President of the RMC Academic Staff Association, Professor Harry Green, told Myers on 19 July 1978 that he was prompted to write by press reports in the *Canberra Times* and the *Australian* on 14 and 15 July claiming that Casey University might be deferred indefinitely.[12] The staff feared there was some substance to the reports and he wanted Myers to know it was damaging faculty morale. He pointed out that many academics had come to RMC believing they would be part of an autonomous degree-granting institution by 1977. By mid-1978 the prospect still seemed a long way off. Green concluded: 'there is a widespread crisis of confidence in the Faculty of Military Studies' and asked for a personal meeting to discuss the situation. Defence tried to reassure the University that there was no cause for alarm. The Deputy Secretary of the Defence Department, Norm Attwood, said the Government had not deviated from its commitment to the Academy despite press reports. The uncertainty, he suggested, was a function of the 'protracted proceedings of the PWC', which was holding up passage of the legislation through parliament. He added: 'because of the delay, it had been decided not to commit any funds to the project in the current financial year. In consequence the opening date was likely to be deferred beyond 1983'. By now the University had a new leadership group. Professor Al Willis had retired in July 1978 and was succeeded by Professor Ray Golding as Pro Vice-Chancellor; Professor Geoff Wilson formally became Dean on 1 November 1978 after Professor Tom Chapman had served as the Acting Dean while John Burns was absent on sick leave. Burns stood down as Dean at the beginning of second term in 1978 after protracted health problems. Thankfully the parliamentary hiatus allowed Golding to become familiar with the range of issues being canvassed; Wilson had been thoroughly acquainted with the challenges for some years.

The inquiry was intentionally slow and laborious, a tactic designed in part to deplete Fraser's personal enthusiasm for the project. One Committee member could not see why the Academy should be pursued when the Government was on an austerity drive and there were other more deserving initiatives. There was also the problem that Government members were opposed to the project and were determined to defy Fraser as a point of principle. They felt democratic processes were at stake and Fraser's insistence on establishing the Academy was dictatorial. The newspapers noted that the 'academy proposal has been criticised by the Vice-Chancellor of the ANU, Professor Anthony Low, who suggested the retention of university-affiliated academies as an alternative to Casey University'. A *Canberra Times* report claimed that the Chief of Defence Force Staff (CDFS), General MacDonald, had overruled the Service Chiefs, who wanted the matter re-examined in principle (a tactic that had also been tried and had failed in Hassett's time as CDFS) while Tange was 'not available' during the inquiry. MacDonald was reported in the *Canberra Times* article as saying:

> Obviously, in the profession of arms, such an education
> [a balanced and liberal university education] must be
> complemented by military education and training – and this
> cannot be provided by other universities. Why, then, does
> the concept of an institution which will aim to meet, at least
> in part, both requirements conjure up 'almost forgotten folk
> tales of sabre scars from Heidelberg'?

This report was inaccurate inasmuch as two of the Service Chiefs, Lieutenant General Sir Donald Dunstan and Vice Admiral Sir Anthony Synnot, were known to be in favour of the Academy project.

The ANU's obstructionist role was having its own effect on relationships between academics at the two institutions. After accepting an invitation to lecture the Masters of Agricultural Economics students without remuneration, Wolfgang Kasper told Dan Etherington of the ANU's Development Studies Centre on 6 June 1978: 'I hope in the presently tense atmosphere regarding Casey University, that my decision to help you will be seen as a friendly gesture from someone in Duntroon and an illustration that the neighbourhood of a second university can have benefits for ANU'. Having had a personal experience with Economics and Social Sciences at the ANU, Kasper maintained some of his teaching and lecturing contacts there and finished publishing a conference volume which he had begun at the ANU. The subsequent dual badging of the volume generated considerable goodwill at the ANU and was a tangible sign of the collaborative spirit that animated the RMC Faculty.

The PWC resumed public hearings in August 1978. The Committee defiantly resisted pressure to expedite its proceedings. The Government wanted to schedule Parliamentary consideration of the Academy Bill in October or November 1978. If there were further delay, the Academy would not be able to open on time. In late August and with no sight of the PWC's report, the planned opening was shifted to at least 1983. On the one hand the protracted Committee proceedings actually helped the Government by delaying the need to spend any money but, on the other, it allowed opponents of the project to rally support. These opponents did not need much encouragement.

The Representative Council of the Federation of Australian University Staff Associations (FAUSA) resolved overwhelmingly to express its opposition to establishing Casey University at a meeting held on 13–14 February 1979.[13] FAUSA's objection focused on Section 5(i)(a) of the Academy Bill: 'the functions of the Academy are to provide, in a military environment, a balanced and liberal university education for officer cadets'. The FAUSA resolution objected completely to the creation of Casey University and 'in particular giving it the title of "University"'. If the 'proposed military academy is established as a university, FAUSA will advise all academics in Australia and overseas not to apply for appointment at that university'. The Association claimed that entry into the 'academy university' was not based on academic merit but other factors. FAUSA also asserted that applicants for Casey University might be discriminated against on political grounds and it was unhappy with the number of military and government appointees on the Council (which numbered 30). It contended that the proposed university was sexist in not permitting the full participation of women, noting that a 'sexist university is a contradiction in terms', although it overlooked the existence of single-sex university halls of residence. It alleged that 'it may not be possible to provide a balanced and liberal university education in a military environment':

> A university education requires that all concepts be

challenged and that a person's outlook within a university should be guided to challenge every concept, to accept nothing until it is substantiated whereas a military environment really requires the opposite, acceptance of decisions by those in authority without question. A defence force in which every member questioned the decision of those in authority would be no defence force at all.

This was very clumsy reasoning. The confusion of education and employment in adjoining sentences is obvious. It is one thing to challenge concepts and another to accept a decision. Concepts and decisions are different things. The RMC Academic Staff Association (which had been a member of FAUSA since August 1970) viewed FAUSA's 'attitude with concern'. It moved a resolution on 15 March 1979 stressing that including the word 'University' in the title of the Academy had been supported by RMC staff since 1972 and called for greater academic representation on the Council of the proposed university.

At the start of 1979, the PWC had still not reported but rumours began to circulate of its likely findings. The Vice-Chancellor of Melbourne University, Professor Sir David Derham, wrote to the PWC secretariat on 13 February 1979 commenting that the reaction of the universities to Casey University was mixed but 'I have been a little surprised at the extent of the adverse reaction. Some of it, of course, comes from people who carry an ordinary human failing with them. Some of the reactions are reactions of envy [...] to many people this institution will look like a gold-plated Rolls Royce'. The wide range of opinion within the AVCC and the extent of the split among its members had alarmed Derham. It was unusual, he thought. Wilson told Myers on 26 February 1979 that the PWC was very likely to recommend against establishing the Academy. He described the

Committee as 'immovable', other than Sam Calder, who appeared to be the most sympathetic. Wilson was worried that FAUSA's resolution would imply that RMC staff were opposed to the Academy when the reverse was actually true. Wilson also said that Defence's own internal and confidential analysis was that:

> ADFA is cost-effective with respect to capital cost and that extensive use of civilian universities involves a lower recurrent cost but is probably unacceptable on other grounds. The idea of a multi-campus Defence university (ie., retaining the colleges but making them part of a single academic entity) was also seen as too expensive.

While the majority of advice tendered to the PWC supported establishing the Academy, 'the long time they have spent on this inquiry plus the variety of anti-ADFA evidence means that they will lose much face unless they come up with a critical report that appears to be helpful'. Wilson, writing to the Vice-Chancellor, then mentioned something that had not been seriously countenanced before:

> I am aware that the FAUSA decision plus the difficulties with the PWC have made Defence think carefully on possibly approaching you on the matter of a continued affiliation – possibly in the context of the planned tri-Service Academy. On this they are extremely embarrassed by some of the rash and unfair statements made to the PWC.

Myers made no comment. There were no conversations with Defence along these lines. Months passed with no apparent progress being made. Behind the scenes, the PWC report was written and re-written until the Committee chair, Mel Bungey, finally tabled it in

Parliament on 31 May 1979.[14] It was the most protracted hearing in terms of sitting hours ever conducted by the PWC and the biggest in terms of the evidence sought and submitted.

As expected, the PWC's conclusion was unanimous: the Academy project should not proceed because the expenditure of public funds could not be justified.[15] The Committee was not persuaded that the number of degree-educated uniformed officers needed to be a third of the entire officer corps. It claimed that there was no clear evidence that a common tri-Service establishment would enhance officer education or that interaction at cadet level would make the three Services any more co-operative. The Committee insisted that 'the situation with regard to the agreement with UNSW is not of itself a singularly strong reason for the establishment of ADFA'. It recommended enhancing single-Service colleges and respecting single-Service education needs, noting 'there is scope for high academic standards by being affiliated with larger institutions which benefit from size and, as present experience shows, from the attraction they have for top quality staff'. The Committee recommended that the ANU should replace UNSW at Duntroon, accepting responsibility for the programs and the staff. The main element in this 'take over' would be to create an engineering school at ANU where one did not previously exist. This was an unexpected and surprising recommendation given that the ANU argued for continuing the existing arrangements and showed no interest in any affiliation with Defence. The AVCC was pleased with the PWC outcome, expressing agreement that 'it is not expedient to proceed with the construction of the proposed work'.

Killen was not surprised by the report, given how much had been leaked before its release. He offered a non-committal response: the Government would examine the report and respond at a later date.[16] Fraser was angry but undeterred. He spoke with Tange, who agreed that the Committee 'might have gone beyond their charter'. Tange remarked that 'they ought to stick to discussing bricks and mortar and not education policy'.[17] While Tange could claim a strong body of support for his arguments and the efficacy of establishing a joint academy, his biographer (Peter Edwards) noted that the strength of this support was 'not sufficient, however, to neutralise' uniformed and political opponents of the idea, who strangely thought the Academy was a Labor scheme that the Coalition ought to reject. It would appear that the main point of contention was not the need for a comprehensive tertiary education but the consolidation of officer education in a joint academy. It was 'jointery', not education, that aroused suspicion and created hostility. Edwards argued that the size of Fraser's parliamentary majority worked against rather than for the establishment of the Academy, as rebel backbenchers felt less constrained to defy their Prime Minister in relation to this project. Tange would shortly retire from the Public Service but remained an energetic supporter of the concept, later being appointed a Visiting Fellow at ADFA, in whose library he lodged a collection of papers relating to higher defence administration.

As the Government pondered the PWC report, backbenchers on both sides of Parliament quietly briefed journalists on their objections to the project and why they had gone beyond 'bricks and mortar' to consider the rationale for the development. Labor had decided to vote against the proposal although it was originally a Labor initiative, while Coalition members were threatening to stage a well-publicised revolt if Fraser persisted in ignoring their views. The only member of Cabinet to refer to the matter publicly was the Minister for Education, Senator John Carrick, who said 'the Government's policy to establish such a tertiary institution still remains', although he conceded that financial priorities had shifted.[18]

The future of the Academy disappeared from view for nearly

six months.[19] Killen's office drafted a Cabinet submission on 23 October 1979. It asked Cabinet to agree to the Defence Minister advising Parliament that:

> the Government does not accept the PWC recommendation that the Academy project does not proceed; and to inform Cabinet of his intention to look into the possibility of an association with UNSW and to examine the possibility of the Academy using RMC land and facilities.[20]

Notably, Killen had not raised the prospect of a continuing UNSW affiliation with Myers and would not mention the matter for another five months. There is nothing to suggest they had a private conversation or whether Killen thought that Myers would move from his previous insistence that the affiliation had to end. Cabinet considered Killen's submission on 27 November and concurred with exploring 'the advantages and practicality of, perhaps, an initial association with UNSW'.[21] This 'exploration' would be confidential.

Killen issued a press release on 7 December 1979 reporting the outcomes of the Cabinet meeting held ten days earlier.[22] This happened to be the day before Carrick vacated the Education portfolio to become Minister for National Development and Energy.[23] If Carrick remained opposed, he could not speak publicly about a matter that was no longer within his portfolio responsibilities. Killen said the Government 'has rejected the conclusions formed by the Public Works Committee. These conclusions are ill-founded'.[24] The Minister went on to say that the Service Chiefs supported the development in part because the existing arrangements were inadequate and 'not economical'. This statement overlooked the lukewarm support of the Chief of Air Staff, Air Marshal Sir James Rowland, and the objections of some retired officers. The AVCC noted that Killen's statement made no reference to Casey University – something it considered a significant omission. It was indeed significant. There was no mention of UNSW in the statement either, as Myers had still not been sounded out and the University Council had not been approached to give its approval in principle.

Newspaper editorials and letters to editors were predominantly against the whole Academy concept, including Casey University. In a letter published in the *Age*, a member of the public named Ross Homes was critical of the expenditure 'at a time of education cuts', noting that the Commonwealth Government's investment in higher education infrastructure was down by 44 per cent since 1975, while the same newspaper's editorial observed that 'champions of the Academy are losing the logical, educational and philosophical argument'. On 15 December 1979, the *Sydney Morning Herald* advocated delaying a decision and initiating more discussion. It noted that Robert O'Neill was in favour of Defence 'having its own university' but was conscious that the ANU Vice-Chancellor had recommended continuing with the Faculty of Military Studies. Of course, this would not help either the Navy or the Air Force. Academics far removed from the conversation continued to argue that the concept of a university was completely incompatible with operating a military academy.

Hugh Smith produced yet another insightful article to clarify the situation and the range of options.[25] He explained that even if $65 million were not spent on building the Academy, approximately $39 million would need to be expended on rejuvenating facilities at the single-Service colleges. This money would be drawn from the Defence budget and not the Education budget. There would also be an annual cost saving of $5000 per student to educate cadets at the tri-Service Academy. He noted that Killen's statement made no mention of Casey University and thought it was 'perhaps too soon

Parliamentary works and political will, 1978–1981

to predict what arrangements will be made for tertiary education at a tri-service institution'. While the cadets at Duntroon were not studying in a university *environment*, they were receiving a university *education* and he observed that the Army was 'entirely satisfied with its arrangement with UNSW'. He thought the idea of limiting recruitment to graduates was an untested and unreliable source of manpower, while fears that cadets would fail to develop Service loyalties were exaggerated.

On 31 January 1980, Tange's deputy, Norm Attwood, wrote to Myers explaining that work was being done within the Department of Defence on new proposals and he would be briefed shortly. No mention was made of a continuing affiliation. There was now no possibility of the Academy opening in 1983. Meanwhile, the mood in the Faculty had changed over the summer break. Writing on behalf of the Duntroon Heads of Department, Wilson told Myers on 27 February 1980 that:

> We have given strong support to the setting up of a
> University/Academy which would be autonomous from
> its inception. However, we now feel that the opposing
> weight of public opinion (expressed by universities, Service
> personnel, newspapers and politicians) makes the setting up
> of an institution through affiliation with UNSW a far more
> realistic goal. The impact of the unanimous PWC Report was
> the last straw which would make it almost impossible for an
> autonomous body to get off to the right start.

In the meantime, Wilson made an urgent plea for funding new facilities, after the shortcomings of existing facilities had been drawn to the attention of staff unions. In November 1979, the Faculty Administrative Officer was approached by the Association of Architects,

Engineers, Surveyors and Draftsmen of Australia and told that technical support officers employed by the Faculty were working in conditions that were 'probably detrimental to their health'. The workspaces were cramped and poorly ventilated, and lacked toilet facilities. The ACT Division Representative said the Association 'expect management to take immediate steps to rectify the situation'. The Association asked for a progress report by 17 December 1979 and urged the Faculty's leadership to grant staff 'special leave' when 'heat conditions become intolerable'. By this time the Navy and the Air Force both concluded they needed change and could not justify the cost of having a small cohort of students in buildings that seriously needed to be restored and expanded. Something had to be done.

Killen went back to his Cabinet colleagues on 2 March 1980 with a submission that critiqued the PWC's report in detail. The Cabinet decided on 25 March that Coalition parliamentarians would be informed of the Government's view of the report and that a meeting of the Joint Party Room would be held before a statement was made in Parliament. This was an opportunity for the Government members of the PWC to put their views, in the hope they would not choose to express them publicly. Fraser was conscious that a minor revolt was gaining momentum within his own party.

Before the war of words began, Killen wrote to Myers on 17 March 1980:

> The question arises whether it is necessary or desirable to move in one step from the present situation of separate Service colleges, each sponsored by a university, to one tri-Service Academy which is an independent university. If a suitable arrangement could be reached with a university, the urgent need to consolidate the education of cadets could be satisfied

by establishing an institution which, suitably governed and developed, could evolve to a point where the ultimate transition to autonomy would be simply the recognition of its effective status. In the light of these considerations, I would be grateful if your University, which is uniquely placed to understand and contribute to our needs, would agree to explore the feasibility of an arrangement whereby the University would ensure the academic integrity of the institution, perhaps by means of the establishment of a college of the University.

Four days later Myers replied to Killen's letter, promising to consider the invitation and pledging an immediate reply.

In the meantime, Professor Don George (Vice-Chancellor of Newcastle University), representing the AVCC, asked Killen to affirm that the Government's preferred option was affiliation with an existing university, something the AVCC strongly endorsed. Nearly two years had passed since the Academy project had been referred to the committee.[26] No-one had expected such a long delay. There were few realistic options open to the Government. The first was to affiliate with the ANU, an unlikely prospect given the continuing disinterest and occasional hostility of successive Vice-Chancellors and Councils towards Defence. The second was to negotiate with other universities, noting that 'this could only take place if there was a clear Government intention to repudiate a formal agreement with UNSW. Awareness among universities of that intention would not provide a good basis for negotiation'. The third option was to extend and enhance the existing agreement with UNSW.

On 26 March, the Government's response to the PWC's report was made public. The PWC members were not, however, prepared to accept criticism of their report quietly or to limit their expression to the Party Room. The chair, Mel Bungey, prepared a paper for tabling in Parliament that sought to 'put in perspective the inferences of bias by the Committee and to correct some of the more major inaccuracies in the analysis' prepared by Defence.[27] The PWC response was critical of Defence's evidence, which it deemed 'unsatisfactory and did not prove the need for this particular project'. The tone then became personal: 'the Committee strongly resents and repudiates [Defence's] inferences that its conduct of the inquiry was biased, dishonest and unfair'. Indeed, the purported grounds for such allegations were 'scant, unsubstantial and dishonest'. There was no criticism of UNSW in Bungey's response before a firm restatement of the original recommendations. The Labor Senator from Queensland, Mal Colston, claimed that 'Casey University is Mr Fraser's pet project. He conceived of the idea and he is determined to give birth to this million-dollar white elephant'. Colston alleged that Fraser 'silenced' his own Education Minister, Senator John Carrick, and called for the project to be deferred until objections to the Academy were adequately answered.

On 26 March, as the Government's response to the PWC's report was made public, Myers wrote to Professor Doug McCallum, the chair of the UNSW Professorial Board, to seek his view and that of the Board before the University Council considered the proposal for a continuing relationship with Defence. Myers was in a difficult position, having assured the Council that the affiliation with RMC was short-term and having made public statements during the PWC hearings that UNSW neither wanted nor was able to continue with the affiliation. McCallum was personally in favour of a continuing relationship with Defence and agreed to persuade his colleagues as to the merits of a continuing affiliation. The Professorial Board discussed the Minister's question on

1 April 1980. It gave in-principle support without hesitation or dissent. Myers prepared a memorandum for the Executive Committee of the UNSW Council on 28 April 1980 – the date that Killen had chosen to circulate another Cabinet submission. The Minister needed the UNSW Council to agree. Myers explained to the Council members that neither the UNSW Staff Association nor the RMC Staff Association wanted the next phase – the establishment of a University College – to be an interim one on the way to eventual autonomy. The University College would be a continuing entity. He concluded:

> The University's Act would seem to make it possible for the Council to approve the establishment of a College in Canberra if it so desired [...] I would support a proposal that the University should explore the feasibility of an association with the Government and believe it would be possible to come to an acceptable arrangement.

Myers told Killen privately that the Council was supportive and the Cabinet submission made mention of 'encouraging indications of UNSW's attitude towards an association with the Academy'.[28] Myers wrote formally to Killen on 12 May 1980, explaining that the University Council had given its approval in-principle to assist the Government with the Academy project by establishing a College of the University in Canberra. He assured the Minister that:

> there is general acceptance of the belief that the experiment with RMC and the Faculty of Military Studies has been very successful [...] it is indeed possible to work cooperatively and fruitfully with both civilian and military officers of the Defence system to develop a first-class education program for

the preparation of Defence Force officers. I am confident that the goodwill and understanding that has been built up during this process and the experience we have gained will stand us in good stead as we proceed to the next important stage.

Three days later Killen made a Ministerial Statement to the House. He admitted that the Government's proposal had indeed changed but affirmed two principles: Defence needed more officers with degrees and academic education needed to occur within a military environment.[29] As Hugh Smith noted:

> For the government the crux of the matter was a tri-Service institution, not an autonomous university. The proposal for a university college would achieve the primary objective while ensuring that cadets received an education from an established and respected university. This outcome appeared to undercut much of the political and academic opposition to the proposal.[30]

Smith had been quietly making this point since 1974.[31] Others had privately maintained the same view but had remained silent because Rupert Myers had been adamant that Casey University was the only option UNSW would support. There was a sigh of relief within the Faculty of Military Studies. The relationship with UNSW, which was tried and tested, would continue. How long was not clear. Would it be decades or in perpetuity?

The critics wondered, however, whether this proposal would lead to establishing Casey University by stealth. A motion commending the statement was carried on party lines, although the three Government members on the PWC abstained from voting, along with the Minister for the Capital Territory, Michael Hodgman.

Parliamentary works and political will, 1978–1981

In reporting these developments to the Faculty, Wilson noted that while the 'Professorial Board gave its strong support for the Minister's proposal, its debate of the matter was vigorous and quite heated'. He urged staff to 'give the highest priority to liaison with members of the Department of Defence and the Department of Housing and Construction over the buildings'. He reported buildings would be ready by 1985, with the first students enrolling in 1986.[32] A new agreement between the University and Defence with a preamble written by John Burns was circulated for comment from the third quarter of 1980.

Fittingly, Myers' last official act as Vice-Chancellor before retiring was to sign an agreement with the Commonwealth to establish the University College in Canberra. He recalled: 'It was a long struggle to get it up and it was finally approved. I'm proud of that'.[33] Burns gave much of the credit to Myers: 'he had such a vast attention to detail and his quickness of thought and his skill in handling [negotiations] was absolutely tremendous'. There were others deserving of praise, especially Professor Rex Vowels:

> He was always admired by everybody at the University. He did all he could for us just as he did all that he could for everybody else. He really was a tower of strength. He frequently came to Interim Council meetings representing Rupert and was a good friend of the place.

Burns also praised the contributions of other colleagues: 'Don Vallentine offered guidance on university procedures as has his successor, John Gannon. Godfrey Macauley was opposed to the agreement but changed his mind and offered his support as did Tom Daly the personnel manager'. Burns thought UNSW had gained 'some sort of moral standing and reputation for having done a national service, perhaps fought in places where it matters, at high government levels [...] it was a good thing to have been involved in'.

Myers displayed no resentment at the outcome. He accepted with good grace that the affiliated University College model was a much better result for Defence than an autonomous degree-granting institution. He also remarked that 'no history of the association should be written which doesn't draw attention to the good fortune we had in the succession of Commandants' at Duntroon. It is truly remarkable that there was no major 'falling out' between successive Deans and Commandants or ruptures between the academic faculty and the military staff over any issue – including occasional cadet misconduct – to derail the proposal and disrupt the relationship with UNSW. In fact, most Deans believed the Commandants were the Faculty's strongest supporters within Defence and the University. Myers also acknowledged the work of departmental officials, particularly Bruce White and Sir Arthur Tange, the latter being 'the most outstanding' because 'he knew how the government system worked and manipulated it for many years, and I thought very well'. Myers regarded the establishment of the UNSW Australian Graduate School of Management and the Australian Defence Force Academy as the foremost contributions he had made to the University and the nation.

The final form of the agreement offered the University favourable financial terms. This was not surprising. The University had extricated Defence from a very deep hole when Casey University became untenable. Had UNSW maintained its insistence that it could not and would not continue with the affiliation, Defence had virtually nowhere to go. It needed an external provider and none of the other universities was experienced or interested in a relationship. The transition to a continuing agreement was not without its own challenges, principally determining the balance between academic education and vocational training, and how the University and Defence staffs would interact. As the University College model had not been tried elsewhere (other than in a limited form at Duntroon), there were likely to be unforeseen problems and unimagined challenges. With sufficient goodwill, both sides believed that compromise would enable collaboration in a manner that would not conflict with the organisational integrity of either the University or Defence. The devil, as always, would be in the detail.

9
From affiliation to Academy, 1981–1986

On 19 February 1981, Prime Minister Malcolm Fraser unveiled a plaque to mark commencement of building work at the Australian Defence Force Academy site in the Canberra suburb of Campbell.[1] There was no turning back. The Academy would be built.[2] The formal agreement between the Commonwealth and UNSW signed on 7 May 1981 was a substantial achievement for both the Government and the University.[3] Arthur Tange and Rupert Myers provided most of the impetus to push on despite obstacles and objections. The Defence Department Secretary was full of admiration for Myer's leadership, remarking that he was 'an excellent bridge between the cultures' of the ADF and the University. Tange thought that he, as 'an administrative reformer', and Myers, as a 'visionary educator', were a formidable team. Myers retired on 31 July 1981 and was made a Knight of the British Empire (KBE) in recognition of his long and productive service to higher education in Australia. His successor was Professor Michael Birt, the inaugural Vice-Chancellor of the University of Wollongong from 1975 to 1981.[4] Birt was familiar with UNSW, as Wollongong had been a campus of UNSW before it became a stand-alone university.

As the decision to proceed with the Academy had been made, the new Vice-Chancellor's role was focused on implementation and administration. This proved to be complicated work but was less demanding than the detailed policy slog that had marked the previous fifteen years. By this time Rear Admiral Dovers had retired as Chief Project Officer.[5] Dovers, whom some university people had initially seen as aggressive and even 'pushy', was completely committed to the project; this was never in any doubt. He was forthright and, at times, forceful. He worked quietly but resolutely against the detractors, he resolved issues with architects, he settled problems with builders and assured the Navy's senior leadership, the source of most of the naysaying, that it need not fear the new institution and its influence on naval culture. It was not surprising that the main administrative building housing the offices of the Commandant and the Rector was later designated the 'Dovers Building'. His contribution deserved such recognition. He was succeeded by Air Vice-Marshal Sam Jordan.

Six weeks after the Agreement was signed, the Development Council was formally disbanded by the Minister for Defence and replaced by the Interim Academy Council. It met for the first time on 28 June 1982.[6] The Council was chaired by Sir Edward Woodward, a widely respected Federal Court judge (1977–1990) and former Director General of Security (1976–1981).[7] Woodward had been approached about the post on Christmas Eve 1981 after a consultation with Michael Birt, who agreed Woodward would be an ideal choice from the University's perspective. Woodward accepted the invitation immediately and was congratulated by Lieutenant General Sir Phillip Bennett, the Chief of the General Staff, who looked forward to his contribution to a 'significant national undertaking'.[8] Bennett considered the Academy to be 'one of the most important long term Defence initiatives taken since Duntroon was established in 1911'.[9] The Council's mandate was outlined in the Commonwealth–UNSW Agreement. It was a document that would attract much attention over the ensuing 20 years.

According to its charter, the Academy would provide 'military education and training (MET) for officer cadets for the purpose of developing their professional abilities and the qualities of character and leadership that are appropriate to officers of the Defence Force [...] [and a] balanced and liberal tertiary education within a military environment'.[10] The Agreement's most contentious section proved to be 3.1. This section effectively created two new institutions pledged to act in partnership – the Defence Force Academy and the University College.[11] Section 3.1 stated that 'the Academy shall consist of the military component referred to in Part 4 and the College of the University referred to in Part 5'. In essence, there were two constitutive elements: a military unit established by the ADF and the University College formed by UNSW. The two relevant sections (3.1 and 5.1) of the Agreement needed to be read in tandem for the nuance to be understood. There is no comparable mention of an 'academic component', but rather of the 'College of the University' which, in section 5.1, is established 'within the ambit' of the Defence Force Academy. Notably, the Agreement did not state that the University College would be part of the Academy. The problem was the wording of section 3.1. To remove any confusion or dispute, had the section simply required Defence to establish the Academy and UNSW to form the College there would have been no doubt that the Academy was not more or less in stature than its 'military component'.[12]

In essence, the Academy was never intended to be the umbrella organisation within which the University College operated. It was a separate organisation with its own chief executive, its own regulations and its own accountabilities. Nor was there any sense in which the 1981 Agreement made UNSW subordinate to the Department of Defence or that made the University College subservient to the ADF. The Academy Commandant had no authority over UNSW staff. That prerogative resided solely with the Rector. The Commandant could not command or direct the academics other than to urge them to comply with the provisions of the 1981 Agreement. Only the Commonwealth, as the principal signatory to the Agreement, could require the University to fulfil its obligations. An agreement that effectively delegated the University's authority to a third party or transferred its prerogatives to another organisation would have been contrary to the *University of NSW Act*. Conversely, the University College was established by UNSW, which exercised complete control of its operation 'within the ambit' of the Academy (section 5.1).

The creation of two distinct and separate entities was intentional. As Hugh Smith explains:

> A compromise was sought which would cover up this dichotomy since there was a desire to create a *sense* of unity at this stage. The phrase 'within the ambit' was accepted because it has a geographical connotation ('within the perimeter of') which both sides could accept, while stopping short of stating that the University College was part of the Academy.[13]

The Agreement reflected and deflected the tensions and the ambiguities that had not been resolved by mid-1981. The ADF wanted to avoid the appearance that it had no control over junior officer education, while UNSW wanted to avert any criticism that Defence was controlling the curriculum. There was, however, trust on both sides and an acceptance that a complex, if slightly contradictory, agreement was the best they could achieve within the available time. Despite previous reservations about the benefits of a tri-Service Academy in some sections of the ADF and the Defence Department, once the 1981 Agreement was signed there was complete commitment on all sides to seeing it built efficiently and on time. The Defence interface would initially be led by Colonel Alan Hodges of the Defence Force Policy Division (representing the ADF) and Rob Tonkin, the CEO of the Defence Training Branch (representing the Defence Department). Defence was pleased to be working with UNSW and did not anticipate any major difficulties during the building phase. The University was highly respected within Defence and well regarded within the higher education community.

The principal challenge, according to Tonkin, was for Defence to be clear about what it wanted and needed from the University.[14] While there was agreement that the undergraduate student body would not include civilians, on the grounds that they would dilute the military culture, there was no vision for postgraduate education and no articulated strategy for exploiting academic research. Defence accepted that postgraduates were needed to promote a research culture. There had been postgraduates at Duntroon, albeit in limited numbers. The postgraduates could use the new facilities beyond 4 p.m., when most cadets were engaged in recreational sport and military training. Defence was yet to appreciate that it would benefit directly from the presence of postgraduates and their supervisors at the Academy in terms of research outputs. In the years that followed and particularly in relation to Masters courses, military students mixed with mid-level public servants and formed relationships across government departments and agencies that would prove invaluable when goodwill and collaboration were needed for a whole-of-government initiative.

Defence certainly wanted to see an increase in cadet retention levels, as the combined resignation and failure rate was then approaching 20 per cent. Tonkin saw the evolving relationship between the two institutions as a diarchy, with the Rector and Commandant reporting to a Council that would ensure the Academy pursued a consistent purpose. The Council had no executive authority over the Academy or the College. It was to serve as a forum to express concerns and exert influence. Behind the scenes, the Chief of Defence Force Staff (renamed Chief of the Defence Force in 1984) would direct the Commandant and the Vice-Chancellor would instruct the Rector. In terms of a wider agenda, the Academy was Defence's demonstration of 'jointery' and a marker of how far the Navy, Army and Air Force had moved towards embracing a tri-Service outlook. It was envisaged that cadets would acquire a cultural familiarity with all three Services and build upon this experience during staff and command courses later in their careers. As the Academy was intended to be a prestigious establishment, the Commandant had to be an officer who was the 'second choice to be Service Chief'. It was not intended to be a retirement post or a position of marginal importance.

The Academy's governance quickly attracted critics, almost all on the uniformed side. An Academy postgraduate student, Major Klaus Felsche, argued in a Masters dissertation that Defence's aspirations would not be realised because the governance was defective. Felsche's views reflected a body of opinion among mid-ranking Service officers who were dissatisfied with the ambiguity of the 1981 Agreement and the difficulty of discerning clear lines of military command and control. He thought the Agreement suffered 'structural deficiencies' as a result of 'political circumstances'. He contended:

> No other nation has attempted to establish a military academy without a mechanism for governing the whole institution internally, and without a formal mechanism to resolve conflicts between the different components of that institution at the highest levels.[15]

Felsche also claimed that 'few military colleges have been allowed the degree of academic independence granted to the University College' and concluded that 'few outsiders would understand how a military academy, fully funded by the Department of Defence, could exist with only minimal input from Defence into the structure and content of the academic courses offered'.

Felsche's claims are contentious. He does not demonstrate Defence's competence to operate a university nor does he show that it had the capacity (or the insight and experience) to organise academic curriculums or to recruit suitable scholars. Like many criticisms of the 1981 Agreement, the problems were potentialities rather than actualities – fears about what might happen rather than laments about what *had* happened. Nor was it fair to blame the Academy's structure or administration for problems that existed elsewhere within Defence or the University, problems that often arose from tribalism and a lack of trust across professions and specialisations.

Felsche contended that the need for an academy had been assumed rather than demonstrated by the Martin Committee in 1970 and asserted that the PWC had highlighted the continuing weakness of the case for its establishment nearly a decade later. He claimed the compromise 'University College' model was 'developed largely in secret and away from public scrutiny' to placate the AVCC, the TEC and the RMC Staff Association. Because the PWC had concluded expenditure on the Academy could not be justified, he alleges that the Government recast the proposal to make it 'little more than an administrative matter for the Department of Defence and UNSW'. Whereas Casey University would have been established by an act of parliament, he lamented that 'the constitution of the Academy is not laid down in legislation' but in a 'conservative' agreement with UNSW, although he does not explain why it is 'conservative'.

From affiliation to Academy, 1981–1986

The result was a problematic institution 'clearly separated from the military and not responsible to the military for its courses'. The 'University College' model was not, according to Felsche, an evolution of the Faculty of Military Studies concept but a departure from that model. Unlike the RMC Commandant, who was responsible to the Chief of the General Staff for everything that happened among Army cadets at Duntroon, the Academy Commandant was not responsible to the Service Chiefs for the academic program provided by the University. He claimed that under the 1967 affiliation agreement, the University provided 'a service to the army, largely on the army's terms'. But this was far from true. Assertions of academic independence and scholarly freedom had shaped the courses taught and the research conducted since 1967. A service had been provided but it was largely on the University's terms. While UNSW degrees were being awarded to RMC cadets it could not have been otherwise. Felsche was also concerned that the Academy had two CEOs and neither were 'directed by the Academy Council. In fact, they are answerable to two different organisations which have few formal opportunities to make coordinated policy decisions'. Clearly troubled by the absence of traditional lines of military command and control across the Academy as a whole, Felsche conveniently attributed any difficulties with co-ordination and co-operation to the University:

> In their eagerness to secure guarantees of academic freedom by forcing a clear separation at the highest levels between the military and academic components of the Academy, the academic world may have caused the creation of an institution which cannot solve its own problems, particularly the problem of officer cadet identity and resource allocation priorities.[16]

He argued that Defence had been obliged to 'surrender' far too much control of the Academy to UNSW in the transition from Casey University to the University College after the PWC complicated the Government's decision making.

To illustrate his point, Felsche noted that academic courses would be approved 'by the UNSW Professorial Board in Kensington, a board which has no Service representatives [...] and is a long way from the in-house mechanisms envisaged for Casey University'. Furthermore, the Academy Council was 'restricted to *advising* Defence and the University', which he thought made it a powerless entity. Felsche wanted 'a carefully structured Academy Council, with powers to direct the Academy's chief executive officers'. He made no mention of the Chief of the Defence Force having command authority over the Commandant and the Vice-Chancellor having organisational authority over the Rector.

Felsche proposed revising the 1981 Agreement to better integrate military training and academic education. His aim was to prevent cadets feeling they were part of 'apparently separate organisations' and to allow Defence more say on the structure of degree programs and course content. Because the Agreement between the Commonwealth and the University 'was not considered to be permanent', he claimed the drafters wanted to 'satisfy the Services that any problems with the hurriedly constructed arrangement could be sorted out in the future'.[17] He assumes, of course, that the Agreement was drafted in haste and that its shortcomings were the result of staff work being rushed. The Agreement actually took twelve months to produce and passed through several iterations. The alleged 'shortcomings' were either deliberate ambiguities or intentional compromises.

While Felsche's critique focuses purely on military concerns and he criticises the Agreement's evident lack of precision and clarity, he and the Academy's other early critics did not have access to Defence documents and had not seen any of the University's internal correspondence. Accordingly, his judgments fail to recognise both the necessity of the compromise and the extent of the goodwill that allowed the two parties to co-operate in the context of a shared aspiration. As Felsche conceded, since 1967 the military and academic staff had 'worked hard at establishing cooperative links between the two bodies'. This was despite 'the struggle between the three Services, eager to retain control over their officers' education, and the Department of Defence with its ambition to rationalise common defence functions'. Felsche also ignored complaints from academic staff at both RMC and RANC that neither the Army nor the Navy was clear or consistent in what they sought from academic education. It is one thing to prescribe the needs of military and naval officers and to ask that these needs be acknowledged and met; it was another for newly posted officers at the Academy without experience of academic education to be authorised to chop and change individual courses, and to have power to review and recast major and minor degree sequences on an annual basis.

Despite its flaws, many of which arose from its brevity and lack of detail, the Agreement would operate for the next two decades on the basis of necessary compromise. The relationship between the University and Defence would rely on the willingness of UNSW to work with Defence and to understand its needs and priorities, and for Defence to trust in the experience and expertise of UNSW in developing and delivering educational programs. Even with a more comprehensive agreement with expansive statements of what Defence wanted and would pay for, the two institutions wanted to build a relationship, if not a partnership, that would transcend the minimalist demands usually associated with a contractual arrangement. In many instances, the University and its staff were willing to give more than Defence could afford.

With the shape of the University's future contribution clearer than it had been for more than a decade, detailed planning for the Academy proceeded with confidence, although it was still business as usual at Duntroon and Jervis Bay. As the Academy would not open for another four years, the UNSW academic faculties at RMC and at RANC continued to teach undergraduate and postgraduate students, although changes in the higher education sector and the employment market were not without their local impacts. In 1982 the engineering courses offered at RMC received only 'conditional recognition' by the Institution of Engineers Australia.[18] The RMC Department of Civil Engineering was said to be the worst equipped in Australia. The shortfalls, which consisted of too few faculty and inadequate facilities, were to be remedied at the Academy. The loss of accreditation was considered a serious matter. It would have an immediate impact on recruiting as the courses had previously enjoyed unconditional recognition. The Institution advised the Army that RMC courses would have all recognition withdrawn if corrective action were not taken by September 1983. The new Civil Engineering building was moved from the back end of the Academy building project to the front.

The Institution had also revised its recognition guidelines in the light of technological change and fresh demands on the profession. These demands included incorporating economics and management courses to ensure that the thinking of newly graduated engineers went beyond technical efficiency to the economic and social dimensions of their profession. Professor Wolfgang Kasper's advice was sought on possibly widening the course offerings for engineers to include economics and management courses; advice that helped to build confidence and secure goodwill within the Institution.

From affiliation to Academy, 1981–1986

The Engineering departments asked whether they could occupy the ADFA buildings in late 1983 or early 1984 because they were to be fitted with equipment not available at RMC, and because the return to unconditional recognition for courses was deemed an urgent priority. Supplementary funding from Defence was sought and obtained. This episode symbolised the essence of the Defence–University relationship. As Defence wanted professional status and public recognition for its future officers, it had to accept some regulation by external agencies with broad authority in a number of disciplines. It also came at a cost.

There were changes in the humanities programs as well. History ceased being compulsory in all RMC courses during 1982. The Executive Committee of the RMC Interim Council noted that:

> [the] Faculty was informed in January 1981 that the Army had reviewed its requirement. It no longer expected that History should be compulsory under the degree rules, though the study of military history would continue to be an essential part of the military curriculum. In reaching this decision the Chief of the General Staff recognised the importance of other subjects taught within the University.[19]

As preparations for the opening continued, the election of the Hawke Labor Government on 11 March 1983 had no bearing on the course or content of the Academy project. The new Defence Minister, Gordon Scholes, quickly affirmed the Labor Government's commitment. The incoming Foreign Minister and former Treasurer in the Whitlam Government, Bill Hayden, was another of Wolfgang Kasper's contacts and supported the expenditure. Hayden had completed his National Service in the Navy during 1951. The title of the new institution continued to vex the planners.

As it would not be the autonomous academic institution proposed in the 1970s, it seemed advisable to drop 'Casey University'. There is no record that the Casey family ever took offence. The title 'University College' was formally adopted to avoid the problems associated with earlier suggestions.

But an old problem resurfaced to distract the attention of academic and uniformed staff. Another bastardisation scandal at Duntroon raised questions about the character of officer education and whether the University was wise to enter into a long-term relationship with Defence. The 1969 RMC bastardisation scandal was so widely known that the popular press continued to cite it as a benchmark for assessing the Army's cultural evolution. Bastardisation was also implicated in the conduct of a former cadet who was sacked from RMC and tried for attempted murder in 1974.[20] A decade on, it seemed as though bastardisation was a thing of the past. It was, many hoped, a dark period through which Duntroon had emerged. The press did not think so.

The Age ran an article on 2 April 1983 outlining serial abuse at Duntroon under the headline: 'Officers and not so gentle men', by reporter Andrew Rule.[21] Three former junior cadets (by now civilians) made a series of anonymous allegations that seemed like echoes of the accusations levelled against senior cadets in 1969. They did not refer to the perpetrators by name but alleged that sadism and brutality were part of the general culture at Duntroon. The Commandant, Major General John Coates, ordered an immediate inquiry and held a press conference on 8 April to outline his response to the claims. Although an initial assessment suggested the former cadets were miscreants and that the relationship between the junior and senior class was 'essentially a comfortable one', some 29 cadets made specific allegations against senior cadets, and 14 cadets were charged with a variety of military offences. Five cadets were expelled from RMC after a more thorough inquiry concluded on 15 April. Another six cadets were severely punished.

General Coates wrote to the Chief of the General Staff, Lieutenant General Sir Phillip Bennett, on 22 April 1983, outlining his actions, defending his decisions and making a series of recommendations which included 'repudiation of generalised sadism and brutality as the code of ethics practiced at Duntroon' and rejection of all accusations of 'coercing witnesses and finding scapegoats'. The Commandant believed the report in *The Age* contained much that was unsubstantiated, exaggerated and lacking in context. His public response was to condemn the reported abuses although privately (and in confidential official correspondence) he thought the press reaction was overblown because cadet behaviour was generally good. He also noted that many cadets were surprised that the allegations were taken seriously or even reported, although the inquiry had led to expulsions. Members of the press were not, however, about to let the matter lie, given that a number of cadets were being shown the door.

General Coates met with academic staff on 19 April and reaffirmed his determination to rid the College of 'irregular behaviour'. During general discussion, the Commandant pleaded with academics not to allow the College's staff to polarise into academic and military camps. Coates conceded that while he could not completely eradicate unacceptable behaviour, he was committed to reducing it to 'the maximum degree possible'. Some academics were still troubled. They had their own views and they did not coincide with those of the Commandant. The Staff Association considered a resolution asking the Government to establish a formal inquiry. According to Dr Peter Dennis, then Senior Lecturer in History, abuse was 'widespread and endemic' and, in his judgment, likely to be 'transferred to ADFA'. This would have been the worst of all outcomes given residual fears in the Navy and the Air Force that their cadets would be exposed to the Army's culture and possibly subjected to mistreatment.

Not persuaded by official assurances, several academics wrote to the Minister for Defence, Gordon Scholes, in a personal capacity. They were concerned that the Army was unable or unwilling to deal strongly enough with behaviour that was petty, demeaning and possibly unlawful. Those who had served at RMC for some time claimed that intimidation, bullying and harassment actually continued after the 1969 scandal although it went 'deeper underground' and persisted because cadets would simply not speak out about abuse. The academics were aware of regrettable episodes and troubling events that had caused cadets to either fall behind in their studies or fall asleep in their classes. A minority within the Faculty thought the problem was overstated and the problem resided with recruiting unsuitable young men. The favoured solution among the majority of academics was abolishing the cadet class system and removing from cadets any authority to punish other cadets.

On 21 April 1983, the Faculty decided to establish a working party to advise the University on its response to the allegations after Professor Ray Golding offered to convey the concerns of UNSW staff to both senior Defence officials and, if necessary, the Minister for Defence personally. The working party consisted of the Faculty Dean (Professor Geoff Wilson), Alan Gilbert, Barry Andrews, John Cashman and Roger Thompson. A number of academics then wrote to Professor Wilson providing evidence of bastardisation and misconduct, attempts to restrain complainants and to curtail dissent, and outlining the mood prevailing among some military staff who thought that academic studies were a burden on officer development rather than a benefit. One academic (Gregory Pemberton), a former RMC cadet and one of the first cadets to graduate with an Arts Honours degree at Duntroon, mentioned that the military always used the term 'academic' or 'acca' pejoratively. He made the general observation that 'many aspects of the military side intrude into the academic side. It is now time to inject some of the academic aspects, for example, free, open and critical discussion, into the military side of the College. The synthesis of this dialectical process can only be a better profession'.[22] Barry Andrews told Wilson that Duntroon cadets were now more racist and sexist and argued that bastardisation contributed to the 'consolidation of intolerance'.

In a letter to Ray Golding dated 10 May 1983, Wilson conveyed the unanimous view of the academics and University administrators that bastardisation was 'contrary to the spirit of a liberal education, and contrary to the best interests of the Army' and that there was an obligation imposed on all academics to 'be concerned about the capacity of cadets to fulfill the intellectual and academic requirements of a full-time university course'. Wilson went on to explain that bastardisation:

> fosters intolerance, anti-intellectualism and general resistance to those attitudes of critical inquiry and cultural pluralism which are essential parts of a 'balanced and liberal education'. Specifically, some cadets exhibit hostility to reading and general study, and there is pressure generally upon them to adopt conformist attitudes.[23]

Once again, an element within the Faculty blamed the cadet class system, believing it responsible not only for bastardisation but for fuelling anti-intellectualism as well. The class system had been indicted in the causes of the 1969 scandal. Some of the academics believed it needed reform because the cadets were not experienced leaders. The majority recommended its abolition before 1986 to 'allow for harmonious interaction with students from the other services'. Long-serving faculty felt vindicated in their belief that the Fox Report reforms of 1969–70 did not go far enough. The latest round of allegations provided a basis for thoroughgoing cultural change that, the Faculty hoped, the Army would not overlook.

Scholes wrote to General Bennett on 10 May and explained that 'having weighed all this evidence carefully, I have concluded that it is not appropriate to set up another inquiry similar to the one undertaken in 1969, but that a number of improvements need to be made'. Surprisingly, the Minister added: 'I have also concluded that it would be inappropriate to punish the five cadets with expulsion, as I believe this would be too severe a punishment, particularly when looked at in the light of what appears to be generally accepted behaviour on the part of the cadets'. It appeared that Scholes was conceding the behaviour was endemic. This decision placed John Coates in a bind. The Commandant's judgment had effectively been challenged and his authority had been publicly undermined by the Minister's refusal to expel the cadets. This appeared to be a clear-cut case of ministerial interference in the Army's internal affairs. Such action seemed out of character for Scholes, who was never considered an 'activist' minister. He had not previously concerned himself with the running of the College and appeared largely indifferent to officer education. Kim Beazley, the son of Whitlam's Education Minister, would replace him the following year.

Some of the more critical academics took more persuading that the matter had been handled adequately. On 12 July, the Minister responded to two letters from John Cashman, the President of the College's Academic Staff Association, by stressing that all forms of harassment and physical and mental violence were prohibited. They would not be tolerated in any context. But Scholes' assurances were lame and some of the academics, particularly those who were active members of the Labor Party, were unsure as to the next step. The Minister affirmed that academic staff would be actively involved in the overall development of cadets and not merely in their intellectual nurture. He directed that the 'RMC Class system is to be de-emphasised as far as practicable' while action would be taken to 'ensure that improper practices are not transferred to ADFA'. This was far from the comprehensive review that some faculty members were seeking.

After the expelled cadets were reinstated, the Commandant addressed the entire cadet corps on *The Age* allegations. Bastardisation, an imprecise term that often conjured up visions of cruelty and sadism, was a waste of time and energy whatever its character, and was prohibited. He outlined what had occurred after the allegations were made, detailed the decisions that had been taken in the light of information received, and made the general point that obtaining a degree through Duntroon was already more demanding than studying at a civilian university because of the military training load. He stressed the complementary importance of academic education:

> I want you to understand clearly the difference between the academic and military approach to problem solving and further action. When you are on the academic side of the College, that is within the Faculty, or even when you are dealing with the theoretical military side, question everything! Probe it, challenge it, examine it [...] Accept nothing at face value. If you do not, you are not getting anything like full value from the course here [...] The military way is quite different from the committee-style system of decision by consensus which is a feature of academic life [...]. I have heard that some of the members of Faculty believe that they have either been black-balled from attending Corps functions or that cadets have been told not to see them. If this has ever

happened, then it must stop. There must be no 'them and us' in relation to the Faculty.

The University's leadership wanted its own assurances from Defence that misconduct would stop. Golding met General Coates and a senior Department of Defence official on 6 June. He was told that Defence warmly welcomed the University's collaboration in dealing with the issues at Duntroon and he was convinced that the Army's proposed action was more than adequate. Golding felt that some progress had been made and a 'line had been drawn in the sand'. To ensure that tangible progress was made, he offered to return to the College on 27 September 1983. Two days later, Coates wrote to Golding with a formal invitation to academic staff whom he hoped might serve on RMC selection boards, such was his positive impression of their contribution.[24] John Coates was a learned man. He had a Master of Arts degree from the ANU and was the author of several books. Coates esteemed academic study and later became Chief of the General Staff. In retirement he would become a Visiting Fellow in the School of Humanities and Social Sciences. Coates respected the views of academics and welcomed rather than resented their perspectives, even if he did not always share them. His foremost aim and that of the Minister was, however, to contain the controversy and ensure that training continued with little interruption. Bastardisation could be addressed as a 'running reform'.

In addition to concerns about cadet behaviour and the attitudes of military instructors, there were also worries about enrolment patterns and vocational needs. The Head of the Physics Department (Dr Don Chaplin) wrote to the Commandant on 18 May 1983 lamenting the small number of students enrolled in Science and Engineering degrees and pledging the assistance of his colleagues to reverse the trend.[25] The problem originated, he suggested, with declining interest in science and technology in the nation's secondary schools and the fact that an Arts degree had fewer contact hours than a Science degree, making it more popular with cadets. He asked the Commandant to target science matriculants during recruiting campaigns to discourage candidates taking the 'early' route into an Arts degree. He proposed offering Honours degrees in Science and Technology and making one or more science subjects compulsory for all cadets, as History had been in the 1970s. He noted, for instance, that all RAAF cadets were required to study physics. But would this requirement continue at the Academy?

The Navy was asking similar questions of the Academy's curriculum in the context of the future remit of the Naval College. Its location on the shores of Jervis Bay made HMAS *Creswell* an ideal site for the study of Oceanography – a discipline in which the University wanted to expand and excel. Dr Alan Carter, Senior Lecturer in Charge of the Department of Oceanography at Kensington, had been involved in teaching marine science at RANC from the early 1970s until the 1980s in concert with a civilian lecturer, Rex Benson, and an instructor officer, Lieutenant Commander John Matthias. Although the University of Sydney offered a marine science major in its undergraduate programs, UNSW hoped to 'capture' the field as the profile of Oceanography at the Naval College increased between 1980 and 1983. It was during this period that Dr Carter had proposed creating an RANC Oceanographic Research Centre that would be available to staff at the RAN Research Laboratory, UNSW and the Academy. It would also be an important way of linking the Academy to critical infrastructure beyond Canberra.

At the RANC Advisory Council meeting held on 7 April 1983, Professor Jack Ratcliffe, Professor of Chemical Engineering and a Commander in the RAN Reserve, proposed that Oceanography become an RAN officer specialisation, not unlike Hydrography and

Meteorology, and that the Navy support the creation and operation of what would become the UNSW School of Oceanography based at Jervis Bay. The new School would enhance these disciplines within the existing Bachelor of Science program and ensure that the Navy had internal expertise in ocean science. In the same way that RMC had attracted outstanding staff by specialising in low-temperature physics, RANC would be bolstered by a similar commitment to Oceanography which, the University added, did not involve the same kind of equipment outlays. Although Alan Carter's vision was never realised, Oceanography remained a research strength at UNSW and later at the Academy by virtue of the connection with Jervis Bay. UNSW became a member of a Marine Science Consortium consisting of Sydney and Macquarie Universities established in 1985. This kind of collaboration would help to transcend the sense of loss the Services felt when the establishment of the Academy was announced.

The Navy was also concerned that the Army was dominating the Academy's development and that the Services were dictating terms to the University. The Navy had respected the University's culture and trusted the judgment of its staff on matters of principle. In a letter to Ray Golding (undated but written in late April 1982) dispatched from Jervis Bay where he was a Visiting Professor, Patrick O'Farrell outlined 'the Navy's strong feeling about its representation [on the Academic Planning Committee (APC)] [...] in regard to what it sees as the disproportionate (in terms of comparative eventual student input) size of the Duntroon contingent. I merely report the resentment and suspicion of this situation is unanimous here, if passively and politely expressed'.[26] The Navy also objected to a motion drafted by the RAAF representative on the committee, a motion considered but not tested with a vote at a meeting of the APC on 20 April 1982:

From affiliation to Academy, 1981–1986

> That we recognise that ADFA is established to serve the needs of the Services, and that any course of study seen as desirable by any of the Services be made available at ADFA provided it has the intellectual content and depth appropriate to a university course and there are sufficient student numbers to warrant it.

O'Farrell, who was not at the meeting and who did not personally support the formation of a tri-Service academy, observed that this motion was:

> variously interpreted at the meeting as a desirable move towards clarification of basic issues, as a sinister invasion of academic freedom and as divisive [...] from what is reported to me, I gather that the Committee discussion moved towards a services versus academics confrontation, rather than towards a clarification of the cooperation necessary to the best interests of this particular tertiary enterprise.

He believed clarity was needed about 'the way in which subjects and syllabuses should originate and be devised in the College'. He continued:

> As I see the present position from impressions formed here, the Services' view is that the College exists within a Defence Academy situation and they are not prepared to accept what they see as a tendency to erect academic dictation in place of assistance and cooperation. The academic positions reported to me suggest that they see no difference in their situation from any other university one, and regard services ideas as to needs and requirements, as tending towards interference. In

both cases, the drift of such attitudes appears likely to lead to difficulties and unpleasantness of an unproductive kind.

O'Farrell's point was not new: there was longstanding disagreement about Defence's role in shaping the curriculum. Service officers wanted some control; some academics were not prepared to concede any. This was, however, overstating the situation. The academics had always acknowledged the need to provide an education relevant to the personal and professional needs of junior officers. O'Farrell was essentially relaying his assessment that disagreement was producing acrimony. The experience of his Canberra colleagues was a little different. Kasper recalled his discussions with Service officers on the structure of management programs and the contents of economic courses. He thought the interaction was cordial and constructive, with agreement on a mix of theoretical and practical subjects that met the needs of Defence and attracted an expanding number of cadets. O'Farrell was, however, an influential figure and his opinions mattered at Kensington.

Golding telephoned O'Farrell on 3 May 1982 to offer some personal reassurance. Golding explained that 'his clear position is (and always has been) that the ADFA arrangements must be on the basis of equality and cooperation, and that there is no necessary presumption that existing Duntroon courses should be adopted. However, he has had continuing difficulty in getting the Duntroon staff to see this [...] he also expressed annoyance that the last meeting [of the Committee] had been held with Duntroon as a venue'.[27] O'Farrell's own hope was that the advent of the Academy would change attitudes to academic education among junior officers. He told the Naval College's Director of Studies that:

> some midshipmen are hostile to study in which they see no practical relevance. They regard such study as frustrating their real ambition – which is to get to sea – and to be undertaken with the minimum effort to secure a bare pass. The result is that such students are personally unhappy and deliberately underachieve. Study, which they have the capacity to enjoy, becomes a burden.[28]

Although this might have been an argument for sending these midshipmen to Kensington (where I happened to be studying when he wrote these words), O'Farrell thought the problem was contextual. He observed that midshipmen were not told why they needed to study nor made aware of the value the Navy attached to their education. 'I personally believe that the academic courses and professional naval training can be readily demonstrated to be a very effective and highly desirable blend, productive of the best potential officers for the future'. While my own recollections of this period are consistent with O'Farrell's, the lack of any opportunity for further study and the absence of any encouragement to pursue a higher degree was also a factor in less than positive attitudes to academic education. In this respect, my cohort were simply echoing the laments of those who had completed the questionnaire distributed by Professor Al Willis in 1974. The Navy had not learned anything.

This disincentive was also apparent to the RAN Liaison Officer at UNSW, Commander Ted Shimmin. In addition to his naval duties, Shimmin was Warden of Basser College between 1980 and 1982. He suggested another strategy to the College's Commanding Officer, Captain Tony Horton, and Professor Ratcliffe. In his five and a half years at Kensington, he observed that 'no ex-naval student has ever returned full-time to this University to do an honours degree, a double degree, or any postgraduate degree or qualification

[...] I believe this is a grossly short-sighted policy'. He went onto explain why education was important, including:

> prestige for the service, improved credibility and consequent better communication with certain civilian members of the Department of Defence, a better sense of achievement for the individuals concerned, etc., but these are essentially value judgements made by society and/or the individual [...] what we need in the services and have the chance to get from degree courses, are people who are capable of adapting to and handling tomorrow's technology as well as today's, and who have the ability to evaluate independently and to criticise constructively their own work and the work of others.

He implored the Navy to offer these opportunities to ensure it retained the best and brightest minds among the officer community. His preference was for Honours degrees as the foundation for doctoral studies at a later point. The following year (1983), the first midshipmen were permitted to complete the new Special Honours program being offered at Kensington. After two years and the graduation of six students with Honours (of which I was one), the Navy decided not to offer further places pending the opening of the Academy. This kind of inconsistency was to mark the Navy's approach to higher education over the next 30 years. Whereas the Army could see the value of an annual cohort of 'Honours' cadets providing the nucleus of an exapnding intellectual community, the opinion of individual naval officers was to determine the Navy's policy – and most had a jaundiced view of what education could offer.

Naval attitudes were also shaped by the numbers of junior officers leaving the RAN College with a diploma (rather than a degree) and positive appraisals of their subsequent service. Captain Peter

Ross, the Commanding Officer of the Naval College from January 1983, wrote to the Chief of Naval Personnel in October 1984, drawing his attention to the educational progress of those who had entered the Naval College in January of that year. There were 49 midshipmen enrolled in UNSW degree programs, of which ten were not expected to proceed to second year based on the failure rate for the previous four years. It was estimated that another 15 to 20 midshipmen could well fail in second or subsequent years.[29] Captain Ross concluded that 'the Undergraduate Scheme does not and will not produce a well motivated naval officer'. Conversely, there were 23 students enrolled in the non-UNSW Diploma of Applied Science then being offered at Jervis Bay. He observed that this course 'has over the last eight years supplied 50 percent of tertiary qualified GL [General List] officers. Of the GL officers who passed out from RANC in 1983, 28 were graduates and 27 were diplomates'. In his judgment, the Diploma needed to be retained after the Academy opened because:

> a large proportion of those officers performing well on the diploma course would be out of their depth with degree studies. Furthermore the diploma course has enabled the RAN to minimise its midshipman degree failure losses. (90 percent of transferees to the Diploma managed to complete their studies). It is clear that the RANC diploma course is a proven and cost-effective method of providing the RAN with a significant number of tertiary qualified officers.

In sum, the Navy would be faced with a critical shortage of officers without the diploma course.[30] There were two options: only accept candidates who were able to complete a degree, or persist with a two-tiered education model. The Naval College preferred the second option and canvassed the possibility of offering a UNSW diploma course at Jervis Bay or an alternative course of study at the Academy. The second option was problematic in a number of respects. Did the Navy actually *need* tertiary-educated officers or simply *prefer* them? Was its vision of higher education aspirational or vocational? Or did it reveal an inability to articulate a clear statement of its needs?

Each of these practical and philosophical challenges remained when Professor Geoff Wilson was formally appointed inaugural Rector of the University College on 2 February 1984 after an international canvas for suitable candidates. He was never promised the job and did not feel entitled to it.[31] Taking nothing for granted but believing his claims were strong, he felt the interview for the position went well and privately conceded he would have been disappointed if not offered the post. The faculty had written confidentially to Vice-Chancellor Michael Birt strongly commending Wilson as their preferred candidate, partly because they wanted to avoid having Ray Golding as their leader.[32] Golding had alienated a section of the faculty by his imperious and at times indecisive manner. He had occasionally disparaged Wilson and could not quite grasp how Defence worked. By way of example, when Defence was considering the allocation of funds to equip various new laboratories at the Academy, Golding opposed the extra funding sought by some Departments. Unfamiliar with local difficulties, he did not understand the extent of the new equipment required to compensate for the paucity of existing equipment. Given the poor state of the facilities being used by the Department of Civil Engineering at RMC, the Acting Head, Dr John Sneddon, was able to negotiate $1.7 million in Defence funding for Civil Engineering compared with the initial offer of $600 000.[33] In a number of instances it was the goodwill of Defence Department personnel with whom the academic staff negotiated directly that achieved some very positive results for teachers and researchers. Hence, the faculty's strong preference for Wilson.

Whatever his strengths and shortcomings, Golding was institutionally too senior for the Academy post and was not, in any event, a contender. The only other possible internal candidate was Professor Alan Gilbert, later Vice-Chancellor of the Universities of Tasmania, Melbourne and Manchester. He was, however, a member of the selection committee and ineligible. Wilson was clearly the more obvious candidate, although a number of his colleagues believed he was considering other academic appointments at the time and feared he might accept another position. They were mistaken. He was offered the inaugural Rectorship and accepted it.

Choosing a title for the academic 'chief executive officer' of the University College had been resolved a decade earlier during autonomy discussions at Duntroon. Wilson explained publicly that the title 'Rector' was a common European equivalent for Vice-Chancellor and 'was meant to indicate a high degree of autonomy and was to be set at the level of Pro Vice-Chancellor of an Australian university'.[34] In Australia, however, the term 'Rector' was more commonly applied to Anglican parish clergy. This led to amusing confusion at times. Wilson would serve as Rector of the University College and Dean of the Faculty of Military Studies concurrently for the next two years as well as being a member of the Vice-Chancellor's Advisory Committee (VCAC) at Kensington. The Deputy Rector was Professor Alan Gilbert, Head of the Department of History.

By this time the ADFA Interim Council had also endorsed a proposal from Ray Golding on academic structures within the University College.[35] The Council agreed to change the name of the Academy's academic units from 'school' (the original designation) to 'department'. The term 'school' was being applied to large bodies within UNSW that comprised several departments. The University College departments were to be similar in size to those at Kensington. As the College developed, there would be scope to establish a school consisting of several departments, most probably in Engineering. It was agreed there would be no military membership of the Assessment Committee, with military input to curriculum development provided through a series of departmental Advisory Committees. The College's senior academic body would be the Academic Board, consisting of academic staff and military members 'appointed by the Vice-Chancellor in consultation with the Commandant and Rector'. The College Secretary would be the College's senior administrative officer and serve as the UNSW Registrar's representative in Canberra.

Two months later, the Academy was established as a Joint Service Unit under Section 32C of the *Defence Act*, with Captain Terry Roach RAN appointed in command.[36] As the 'senior Service' was least persuaded by the need for its establishment, there was much sense in bringing the Navy onboard early. The first commandant was Rear Admiral Peter Sinclair. His appointment took effect from 9 July 1984. Sinclair had joined the Navy as a thirteen-year-old cadet midshipman in 1948 and had served at sea and shore, including an appointment as Executive Officer of the Junior Recruit Training Establishment, HMAS *Leeuwin*, in 1972. Both Wilson and Sinclair reported to the Interim Council, which consisted of 22 members including, among the academics, the Vice-Chancellor of UNSW, Professor Michel Birt, the Chair of the Professorial Board at UNSW Kensington, Professor Douglas McCallum, and four academics from Kensington and Canberra elected or appointed by their colleagues at the respective campuses. The Interim Council also consisted of the Service Chiefs, the

From affiliation to Academy, 1981–1986

Secretary of the Department of Defence and two students (a postgraduate and an undergraduate).

The brief handed to Sinclair was complex if not contradictory. He had succeeded Air Vice-Marshal Sam Jordan as the Academy's Chief Project Officer. This made him responsible to Defence for constructing the Academy within the design cost limits. He was also the inaugural Commandant charged with achieving the best operational outcomes. One side of his brief required financial frugality; the other side sought a generous provision of resources. Sinclair's position was unenviable. Part of him needed to save money and another part wanted to spend it. His task was made even more complicated by the simultaneous construction of the new Parliament House. This much larger project was managed by the same prime contractor and would lead to directly conflicting priorities. Sinclair later noted: 'the Academy always came off second best when there were competing demands'.[37] Additionally, he wanted the new-entry intake to be the only student cohort at the Academy when it opened in 1986. His strong preference was to establish a new culture at the Academy among a group of students who had not experienced and would not want, therefore, to replicate at the Academy what they had experienced at the single-Service Colleges. He was over-ruled. The first year of the Academy would have first-, second- and third-year cadets. The third-year Army cadets were allegedly the most fractious. This decision was to have serious and unforeseen consequences in the first decade of the Academy's existence.

While the formal arrangements were set out on paper, it fell to Wilson and Sinclair to make the relationship work – and they did. Overcoming any personal reservations about the wisdom of establishing the Academy as a tri-Service venture, Sinclair embraced the challenge of creating a new institution and reached out to senior academics with generous personal hospitality. His determination to overcome every hurdle and obstacle instilled great confidence in both the continuing and newly appointed UNSW staff, who feared they were being drawn into an entity that would diminish the importance of academic education and relegate them to the margins of the enterprise. Wilson would enjoy good relations with the first three commandants, demonstrating that there were no institutionalised impediments to a workable diarchy. Neither Wilson nor his uniformed counterparts ever sought a revision of the 1981 Agreement. There is no recorded instance in which either Wilson or Sinclair sought or needed external assistance to resolve an internal dispute.

By late 1984 and with the senior leadership in place, those academics and administrative staff who were to be offered places in the new Academy had received letters of appointment. The Finance and Administration Office (FAO) at Kensington noted that RMC staff apparently believed their conditions of employment ought to be aligned with those at the ANU.[38] The staff administrative officers from Kensington explained that while 'UNSW will continue to have regard for and keep in touch with situations at ANU, the policy of UNSW is the one that will be followed and UNSW will not adopt, for introduction in the FMS, every change in conditions and practices made by the ANU'. But on reflection and after further discussion, the FAO at UNSW agreed to model some FMS support staff salaries and conditions on those offered at the ANU. It stated 'we cannot escape the consequences of the undertaking to match the ANU' but counselled against making longer-term commitments, given the advent of the Academy. It was noted that the ANU tended to pay some staff categories, such as Research Assistants, at a higher rate than anyone else in the tertiary sector.

Many of the key UNSW administrative staff at the Academy began work in early 1985. With much of the Academy under construction, they were located initially in what was to be the Academy Main Store, where the Commandant and Rector were also temporarily accommodated. One of the leading figures was Trevor Short. He served as the University College Secretary from January 1985 until September 1990. A graduate of Melbourne University and with National Service experience in the Army, he had gained considerable expertise in university administration at the Monash Teaching Hospitals and then as Executive Secretary to the Victorian Universities Admissions Committee, a post he held concurrently with being Executive Secretary of the Victorian Vice-Chancellors' Committee. His principal task was financial management. He liaised with heads of academic departments to ensure the buildings they would occupy were fitted out with the necessary equipment, in addition to ensuring that existing and new staff were paid the correct salary and proper allowances. There were the usual arguments between departments over staffing and budgets, which were made more complex by the influx of new appointments and the transfer of lecturers from Jervis Bay and Point Cook who had their own particular wants and needs. The new buildings, the presence of Navy and Air Force cadets and the former faculty of Jervis Bay and Point Cook combined to create a very different mood than the one pervading the Faculty of Military Studies at Duntroon. Many of the women employed by the University felt the emerging Academy culture was less 'blokey' and slightly more inclusive, although there would be a series of necessary reforms over the ensuing 15 years that made the campus more welcoming and more encouraging to women.

The relations between the UNSW and uniformed staff were by all accounts constructive and respectful, although some of the new UNSW people were confounded by the organisational complexity of Defence and sometimes struggled to find the right person, whether civilian or uniform, to answer a question or to make a decision. There was generous assistance from the administrative staff at Kensington, foremost the Registrar, John Gannon, who had a military background as well. He was acutely aware of the size of the task being undertaken at the Academy. While Short was conscious that individual departments found ways of retaining unspent funds, the Academy Librarian, Lynn Hard, made the most of the rare opportunity to establish a new Library. Hard had served in South Vietnam as a helicopter door-gunner with the United States Army and was known for progressive thinking. He was working at Melbourne University when he applied for the position of Academy Librarian in 1981. Initially located at Duntroon, he had authority to examine the single-Service college libraries for items that could be transferred to the Academy collection. He encountered strong resistance at both Point Cook and Duntroon to the cannibalising of holdings that included rare manuscripts and historic objects better suited to display cabinets than library shelves. The tensions were resolved and the Academy collection developed. He, too, noted the difficulties that arose from the frequent rotation of Defence staff and the almost constant need to deal with new members on advisory and executive committees.

Deemed a priority facility, the Library was operating before the Academy formally opened. It soon became known nationally and internationally for the depth of its collection in certain specialist areas, including a valuable collection of personal papers belonging to leading literary figures, and for the large number of inter-library loans. Hard created computer classification systems that made it much easier for staff and students to find material relating to their specialist subject area. The Library drew uniformed people and civilians to the Academy who would otherwise have no reason to visit.

There was one practical challenge that no-one had anticipated.

From affiliation to Academy, 1981–1986

The Academy's location was unknown to most Canberrans, including members of parliament and the press. Unlike the ceremonial entrance to Duntroon on the busy Morshead Drive in Canberra and the iconic clocktower at the Naval College, which was visible for miles across Jervis Bay, the main entrance to the Academy was on nondescript Northcott Drive. It was not one of the national capital's major thoroughfares. In fact, Northcott Drive was nothing more than the connecting road between the two main Defence office complexes – Russell Hill and Campbell Park. It was only used by those with business at either place. Residents and tourists had no reason to use it. To make the location of the Academy even less obvious, it was enveloped by Duntroon on two sides, leading many Canberrans to think the Academy's buildings were actually part of RMC. The Academy was not visible from the Parliamentary Triangle or from the Black Mountain lookout. Road signage was minimal. It was almost as if the Government hoped its whereabouts would remain a secret. That it was also a campus of UNSW was a mystery to many and would remain so.

What happened in Canberra had little significance for UNSW staff at Kensington. For the first half of the 1980s, neither the University Council nor the Professorial Board showed much interest in the establishment of the Academy. In this respect, the Academy was unlike the former University Colleges in Newcastle and Wollongong before they became autonomous universities. The University College would not extract any funding from the main budget or require much input from the faculties. There was an expectation based on the RMC experience that it would have good academic standards. The staff–student ratio was enviable. The Canberra campus offered facilities for visiting Kensington staff.[39] Outwardly, the Academy was progressing well although slowly. Buildings were going up and staff were coming in. There were no acute differences on principle or procedure and Defence had confidence in the competence of the University's staff – academic and administrative. Much of the competence had, of course, been acquired from the RMC and, to a lesser degree, the RANC experiences. The only area of continuing concern for Defence was the poor grasp that some University staff had of forward financial planning, mainly budgets and estimates. Rob Tonkin from Defence observed a 'hand to mouth' attitude within the Faculty of Military Studies. The Faculty was, of course, an 'interim' entity for its entire existence and this naturally promoted a short-term view. Defence realised the need to rethink some of its initial policies on establishing the Academy. Norm Attwood, who deputised for Tange during the establishment phase, arranged for the University to receive a block allocation of funds to avoid processing a multitude of separate requests for payment as the Academy was built.[40]

Throughout this period the Academy Advisory Council reflected the intention of its title: it advised. Although its meetings were very formal and meeting discipline was strictly observed, it did not direct the Academy's activities or determine strategic policy. Its role was to reassure stakeholders that their interests were being served and that the Academy's leaders were responsive to advice. The meetings were essentially forums in which to raise concerns and acknowledge successes. Occasionally, ideas and proposed initiatives would be tested against prevailing opinion but the outcome would be guidance rather than direction.

As expected, the ANU was neither alarmed nor even much concerned by the opening of a new university campus in Canberra. The ANU was not threatened by the University College, whose remit was narrow and whose initial postgraduate cohort was small. Wilson had positive relations with his disciplinary colleagues at the ANU and was invited to its graduations in addition to regular social meetings with the Vice-Chancellor, Professor Peter Karmel. In time, the net flow of academics would be from the ANU to the University College. This essentially continued a pattern that had begun in the late 1960s, when the Faculty of Military Studies had become established, gained momentum and acquired a reputation for expertise in areas of inquiry that were non-existent or poorly supported at the ANU. The Academy was considered a much happier place to work than the ANU, which seemed perpetually beset by tribalism and factional strife.

The University College certainly had no difficulty attracting talented staff. The number of academics willing to live in Canberra was smaller than those who preferred to live in Sydney but there were many compensations for living in the national capital and working at the College, including its promixity to a number of major cultural institutions. In sum, there is no evidence that the relative isolation of the College (when compared to the location of the main campus in Sydney) had any impact on the quality of staff. But the size of the College definitely had a direct bearing on what it could do. It was nowhere near as large as the main campus and could not purchase equipment on anything like the same scale as Kensington. Defence often provided generous support, recognising the challenges the University faced and honouring the devotion of academics to their teaching and research – both of which benefitted Defence. The small size and multi-disciplinarity of the Faculty of Military Studies had also been an asset. There were none of the ideological feuds or practical turf wars that plagued some universities. The gathering of academics from across disciplines at morning tea in the Officers' Mess at Duntroon House had worked against the formation of intellectual silos that can fracture a scholarly community. There were fears that these demarcations might emerge at the Academy as each department would have its own staff room and all-discipline gatherings at the Officers' Mess would not be as well attended. The size of the Academy and the much greater physical separation of staff meant there was less contact between personnel from various departments. Those transferring to the Academy from RMC noted that the kinds of friendships they had enjoyed at Duntroon were not replicated at the Academy. This was a source of widespread regret. But the dreaded silos did not eventuate; faculty just worked on fewer related problems.

Defence also contributed to the Academy through the Visiting Military Fellow Scheme. This was another important element in the evolving relationship between UNSW and Defence. The concept began at Duntroon when competent officers serving in a range of corps were involved in delivering undergraduate education. They connected the cadets to the professional communities they aspired to enter. At the Academy, the scheme was larger and formalised. In brief, mid-ranking ADF officers were posted to the Academy staff as lecturers in various departments. The idea had originated with Lieutenant Colonel David Horner, an RMC graduate who later obtained an ANU doctorate in military history, who travelled to work with Bill Coburn, a highly influential Defence civilian in the personnel area. The two discussed the possibility during 1984 and the first half of 1985, focusing on the benefits of Defence posting 'Military Fellows' to the academic faculty at the Academy. The idea was eventually accepted and Horner was posted to the Academy's Department of History in 1985 – the semester before the formal opening. Another of the small first group was Lieutenant Colonel Dick Warren, an engineer officer who went to the Department of Civil Engineering.

Shortly after he arrived at the Academy, Horner was to experience personally the Academy's complexity or ambiguity. Dr Roger Thompson, a senior figure in the Department of History inclined to occasional hostility towards the military, insisted that Horner wear civilian clothes because he was part of the academic faculty. The Commandant had insisted that he wear Service uniform because he was an Army officer. This episode reflected different approaches between the two sides at the Academy. Horner decided the discretion rested with him; he would decide what clothing was appropriate and when. The 'VMFs' would make an important contribution to the Academy over the first five years when commitment to the scheme began to languish. Indeed, a number of departments relied upon the provision of VMFs to meet their core teaching commitments. The reason for the decline, according to the ADF, was the difficulty of finding suitably qualified officers who could be spared from other duties. There was no opposition from the academic side to appointing VMFs but there was some resistance to using qualified Service officers as tutors when the enrolments exceeded expectations. Fears that the uniformed tutors would dilute the academic content of College programs subsided as the officers who provided this assistance proved to be a valuable conduit to the ADF and its evolving educational needs.

To prepare for the first Academy intake of cadets (now set for January 1986), Professor Wilson and Rear Admiral Sinclair visited 14 Defence academies in six countries to assess alternative models of organisation and management. Their first visit was to India. They met with senior uniformed staff and the Vice-Chancellor of the University of New Delhi, which offered some of the courses conducted at the Indian Military College at Pune. They then proceeded to Munich and Hamburg, where their growing conviction that tri-Service education did not necessarily lead to the loss of single-Service traditions was confirmed. These visits were followed by similar experiences at Plymouth and Shrivenham. The United States Naval Academy at Annapolis, the United States Military Academy at West Point and the United States

WIDENING MINDS

Air Force Academy at Colorado Springs led them to believe that women could and should be fully integrated into the education and training system. After the United States, they visited the Royal Military College at Kingston in Canada and its partner institution at Saint Jean in Quebec. On the way home, they toured the Japanese Defence Academy at Yokosuka. The main lesson they took from this trip was the international movement towards tri-Service or common curriculum arrangements. They reported:

> Four of the six countries have tertiary education and training systems within a tri-Service environment. Of the other two, the United Kingdom would seem to be moving slightly in that direction (at least in the case of Shrivenham) and the United States is sympathetic towards the concept but is constrained by history and the scale of their operations [...] Shrivenham probably goes closest to our Academy in respect to association with a civilian university. Shrivenham quite recently entered into a contractual arrangement with the Cranfield Institute of Technology and this appears to have given the establishment renewed vitality and, dare we say it, credibility.
>
> There is no clear pattern with the composition of academic staff in overseas institutions. Some have all civilian staff and others all military. There are advantages and disadvantages with each but our policy of having predominantly civilian staff with a small but increasing number of Service officers spread throughout the Faculty should achieve the best of both worlds [...]
>
> The academic standing of some postgraduate courses and teaching facilities overseas was often disappointing and there appears to be a trend towards the lowering of standards in the

interests of achieving narrowly based professional relevance. We believe we must avoid falling into this trap at all costs. However, the conduct of postgraduate degree courses of high academic standing should not preclude other means of providing higher education for Service officers in various professional areas of application.[41]

After the Academy opened, the Dean of Shrivenham, Professor Frank Hartley, visited Canberra. His visit was prompted by the Thatcher Government's decision to seek tenders from British universities to deliver higher education at the Royal Military College of Science. The contract was subsequently awarded to Cranfield University. It had recently been upgraded from an institute of technology. The agreement between the UK Government and Cranfield was modelled on that between the Commonwealth and UNSW for the Academy. [Hartley later became Cranfield's Vice-Chancellor and was instrumental in promoting its rapid rise in the standing of new British universities.] Other Canberra-based academic staff were also given an opportunity to visit the single-Service colleges and professional military training establishments before the Academy was officially operational.

As the opening date loomed, the UNSW Professorial Board approved the proposed suite of courses in Arts, Science and Engineering with minimal amendment. Wilson noted that:

> Despite the probable views of some of those in academia that our courses would be to the ultra right, one small subject in the History 3 offerings was entitled 'Australia: From a British colony to an American appendage'. This was too much for the Board and we had to resubmit with the change to 'Australia from 1900–1970'.

Before the Academy opened, there were even suggestions its remit might change. Rear Admiral Sinclair wrote to the Rector and the Heads of Department on 16 August 1985 about the possibility that the Academy could incorporate or subsume the functions of the Service staff colleges or that the University might accredit staff college courses. Hugh Smith was against both proposals because he felt the development of the Academy would become 'far too closely linked to support of the Defence Force' and that academic independence was needed. If the University did not devise, develop or deliver the course, it should not be considering its accreditation. The suggestion did not proceed beyond the 'idea phase'. There were, however, other small changes that made sense. In due course and after some debate, credit towards coursework masters degrees was given by the University College to officers who had completed selected staff college courses.

The 1981 Agreement was amended slightly on 26 June 1985 to change the composition of the Academy Council to include an additional number of academic and general staff from the University. There was also a need to determine payments to the University in respect of its operations at Duntroon and those associated with preparations to occupy the new buildings at the Academy. While the negotiations were detailed and complex, there was goodwill between the parties, and neither accused the other of failing to meet obligations. The contribution of former uniformed men and women now employed by the University helped to resolve Defence's concerns while they were integral to the operations of the University College. The Faculty Administrative Officer at Duntroon, Dennis Harverson, became the Deputy College Secretary (Personnel and Finance) when the Academy opened. He had more than two decades of service in the Australian Army and was well known and widely respected.

As preparations to open the Academy continued, the Naval

From affiliation to Academy, 1981–1986

College staff prepared for the final cohort of midshipmen undertaking degrees to depart. Most would relocate to the Academy to complete their studies. Textbooks were transferred from Jervis Bay to the Academy Library along with 'second' copies of naval and maritime books. The University offered its expertise to the Naval College on the management of its rare books and manuscripts collection.[42] In Canberra, the RMC Interim Council held its final meeting on 7 November 1985. The Council had existed for 18 years, considerably longer than imagined when it was established. Only Professor Douglas McCallum and Roy Pugh, its Secretary, remained of the original membership. The meeting was devoted to reports of transitional arrangements and words of appreciation for the goodwill and generosity of the Interim Council's members. Ray Golding spoke on behalf of the University and noted there were mixed emotions. It was the end of 'a most successful joint venture' and 'the culmination of probably one of the most successful partnerships in Australia'. He compared the first year's enrolment of 68 students with the 503 enrolled in 1985, and the increase in staff from 26 in 1968 to 110 in 1985. The first PhD was awarded in 1974 and many more followed. Golding stressed that there were problems to resolve and challenges to be overcome but every issue had been handled through negotiation and friendliness. The Council's final resolution was that 'the entry standard for the University College should be set at a HSC aggregate of 300 (or the equivalent in a state other than New South Wales) but that there should be discretion for the acceptance of lower aggregates in particular cases'. The dissolution of the Interim Council did not mark the demise of the Faculty of Military Studies. It continued until 30 December 1986 to enable the final class to graduate from the Faculty on 9 December 1986.

To mark this new beginning, there was a special edition of the UNSW Alumni Association magazine focusing on 'Educating the

Military' in November 1985. Commodore Jim McKeegan, who gained his PhD at UNSW in the mid-1970s while seconded to the School of Mathematics at the Kensington campus as a naval liaison officer, explained that an organic connection between uniformed people and civilians was a 'sociological safety factor'. In an interview for UNSW *Alumni Papers* he said that 'the involvement of UNSW with the education of military personnel means a healthy level of contact between the military and civilians. That outside influence is a very big advantage'. He thought the absence of midshipmen on the Kensington campus after 1985 would be 'a bit of a pity'.[43] Ray Golding described the University's partnership with Defence as:

> one of our great success stories. I think that we are demonstrating that our experience and knowledge can be used in a number of ways, and one of them is in training officer cadets [...] there are enormous advantages in a training program which looks back to the resources of a larger organization, such as the Kensington campus. If the Academy had been fully autonomous that would have been impossible.[44]

Professor Harry Heseltine, head of the Department of English at Duntroon, was typical of the Canberra faculty in looking forward to the opening of the Academy because it afforded expanded teaching opportunities. He stressed that the Department of English was committed to ensuring cadets had 'a sense of the values of the culture they live in [...] we try to impart the same sets of human values as any civilian university. In effect we teach them the arts of peace [...] in the tradition of the soldier-scholar'. He thought the Academy's long incubation was a positive experience in that it led to a continuing collaboration with UNSW and demonstrated Defence's commitment to higher education.

From affiliation to Academy, 1981–1986

No-one rose to defend the Casey University concept. The continuing link with UNSW was deemed the Academy's foremost attribute. The greatest advantage that the Academy had over the single-Service Colleges was that it could do things on a much larger scale. The Academy had cost $130 million, much more than Prime Minister Fraser had ever imagined. Its annual operating bill was $47 million. The University College would consist of 150 academics, 130 technical and general staff and another 100 staff engaged in administrative and support tasks. The initial student cohort was expected to be 700 undergraduates. Within five years, that number would increase to 1300 students consisting of 276 undergraduates from the Navy, 385 undergraduates from the Air Force, 490 undergraduates from the Army, 50 international students and 100 postgraduates. The inclusion of women was perhaps the most significant personnel innovation to accompany the advent of the Academy. The presence of female cadets would work against the excesses of the 'alpha male' culture that had predominated at Duntroon and, according to those academics who made the transition from Duntroon to the Academy, brought a new level of very welcome civility to the classroom. Women certainly enhanced the environment associated with academic field trips. They were generally more diligent professionally and more disciplined personally in their approach to learning. Many women were simply better students, something that irritated some of their less capable male counterparts.

The first intake of Academy cadets would arrive on 17 January 1986. The new institution was becoming a hub of activity, establishing organisations to enhance the new forms of community life that were taking shape. In addition to the opening of the 'Thesaurus Bookshop', a commercial bank, a credit union and a hairdresser opened for business. There were sporting competitions between

Academy cadets and local teams, and social games between staff and students. The management of the Academy Staff Club was finalised and an Academy Amenities Committee was recruited. A University College branch of the Australian Association of University Staff (AAUS) was inaugurated. An Academy newsletter known as *Acadfinitas* was launched. Difficulties with the new buildings were identified and eventually rectified. Staff seminars were announced, the Library was opened to faculty and students, and clashes in class timetables were resolved. The Commandant attended the occasional class or seminar by invitation and his interest was warmly welcomed by the academics. Rear Admiral Sinclair had provided the necessary oil to ensure the Academy 'machine' functioned smoothly and efficiently. He had gained the respect and the loyalty of the vast majority of academics.

The Academy staff and students would benefit from new, well-equipped buildings, notwithstanding the antics of several obstructive trade unionists who tried to exploit the tight construction deadline to benefit their members, a less than efficient Department of Housing and Construction, and the mistakes of several architects who failed to read the design brief. The Academy was also unique in generously providing personal computers for all academics. This was an indication of things to come. Defence had led the way. In sum, there was a good deal of optimism about the Academy's potential. It would not benefit in the short term from alumni or benefactor generosity because it lacked a cohort of former students keen to 'give something back'. Those who had graduated from the single-Service colleges were without any loyalty to the Academy. Few had much affection for the new institution. The Academy was not even associated with their Service. It was a hybrid; an indicator of things to come as a Joint Service Unit. But these were early days and hopes were high within both the University and Defence.

Professor Sir Philip Baxter, the Vice-Chancellor of UNSW, at the opening of the new Science Block at the Royal Australian Naval College, HMAS *Creswell*, in 1968. *UNSW Sydney Archives*

Signing the 1981 Academy Agreement, (standing left to right) Jim Killen (Minister for Defence), William Pritchett (Secretary of the Department of Defence), Admiral Sir Anthony Synnot (Chief of Defence Force Staff); seated, Professor Rupert Myers (Vice-Chancellor, UNSW). *UNSW Sydney Archives*

Above The Prime Minister of Australia, Malcolm Fraser (right), unveiling the plaque to mark the commencement of construction work on the Academy site, February 1981. *UNSW Sydney Archives*

Right Mr Roy Fisher, one of many construction workers employed at the Defence Academy site, doing work on the steel reinforcing for the concrete floor of the Academy Library. *University College, Australian Defence Force Academy*

Opposite A flag-raising ceremony held in 1984 attended by representatives of the three Services and UNSW, proclaiming the Defence Academy as ADF joint service unit. *University College, Australian Defence Force Academy*

The Defence Academy engineering buildings under construction, February 1984.
University College, Australian Defence Force Academy

Professor Geoff Wilson, the first Rector of the University College, in 1984.
University College, Australian Defence Force Academy

Opening of the Defence Academy in 1986, (left to right) Mrs Jacqueline Samuels, Justice Gordon Samuels (Chancellor, UNSW), Rear Admiral Peter Sinclair (Commandant) and Professor Geoff Wilson (Rector, University College). *University College, Australian Defence Force Academy*

The Defence Academy shortly after its opening in 1986. The parade ground and the Dovers Building (central administration block) are in the foreground. *University College, Australian Defence Force Academy*

The first Academy Council in 1986, (top row, left to right) Major General Peter Day (Acting Chief of the General Staff), Cyril Streatfield, Dr Ray Watson, Colonel Ron Boxall, Dr Don Chaplin, Professor Ray Golding, Professor Michael Birt, Sir William Keys, Malcolm Kelson, Professor Noel Svennson, Ken Dean. (Bottom row, left to right) Professor Douglas McCallum, Air Marshal Jake Newham, General Peter Gration, Professor Geoff Wilson, Sir Edward Woodward, Rear Admiral Peter Sinclair, Sir William Cole, Elizabeth Alexander, Vice Admiral Michael Hudson. *University College, Australian Defence Force Academy*

Drill practice in the snow, October 1987. *University College, Australian Defence Force Academy*

The first Academy Graduation, 1987. *University College, Australian Defence Force Academy*

The Vice-Chancellor of UNSW, Professor Michael Birt, and the second Academy Commandant, Major General Peter Day, 1987. *University College, Australian Defence Force Academy*

Her Majesty Queen Elizabeth II with Academy Cadets during the Royal Tour of Australia, May 1988. *University College, Australian Defence Force Academy*

Governor General, Sir Ninian Stephen, inspects Graduation Parade, 1988. *University College, Australian Defence Force Academy*

Academy cadets undertaking their undergraduate studies in technical disciplines.
University College, Australian Defence Force Academy

Launch of the *Oxford Companion to Australian Literature* held at the State Library in Melbourne, 1985. Produced by faculty from the Department of English, Faculty of Military Studies, Duntroon. (Left to right) Bill Wilde, Gough Whitlam (former Prime Minister of Australia), Barry Andrews and Joy Hooton. *University College, Australian Defence Force Academy*

Tossing of the coin, annual Rector's XI vs Commandant's XI cricket match Professor Harry Heseltine and Air Vice-Marshal Richard Bomball, 1992. *University College, Australian Defence Force Academy*

Annual Rector's XI vs Commandant's XI cricket match, Rector's team, 1992.
University College, Australian Defence Force Academy

Inaugural University College Lecture delivered by the Honourable Kim Beazley MP, September 1992. At left Air Vice-Marshal Richard Bomball and Chancellor of UNSW, Justice Gordon Samuels at right. *University College, Australian Defence Force Academy*

Visit to the Defence Academy by Kim Beazley (former Minister for Defence), 1992, (left to right) Kim Beazley, Air Vice-Marshal Richard Bomball (Commandant), Professor John Niland (Vice-Chancellor, UNSW), Justice Gordon Samuels (Chancellor, UNSW) and Professor Harry Heseltine (Rector). *University College, Australian Defence Force Academy*

Visit to the Defence Academy by Alexander Downer (later Foreign Minister in the Howard Government), 1992, (left to right) Mr Peter Jennings, Professor Harry Haseltine (Rector), Alexander Downer and Air Vice-Marshal Richard Bomball. *University College, Australian Defence Force Academy*

Visit to the Defence Academy by Ms Quentin Bryce, Federal Sex Discrimination Commissioner (later Governor General), with Professor Harry Heseltine (Rector) and Rear Admiral Gerry Carwardine (Commandant), 1993. *University College, Australian Defence Force Academy*

UNSW Faculty as part of the Academy's Freedom of the City parade in 1995. *University College, Australian Defence Force Academy*

Above Visit to the Defence Academy by Gareth Evans (former Minister for Foreign Affairs), 1996, (left to right) Professor John Richards (the third Academy Rector), Group Captain Mark Lewis RAAF, Gareth Evans and Associate Professor Anthony Bergin. *University College, Australian Defence Force Academy*

Right Professor Robert King, the fourth Rector of the University College. *University College, Australian Defence Force Academy*

Opposite Professor John Baird, later fifth Rector of UNSW Canberra. *University College, Australian Defence Force Academy*

The School of Geography, Fieldschool, 2006.
University College, Australian Defence Force Academy

10
Reviews, reforms and restructures, 1986–1995

On the morning of 17 January 1986, 351 new-entry officers under training, including 52 women, arrived at the Academy site.[1] With those who had transferred from the single-Service colleges as second- and third-year cadets, the student body now consisted of 657 undergraduates and 90 postgraduates – already exceeding the 700 students expected. The Academy opening attracted all the usual expressions of delight and hope, pride and satisfaction. The foremost personal sentiments were relief – that it was open and operating after such a long and controversial gestation. The UNSW component was known as the 'University College'. It appeared as 'UC' in correspondence until the University of Canberra was established in 1990, which led to fears of confusion between the two entities. A later Rector, John Richards, had suggested to Canberra's Vice-Chancellor, Professor Don Aitkin, that his institution be referred to as 'UCAN' to avoid any confusion with 'UC UNSW'. Aitkin feared the abbreviation would be mutated to 'UCAN'T'. The 'UC' acronym then fell into disuse within UNSW. The new institution was very much a work in progress.

The initial focus on undergraduate teaching was soon supplemented by the establishment of a number of specialist research and service entities, including the Australian Defence Studies Centre, the Audio-Visual Centre and the Computer Centre. As academics began to occupy the buildings and conduct classes, there were all the standard complaints from staff and students about the facilities being poorly designed or badly constructed. There had been significant delays in completing some buildings in what was the largest Defence facilities project ever undertaken. The entire cost of the Academy was, however, less than the 'Fittings and Furniture' budget for the new Parliament House that would open in 1988. The builders remained on site until 1987, with the Officers' Mess and the Fitness Centre not available until the middle of that year. The librarian, Lynn Hard, contended that the Library could only contain half of its collection, while proposals for a chapel and a commandant's house had been shelved. The second commandant, Major General Peter Day, recalled that 'the attitude of the Defence Department was that so much money had been spent on the facilities already, any more should await some proof that the Academy was viable'.[2] But the Academy represented a major step forward from the sub-standard conditions endured by academics at Jervis Bay, Duntroon and Point Cook. If new equipment was needed, it was acquired with alacrity. In fact, almost no expense was spared.

In the first few years of the Academy's life the difficulties of adjusting to the new arrangements and addressing facilities problems were matched by a strong desire on the part of the University to see the venture succeed – and to prove the naysayers were mistaken. Most of Defence's senior leaders were actively committed to the Academy. Several of the Service chiefs were yet to be persuaded but they were outwardly supportive. The Chief of the Defence Force, General Peter Gration, was always solidly in favour, as was the Chief of the Air Staff, Air Marshal Ray Funnell. Gration's successors over the next decade, Admiral Alan Beaumont and General John Baker, were very pro-Academy and never wavered in their support. Tony Ayers had no views on the Academy before becoming Secretary of the Defence Department in August 1988. He spoke with both Malcolm Fraser and Sir Arthur Tange, who persuaded him that its establishment made sense and was necessary to the larger reform agenda being pursued.[3] As one of Tange's successors, Ayers thought he might need to protect the University's standing from Defence and the Government more generally. In the years that followed, Ayers never felt the University's independence needed protection, although he asked that the University hear, and when necessary heed, Defence's concerns about its operations. Ayers did not think any other university was better placed to deliver education to Defence and felt UNSW's loyalty to Defence ought to be honoured if and when the 1981 Agreement was renegotiated. His backing would be important throughout the next decade.

The working relationship between the academic and military sides was, for the most part, happy and healthy. There were the inevitable stresses and strains associated with competing for cadet time but these were, for the greatest part, managed at a working level and never escalated to become an institutional problem. External relations were a little more complicated, especially the place of the Academy in the national capital's education landscape. In response to the Stephen Report into higher education for the Australian Capital Territory completed in 1990, Professor Ian Chubb, Head of the Commonwealth Government's Higher Education Council, noted the benefits of creating a new university in Canberra by amalgamating the ANU Faculties, the Canberra Institute of the Arts and the University of Canberra. There was a sly footnote to Chubb's observation:

> I have not had the opportunity to explore this [recommendation] with relevant staff of the Academy or of the University of NSW. It does, however, have the support

of academic staff in the other institutions. Higher education in the ACT would also be more logically organised if the Academy were affiliated with an institution located in the ACT and not a university based in Sydney.

The Deputy Vice-Chancellor at UNSW, Professor Jarlath Ronayne, wrote to the Minister for Higher Education, Peter Baldwin, stressing that Chubb's statement demonstrated 'a lack of understanding of the nature of the relationship between the Academy and the University of NSW'.[4] He detailed the 'remarkable success' of the arrangements, the building of trust and goodwill over the previous two decades, and the Academy's anticipated 'rise as a national institution that was not in competition with any ACT institution'. The fact that the main campus at Kensington was a 'distance of some 250km' should have led Chubb to condemn all the Government's reforms designed to link otherwise disparate campuses around the country at that time. Ronayne was astounded that Chubb did not speak with anyone in the University College about his recommendation and confidently asserted that:

> an amalgamation of University College with the new university [i.e. the university formed from all of the undergraduate education providers in the ACT] would have little support from the academic staff of the College; and none from the University, the Department of Defence or the three Services.

He warned that UNSW would fight any attempt to amalgamate the Academy into a new entity or to alter the extant arrangements. On February 1991, the chair of the Academy Advisory Committee, Sir Edward Woodward, told Ronayne that it was 'hard to believe that this ill-considered suggestion will be taken seriously'. It wasn't. But this episode revealed the extent to which the University jealously guarded its place in the national capital.

Having set such a firm foundation over the previous decade, the Rector, Professor Geoff Wilson, accepted the Vice-Chancellorship of the University of Central Queensland based in Rockhampton. He left Canberra on 13 March 1991.[5] The debt owed to Professor Wilson by all stakeholders was a significant one. It began during his time as Dean of the Faculty of Military Studies. As Rector, he had personally conducted most of the negotiations and carried the bulk of the relationship with Defence. If the Rector and the Commandant could not work together, the tone and tenor of the University's interactions with Defence at all levels would be adversely affected. He retained the respect and the goodwill of the Faculty throughout the protracted Academy negotiations and was rightly praised for his calm approach and consistent demeanour. His scholarly abilities were first-rate. Unlike some academic administrators, Wilson remained active in his own discipline and continued to publish research papers. As Rector he had brought people together and made disunity a stranger. He was a risk-taker without being reckless. Academic and administrative staff appreciated his easy manner and friendly nature, although some believed he was uneven in distributing funds across the academic departments. It was unsurprising that some thought the allocation to Physics was overly generous. Wilson knew how to 'manage' expectations and obligations at Kensington, so that the main campus was neither too close nor too far to play a part. He enjoyed the goodwill of senior staff at the ANU and managed to avoid any outward antipathy or unfriendly rivalry between the campuses. Within Defence, he was universally admired and managed to find ways of extracting additional money for facilities without attracting the charge that the University was engaging in special pleading. A man of stable temperament who never appeared ruffled, he accepted the complexities and the controversies of the UNSW–Defence relationship when others might have descended into despair. After a short interregnum, the University announced that the new Rector would be the Professor of English and Deputy Rector, Harry Heseltine.

Heseltine commenced his duties on 2 July 1991 with the expectation that he would serve for five years. As with all but one of his successors (and Professor John Richards had been a member of the senior leadership team for a decade before becoming Rector in 1996), Heseltine's term of office would be longer than that held by any of the Academy Commandants who have ever served in the institution. Over the next five years there would, in fact, be four commandants: Air Vice-Marshal Richard Bomball, Rear Admiral Gerry Carwardine, Major General Frank Hickling and Air Vice-Marshal Gary Beck. The short tenure of commandants was replicated in the rapid turnover of less senior uniformed staff with whom the University was obliged to deal. Officers familiar with the challenges of educating cadets and acquainted with the complexity of University administration were often posted elsewhere before the experience they acquired could make a difference to the Academy's structural arrangements. Defence's own interests were not well served by a lack of continuity among those charged with achieving the best outcomes. Despite the difficulties, Heseltine told the Vice-Chancellor (Professor Michael Birt) not long after assuming office as Rector that 'relations between the academic and military elements of the Academy have never been more harmonious and positive'.[6]

Professor Heseltine had enjoyed a distinguished academic career before starting work as Head of the Department of Language and Literature within the Faculty of Military Studies at Duntroon in early 1982. He had been in the Army cadets at school in Western

Australia, but was too young for service in the Second World War and too old for the national service scheme associated with the Vietnam War. Heseltine had been the final chair of the Faculty of Military Studies before it passed out of existence in late 1986. He became the Deputy Rector in 1988, succeeding Professor Alan Gilbert, who was appointed Pro Vice-Chancellor at the Kensington campus. Heseltine believed that Geoff Wilson appointed him Deputy Rector because he (Wilson) was a physicist and Heseltine a humanist, and the mix of disciplines would provide balance in the College's senior administration. Wilson simply thought Heseltine the better of two internal candidates for the position.

Heseltine was certainly experienced in university administration. He had participated in planning for the new Academy and was familiar with the challenges of working with Defence, including the too-frequent turnover of commandants – something that Heseltine also felt bedevilled the development of the institution. He was widely esteemed as a scholar of Australian literature and universally respected as a man, a respect he retained after dealing with several acute conflicts within and among the College staff. He was later made an Officer of the Order of Australia (AO), recognition that was universally acclaimed. The Rectorship would be his final academic post but he always exuded vigour and enthusiasm for the task at hand. Although the Vice-Chancellor was being slowly debilitated by dementia and would shortly resign, Professor Michael Birt had shown considerable trust in the University College's senior staff and felt no need to intervene in the operation of the Canberra campus. His successor at Kensington felt similarly.

The appointment of John Niland as Vice-Chancellor in late 1992 also brought a new mood to the University as a whole, with an emphasis on eradicating gender bias, closer attention to staff development, a renewed focus on information technology and enhancing the Kensington site. Niland sensed the strategic value of the Canberra campus and took no persuading as to its value.[7] Educated at Lismore High School and recipient of the Brigade Prize for the Best Cadet at the Potential Officers' Course held in May 1957, Niland had briefly considered a career in the Army and had thought about applying for Duntroon. He was aware of activities at the Academy campus while he was Dean of Commerce and Economics and attending VCAC meetings with Geoff Wilson. He was comfortable with the governance oversight provided by Sir Edward Woodward and the institutional leadership exercised by Harry Heseltine. As Niland was committed to a campus-wide facilities redevelopment and to substantially restructuring faculties at Kensington, he did not need to focus closely on the University's operations in Canberra and did not regard the Academy as another source of revenue (something precluded by the 1981 Agreement in any event). He attended graduations and major events in Canberra but did not personally intervene in Academy affairs other than when his advice and guidance was requested by the Rector.

The incoming Rector found the circumstances much changed from those enjoyed by his predecessor. The first he encountered was the development of the inaugural UNSW strategic plan. This was an initiative of Michael Birt, who felt the university's evolution needed to be directed and disciplined. The University College would develop its own plan, in addition to identifying where and how it contributed to the wider objectives of the University. Heseltine invited the Commandant, Air Vice-Marshal Richard Bomball, and his deputy, Colonel Roger Powell, to a weekend planning retreat with his departmental heads at Milton Park. The University College plan was completed by the end of 1991 and was handed to the Vice-Chancellor at the graduation ceremony that year. This was the first exercise of its kind for the College and is notable for including uniformed personnel in the drafting process. There was no such reciprocity in the military's policy-making committees. Whereas the military staff were present at faculty meetings and at end-of-year examiners' meetings, the civilian academics were not involved in meetings relating to the performance of cadets in professional training. The organisational segregation reflected the physical separation of the two staffs on campus, with the academics on the western side of the boulevard leading to the main administration building and the uniforms on the eastern side. However, the decision of the Commandant (Richard Bomball) to expand the name of the Officers' Mess to include the words 'Senior Common Room' was a significant and generous gesture.

The principal variation in Wilson's and Heseltine's tenures was, however, a sudden shortage of money. As Heseltine later explained:

> Geoff Wilson had always had a very good personal relationship with Ron McLeod [Deputy Secretary Budgets and Management in the Department of Defence] [...] and every year Geoff would go to Ron and say: 'what about some additional estimates?' and he'd get them. So Geoff always banked on having additional monies coming in around February. In 1991 [...] just about when he's due to go, he went to Ron and asked: 'what about additional estimates?' 'Sorry', said Ron. So Geoff left this institution with a potential deficit of $2.6 million staring me in the face [...] In my first day as Acting Rector, at 4pm in the afternoon, I sent out a fax to all department heads that there was going to be an absolute stay on all spending until the end of the financial year.[8]

Times had changed and Defence was looking to reduce costs across its establishments. The special relationship that the University had enjoyed with the Department in terms of finances and its foundation had ended. Because Heseltine had been able to reduce the deficit by $800 000 while he was the Acting Rector, Defence agreed to 'wipe the slate clean' and cancel the deficit when he was appointed Rector in mid-July. The mid-1990s also marked a change of climate for the universities nationally in terms of their operations and funding. They needed to be far more entrepreneurial, to find alternative sources of funding and to exercise greater control over their expenditure. By the end of the year, the Rector was able to advise Faculty that he was:

> confident that, in spite of the difficulties of the early months of 1991, financial year 1991–92 will signal the beginning of a new period of more orderly economic planning for the College, of fuller and more sympathetic negotiations with Defence, and consequently a more stable and predictable financial situation for us all.[9]

Many academics believed the allocation of funding between departments was fairer and more transparent under Heseltine's leadership.

In addition to recasting the budget and reviewing expenditure, Heseltine began to make his mark on the institution. As the establishment phase had ended and excuses about the newness of the University College were no longer relevant, UNSW needed to consolidate its position within the national capital. In August 1991 a Media Liaison Unit was established to 'raise the level of public awareness of University College, particularly in the Canberra community'. The first two annual 'Open Days' in 1992 and 1993 were moderately successful, with between 400 and 600 people attending. (They now attract 8000–10 000 people.) The 1994 Open Day benefitted from advertising sponsored by

Kensington; the attendance figure was estimated to be more than 2500 people. The fourth Commandant, Rear Admiral Gerry Carwardine, created the Defence Academy Community Relations Committee (DACRC) in November 1993 to secure positive press for the Academy and find ways to enhance its public image. The Committee's membership included representatives of the University College. The intellectual life of the University College also continued to expand and deepen. Sir Arthur Tange was invited to join the Department of History as an Honorary Visiting Fellow and the former Minister for Defence, Kim Beazley, delivered the Inaugural College Lecture.

Against this backdrop, the question was being asked, perhaps prematurely: had the University fulfilled expectations? The main focus was on undergraduate education as this was the area of priority for Defence. The statistics for the first six years of the Academy's operations (1986–92) were a revelation. In terms of degree choices, the change in student enrolment patterns was marked and substantial: Bachelor of Arts: Army from 123 to 195, Navy from 33 to 104 and Air Force from 14 to 53; Bachelor of Science: Army from 71 to 105, Navy from 46 to 101 and Air Force from 56 to 135; Bachelor of Civil Engineering: Army from 34 to 41; Bachelor of Electrical Engineering: Army from 17 to 24, Navy from 26 to 35 and Air Force from 18 to 55; Bachelor of Mechanical Engineering: Army from 8 to 9 and Navy from 13 to 20; Bachelor of Aeronautical Engineering: Air Force from 17 to 66. The number of undergraduates had increased from 770 to 969 with an additional 365 postgraduates in 1992, of whom 116 were uniformed Defence, 37 civilian Defence, 32 UNSW staff and 180 civilian. More civilians were being admitted to postgraduate courses, with many travelling from overseas to study at the Academy. As they were postgraduate students, this trend also influenced the collective mood. International students increased from 42 in 1986 (Arts 14, Science 12 and Engineering 17) to 57 in 1989 (Arts 19, Science 14 and Engineering 24).

In terms of the undergraduate full-time student load across the twelve departments between 1988 and 1992: Chemistry remained unchanged at 41, Civil and Maritime Engineering increased from 19 to 27, Electrical Engineering from 51 to 61, History from 60 to 93, Mechanical Engineering from 21 to 45, Computer Science from 101 to 114, Economics and Management from 87 to 89, English from 44 to 49; Geography and Oceanography declined from 147 to 139, Politics from 85 to 83, Mathematics from 132 to 116, and Physics from 76 to 72. The postgraduate student load from 1988 to 1992 showed growth in all disciplines: Civil and Maritime Engineering from 15 to 23, Electrical Engineering from 11 to 22, Mechanical Engineering from 2 to 9, Economics and Management from 0.5 to 19, Computer Science from 1 to 25, English from 7 to 12, Geography/Oceanography from 6 to 7, History from 9 to 11 and Politics from 11 to 16. Most notably, research grant income increased from $223 000 in 1986 to $2 773 000 in 1992. Grant income declined slightly in 1993 ($2 429 000) and 1994 ($2 092 000) but rebounded strongly, with the single biggest increase in 1995 when the total reached $3 187 000. Not surprisingly, staff publications more than doubled in the same period from 202 to 476, with increases in all departments.

The increased teaching and supervision load when coupled with research grants, saw the academic staff increase from 112 in 1985 to 183 in 1994 and general staff from 192 to 300 over the same period, with the majority of staff aged between 35 and 44 years. In terms of comparisons with the main campus at Kensington, the ratio of academic to general staff was the same (1:1.6) but the percentage of female staff in Canberra was lower (13 per cent compared to 24 per cent at Kensington), the percentage of tenured staff was higher (77 per cent compared to 66 per cent at Kensington), and the number of staff aged over 45 was higher (53 per cent compared to 47 per cent at Kensington). Postgraduate study was proving to be the area of unexpected growth and the increase was marked. After 1987, enrolments in Masters programs accounted for one-third of total uniformed enrolments. This student cohort was not anticipated when the 1981 Agreement was signed and the University College was being planned. Conversely, the number of new-entry officer cadets declined. The initial estimate of 1100 undergraduates would, in fact, never be reached as the ADF's workforce contracted rather than expanded. Fewer new officers were needed. Officers for the permanent forces were also being secured by other means. With increases in the transfer of reserve officers to regular commissions, the transition of more 'other ranks' personnel to commissioned service, and larger numbers of overseas recruitments and graduate entrants, the Academy was providing only one-third of officers for the Services.

Notably, the proportion of ADF officers with degrees doubled between 1977 and 1992. The culture had changed across the Services. Uniformed people not only wanted to reflect critically on past service and to better prepare themselves for future service, there was also an element of 'bracket creep'. There was a growing feeling that all officers ought to have degrees and, in time, postgraduate qualifications as well. Like the United States, where a number of Service chiefs had doctorates and most had postgraduate qualifications of some kind, educational attainments would influence promotion prospects and posting opportunities. The demand for coursework Masters degrees had a direct and substantial impact on staff workloads and on priorities for curriculum development. It was a need generated principally by Defence; the University College was obliged to respond positively, given that the staff–student ratio compared favourably to Kensington. Creating the Master of Defence Studies in very short order during 1986–87 was not simply a positive response but an enthusiastic one; the academic staff involved were keen to offer courses in their specialised areas to students who would be mature, already equipped with bachelors degrees, eager to learn and volunteers for study – the opposite of many cadets, who clearly begrudged having to study or were content with a bare pass. The value of a Masters degree for a military career was certainly recognised by ADF personnel who enrolled as students. Their motivation often included learning for its own sake. It also made political and economic sense to support the Academy's long-term growth. By April 1992, the University College had more than 360 postgraduates. The majority of postgraduates were part-time but 98 were full-time, of which 43 per cent were serving military personnel or officers of the Department of Defence. In addition to student research, the research capacity of the staff was also expanding.

To make the most of this, the Rector and the Commandant met with Dr Richard Brabin-Smith, the First Assistant Secretary of the Strategic and International Policy (IP) Division of the Department of Defence, to develop closer links between the Academy and Defence in July 1992. The creation of an 'ADFA Users' Committee' was intended to promote the continuing work of academics and to receive notice of projects of interest to Defence that academics might consider undertaking themselves. The University College had also established a number of important institutional affiliations with overseas academic institutions – a development of interest to the IP Division. The College's 'External Affairs Committee' facilitated and recorded these interactions, which included the United Kingdom (particularly Shrivenham) and the United States, Indonesia and China, Russia and Canada, South Korea, the Philippines, Singapore, and Fiji. The Australian Defence Studies Centre was also active in cultivating international contacts with these countries and Taiwan, Israel and Vietnam. In a follow-up letter, Brabin-Smith explained:

> The thinking already underway in the Civil and Maritime Engineering Department on 'appropriate technologies' for Defence construction activities in the Pacific is an example of the kind of research that could be directed to activities of common benefit to the College and the Department.[10]

Conversely, there were some nations and regimes with which the Australian Government had concerns, especially in relation to human rights abuses. The Government preferred that the University also suspend its contact with these countries, if only for a short time. Was this a form of censorship or an exercise in propaganda? Brabin-Smith explained that government concern did not 'automatically put individual countries on a proscribed list but it does indicate where care should be exercised on the intensity of interaction'. As the Department was only able to give 'general guidance in priorities and proscriptions, which in any case vary over time', he highlighted the 'importance of consultation on specifics with IP Division'. This was a polite way of reminding academics that they needed to be mindful of sensitivities.

The persistent problem for Defence was identifying and articulating its educational needs. For instance, the Maritime Engineering program was introduced at a substantial cost to Defence before being abandoned when student numbers were insufficient to justify its continuation. As he prepared to retire from the RAAF, the third Commandant, Air Vice-Marshal Richard Bomball, produced a 'Review of Tenure' dated 1 March 1993 in which he chastised the single-Service personnel divisions for not knowing what they wanted from the academic curriculum and making too many changes after too little thought. 'A more consistent and coordinated approach could result in a better product and better utilisation of College resources without affecting the liberal nature of the education.' He highlighted a lack of motivation among cadets, deficient military standards and a culture that did not reflect the wider ADF mood. He would not be the last to identify these pressing issues.

As the University College tried to more closely align staff research activities with Defence priorities in the hope of securing financial support where interests converged, the College strived to be more attentive to the changing educational needs of the ADF as far as they could be detected. In March 1994, a new aeronautical engineering degree course (the Bachelor of Technology (Aero)) was designed and introduced at the Academy to provide a technology option for prospective pilots and navigators. This was an important innovation, welcomed by Defence, for which Professor John Baird deserved great credit. If the BTech (Aero) graduates happened to be among the 60–70 per cent of students who later failed to complete flight training, they could convert their existing qualification to a four-year aeronautical engineering degree.

On almost every front, the College was expanding. It began with an extension to the library building early the following year, in tandem with refurbishing the existing building. The work was to be completed by March 1996. The aim was to provide capacity for 425 000 volumes and seating for 150 library users. This would allow all holdings to be housed on the Academy site rather than in warehouse storage at nearby Fyshwick. Before the project was completed, 70 000 volumes and 95 per cent of the Australian Special Research Collection were held off site. In the meantime a dedicated building for the Australian Defence Studies Centre was opened in February 1995. These had been the first major building projects at the Academy for a decade. They were to be the last for some time as Defence showed consistent reluctance to spend more money on the Academy despite the greatly enlarged student body.

There were other creative initiatives between the University and Defence. The School of Politics, led by Professor Ian McAllister and with the co-operation of the ADF, initiated a longitudinal Survey of the Military Profession which involved administering questionnaires to cadets and midshipmen in first year in 1987–1989 and following up those individuals five years and then ten years later. The School of History, which had been built upon core expertise inherited from RMC Duntroon, lacked any specialised capacity in naval history and the study of sea power. By the early 1990s both academic staff and students saw the need to fill this void. Key players in filling the gap were Professor Peter Dennis, the Head of School, and Rear Admiral David Campbell, who had won the Graham Naval History Prize at Dartmouth three decades earlier and was then Deputy Chief of Naval Staff. He was told the School of History's structure precluded another professorial position but a senior lecturership would be possible. The University and the Navy signed a contract in July 1995 to fund the appointment of a naval historian for ten years, to be known as the 'Osborne Fellow in Naval History', together with associated research and library costs. The position's title was drawn from the 1913 location of the Interim Naval College at Osborne House in Geelong. The post was advertised in late 1996 with the incumbent to begin in July 1997. The appointment committee included Captain (later Commodore) Jack McCaffrie RAN as the Navy's representative. The appointee was Dr John Reeve, an Australian graduate of Melbourne and Cambridge Universities with extensive international experience, who was then at Sydney University. The Navy later extended its funding for a further two years until 2009, after which the position was funded by the University. A popular suite of courses was established at senior undergraduate and postgraduate levels covering naval history and maritime strategy from 1500 AD to the present day, with further courses dealing with amphibious warfare later created in response to the new strategic trends of the early 2000s. Dr Reeve also contributed to the national strategic policy debates surrounding the 2000 and 2009 Defence White Papers and during the Parliamentary Inquiry into Australia's Maritime Strategy in 2003–2004. The establishment of this position was a welcome initiative (despite some serious unease among some members of the School) and one that showed the capacity of both organisations to collaborate effectively to achieve positive outcomes.

There were, inevitably, some tensions between the outlook of academics and the mindset of military officers. For instance, University staff did not always appreciate the acute sensitivity of the Services to public controversy, especially if that controversy involved their corporate reputation. One episode illuminates the potential for tension. Ahead of the Academy's opening in 1986, the Head of the English Department at Duntroon, Professor Harry Heseltine, proposed producing a lexicon of the distinctive cadet slang that had developed at RMC over the previous 70 years. A survey questionnaire distributed in late 1983 with the approval of the RMC Commandant, Major General Barry Hockney, elicited very little material. Dr Bruce Moore then took up the project. The cadets responded diligently thanks to the 'active encouragement of the Director of Military Art' at RMC. The lexicon was to be published as a very substantial book of 460 pages in 1992. The cover featured a recent Queens Birthday Parade held at the College. The original imprint page stated it was published by the 'Australian National Dictionary Centre, ANU' and the 'English Department, University College, Australian Defence Force Academy'. The lexicon included hundreds of terms that dealt with all aspects of cadet life, including the relationship that existed between the junior and senior classes at the College. Some were humorous and benign. Others were offensive and crude, conveying racism, sexism and bigotry. At the time of the book's intended release, Heseltine was the Academy Rector and Moore was the Acting Head of English.

Given the possibility of controversy, an advance copy of the book was provided to the Commandant of RMC, Brigadier Rodney Curtis, ahead of wider distribution. Shortly afterwards, Heseltine and Moore were asked to meet the Academy Commandant, Richard Bomball, and his Deputy, Colonel Roger Powell, and the RMC Commandant, Brigadier Curtis. The lexicon was indeed controversial. In sum, the Army could not countenance the book's release because it would damage its reputation and that of RMC. It was alleged that Moore had been 'conned' by some mischievous cadets and that the lexicon presented an unfair and distorted view of RMC and its values. Many of the colourful terms appearing in the book had, it was asserted, never been used or gained currency. While Moore was prepared to concede there had been problems confirming the veracity and accuracy of some slang terms, he defended the lexicon's scholarship. The intended timing of the book's release was another critical issue. A number of cadets were then being court-martialled in the Academy Assembly Hall for misconduct. They had been accused of 'woofering', a form of humiliation in which a vacuum cleaner was applied to the genitals of a cadet who was restrained.

Heseltine agreed to pulping the entire print run with Defence paying for a new edition with a plain front cover, the words 'in the period' inserted into the sub-title to illuminate the historical nature of the study, and publication attributed solely to the ANU's Australian National Dictionary Centre. He also agreed to withhold release of the revised edition for six months to allow the court-martials to conclude. The replacement volume appeared with a revised preface that included the statement:

> Some of the terms in this Lexicon may cause offence to readers [...] the attitudes reflected by such terms, and by the examples of usage, are not those of myself, of the publisher, or of anyone involved in the various processes of the production or distribution of the book.[11]

Moore noted that some of the terms were 'not used by, and in some cases were not known by, all cadets'. He went on to emphasise that 'since the 1983–85 period Commandants at both Duntroon and the [Academy] have introduced programmes which seek to eliminate those very attitudes which some readers may find offensive. Some parts of this book, therefore, play the historical role of indicating why those attitudes had to be changed'.

There are several ways of interpreting this regrettable incident. From the University's perspective, there was a satisfactory outcome: censorship was resisted and the book appeared. On the other hand, there was every possibility that the Army had an enduring attitudinal problem if the RMC slang had persisted or penetrated the Academy. The crudity of cadet slang could not be excused as an inevitable or unavoidable byproduct of preparing young men for uniformed service and armed conflict. The University did not create the problem but it had revealed the difficulties that could arise when disinterested academic research was conducted in a military environment. This episode also revealed the extent of what might be considered academic innocence. That the book originated from 'within' the Academy made denying its contents next to impossible for the Army if the media wanted a pretext for a controversial story.

Reviews, reforms and restructures, 1986–1995

The ADF was to be completely dissociated from the book, hence its publication by the ANU. This achieved little and proved nothing because the lexicon itself was unchanged. Not a word was altered. Even with its imperfections, the lexicon provided a compelling catalyst for cultural change. Although this episode was dealt with quietly, it exposed the gulf that existed between the University and the military on matters with a 'public relations' dimension, especially in relation to the Academy, which had already endured a good deal of bad press. Within a few years the need for cultural reform would become irresistible. In the meantime, the Academy was fulfilling its core function, according to a range of indicators.

By the end of 1995 and the Academy's first decade of operation, 1500 students had received degrees and other qualifications at undergraduate and postgraduate levels. The undergraduate programs were expanded to cover Management, Oceanography, Information Systems and Aeronautical Engineering. The postgraduate growth continued as well. Admissions on the tenth anniversary were the highest on record, with 990 cadets studying for undergraduate degrees and 450 new postgraduates, taking total student enrolment numbers beyond 1500. The best news was probably the increase in Civil Engineering students, with 27 new-entry cadets enrolling in 1995, a substantial increase from the annual average of 15 over the previous few years. Collaboration between the Army and the University to raise the profile of Engineering was nonetheless making a difference to career choices.

The tenth anniversary was a marker for UNSW to complete its corporate image renewal project. This project included the University College and led to an agreement between the Vice-Chancellor, the Commandant and the Rector on a single approach to 'corporate branding' in order to 'cement the unique partnership' between UNSW and Defence 'in a very visual way' by combining the ADFA badge and the University logo with a common typeface – Optima. In releasing a suite of new images and protocols for their use, both 'partners' hoped that the 'new image will go some way towards educating the community about the relationship between UNSW and Defence'. The first application was on stationery and signage. While the focus on symbols was not unimportant, there were matters of substance that the symbols could not obscure. The 'dual badging' did not last long, with Defence wanting to distance itself from potentially controversial activities conducted within the University College and the University College concerned that media reporting of cadet misconduct would tarnish its reputation.

Although cadet misconduct was a persistent problem, the most pressing issues were academic results and retention rates. Had they improved at the Academy, as promised by its advocates? As early as 1990, the Academy Council had expressed concern that the undergraduates were not making the most of the opportunities presented by the Academy and their academic performance was not meeting expectations. At the Council's request, Emeritus Professor John Burns and Lieutenant Colonel Bob Bradford were asked to examine the cadet experience and to make recommendations for change.[12] The aphorism: '51 per cent wasted effort; 50 per cent sufficient effort; 49 per cent wasted year' was widely known within the undergraduate student body. The aim of many was to secure a bare pass. These students believed nothing valuable was achieved by doing better but doing worse meant the whole year would need to be repeated. It was an attitude that members of the military staff were concerned was influencing the approach of cadets to their vocational training as well.

Burns and Bradford received submissions from staff and interviewed former and current students. After noting that the Academy was effectively 'out of sight and out of mind' for most uniformed personnel and the Canberra community, they believed the Academy needed to improve its profile within and beyond the ADF to overcome an element of isolation from both the Services and from the civilian population which, they asserted, had led to an ethos of mediocrity. Suggestions included promoting open days, a new format for the Academy annual report, and forming a 'Friends of the Academy' organisation. They were disappointed on the need to report: the 'demeaning' treatment of first-year cadets (particularly women) by third-year cadets; that few cadets were striving for academic excellence; the apparent lack of interest in degree structures and course content within the Services; the posting of less than suitable military staff to the Academy; and poor liaison between the military and academic staff, notwithstanding the mutual respect that existed between the two.

The principal challenge was improving cadet motivation, especially in relation to the academic program. Plainly, there were some Academy staff and some notable senior officers in Service headquarters who were denigrating the importance of education, and this eroded the sense of purpose vital to good undergraduate results. Achieving a consistent message would be helped, it was thought, by raising the profile of the Visiting Military Fellow program and providing more full-time study places for serving officers. Attitudinal change was needed, but this would not happen quickly. The academic staff would assist in this process by paying closer attention to the development of study skills (which were assumed but not assessed in first year), by more overt descriptions of course objectives and outcomes, by ensuring that course workloads and marking criteria were consistent, and avoiding any sense in which some courses were soft or easy options. A unit tasked with assisting academic skills, including English language skills for postgraduate overseas students, was established in response. Separately, cadets were entitled to complain that the documentation accompanying some courses was sparse and that the 12 departments had different expectations and requirements of their students. It was clear that the University College had its own contribution to make to improving student attitudes. The Academy Council welcomed the report and recommended a series of actions for both the military and academy staffs. The most significant outcome over the next few years was a halving of the first-year resignation rate.

Not long after this internally generated review was completed and changes made, the Academy was subjected to its first external review by the Inspector-General (IG) Division of the Department of Defence. The review commenced in early 1992 and was released publicly in February 1993.[13] While a number of Defence and University officials believed it was too early to review the Academy, there was pressure on the Inspector-General's Division to conduct a program review of a major Defence institution. The Division had previously reviewed the Aeronautical Research Laboratory, the RAN's Hydrographic Service, the Army's ammunition management, Defence Force Recruiting, Natural Disaster and Civil Defence and the Army's Land Surveillance Forces. The Academy seemed like a Defence entity that lent itself to a controlled and manageable review. The review was conducted over a ten-month period and involved a very large steering committee chaired by a former Vice-Chancellor of Queensland University, Professor Brian Wilson, that included representatives from the Services and the Department of Defence, the Department of Employment, Education and Training, and the Australian Vice-Chancellors' Committee. More than one hundred interviews were conducted in what would prove to be a most thorough stocktake of the Academy's first five years.

Many academics thought Professor Brian Wilson was 'a bit hostile' towards the College and that the review team was predisposed to finding fault.[14] The only stated internal impetus for the review was an unattributed concern that Defence had 'tended to adopt a hands-off approach to allow this tri-Service establishment freedom to develop. There are a number of instances where greater Service input would have assisted the development of the Academy'.[15] It emerged that these inputs included more specific advice to UNSW on what Defence sought from both the undergraduate and postgraduate programs, a detailed breakdown of the funding provided to the University, the apparent absence of performance standards by which the University's operations could be assessed, and the lack of any transparency in the University's postgraduate enrolment policies and teaching practices.

The IG's review was foremost an exercise in determining whether Defence was getting value for money from funding the Academy. While the review team conceded that it was 'too soon to assess the long-term effectiveness of graduates' and a series of subsequent reviews were to find it was an almost impossible exercise, the final report could not come to a firm view on how to determine cost effectiveness because so much information was either unavailable or recorded in ways that did not allow efficiency to be readily assessed. Because the ADF lacked a coherent statement on the purposes and priorities of higher education in academic disciplines other than Engineering and Applied Science, it was difficult for the review team to make any judgments about value for money. There was also the rapid rotation of uniformed officers who could be consulted on curriculum questions. Once the ADF settled on what it wanted from the Academy, the review team felt that critical evaluations could then be made.

The 1981 Agreement was, however, considered the root of several administrative and financial problems. While there was a clear understanding that Defence was paying more to educate its students at the Academy than at the main campus at Kensington, it was not clear what this 'premium' covered and whether it was justified. Plainly, there were critics of the Academy who believed that UNSW was either exploiting Defence or concealing its true costs to undertake functions not prescribed in the Agreement. I recall from my time as a doctoral student at the Academy (1990–1991) that some middle-ranking naval officers contended the University saw the Academy as 'rivers of gold' and that Kensington was profiting from the agreement. This was far from the truth, although the Academy was better funded at that time than most comparable tertiary education institutions. The paucity of funding for other institutions did not mean, of course, that the Academy was over-funded. The Academy was new and it needed to establish its own reputation for quality teaching and research excellence – and excellence would not be achieved without investment.

The Academy also benefitted from Defence support for a range of teaching and research activities that were simply unavailable to other institutions. For instance, the Department of Civil Engineering offered a final-year subject named 'Civil Engineering Project'. It was usually based on a scenario that involved establishing a new Defence logistics base to support an exercise or operation. The first site was Wewak in Papua New Guinea. Later sites included the Shoalwater Bay Training Area. Part of the project involved a site investigation utilising equipment transported by RAAF C-130 Hercules transport aircraft, locally allocated fixed and rotary wing aircraft, in addition to Land Rover vehicles and tented accommodation. Civilian universities could not call on support of this kind or this magnitude for an undergraduate program.

After urging Defence to develop a comprehensive education policy, the review team called for the 1981 Agreement to be replaced with 'an arrangement more appropriate to modern public sector practice'. The University agreed and believed that the existing agreement could be modified through negotiations conducted under Part 9 of the extant document. The most pressing issue for the IG was the absence of accountability mechanisms with respect to performance and expenditure, although the final report would stress that UNSW had performed entirely to the satisfaction of Defence since 1981. It had also dealt with the number of undergraduates increasing from 770 in 1986 to 1034 in 1991 and the number of postgraduates increasing from 90 to 356 in the same period. The size and the nature of the task confronting the University College had changed substantially during the first five years of the Academy's operations.

Notably, Defence had no desire to seek another partner because UNSW had under-performed in any way or failed to meet expectations. But the University College had not reported to Defence 'in a format consistent with performance monitoring in the United (National) University System'. Nor had it kept Defence sufficiently informed of its activities in postgraduate education or detailed how these activities were consistent with Defence's overall objectives. There was no evidence that the University College had explored the possibility of using academics based at Kensington to deliver courses or whether too many courses were being offered to too few students. The University had also been reactive rather than proactive in defining what it considered to be acceptable influence on undergraduate programs. Rather than outlining the guidance it welcomed, the University had simply complained about pressure it resented. As both the University and Defence would be helped by developing the curriculum as partners, the University needed to be more open to Defence's desire to have input. This was agreed. But the University would not accept the review team's suggestion that military training be accredited within the undergraduate degree program. Nor would the ADF accept the suggestion that full fee-paying civilian undergraduates might be permitted to enrol in courses at the Academy.

Despite reservations about whether such an early examination of the Academy was worthwhile and what the report could achieve, its tone was reasonable and its approach was sensible, leading to dozens of findings that were insightful and timely. The IG even provided the Government and the University with a timetable for implementing its recommendations. At the first meeting of the Academy Council after the report was released, both the Chief of the Defence Force (General Peter Gration) and the Departmental Secretary (Tony Ayers) expressed their satisfaction with the report and their general willingness to accept its recommendations. A joint University–Defence 'Academy Review Implementation Committee (ARIC)' was established, with Professor John Richards representing UNSW and Colonel Chris Hunter leading the ADF element. Richards later remarked that Hunter 'was excellent to work with and the fact that the Academy didn't suffer as much as it may have was due in part to his understanding of the nature of the University and the importance of the Academy'.[16]

But the IG's principal recommendation was ignored. The 1981 Agreement was not revised largely because the Deputy Vice-Chancellor, Tony Wicken, and Niland had quietly concluded that it operated in the University's favour and did not want to see any changes to the arrangements. Apart from the Agreement, some of the more complex and demanding accountability mechanisms proposed by the IG were ignored. Why? Because those who received the report stopped reading the text after noting that the Academy was generally performing well, that the cost was within Defence's general expectations (and budget) and that change was needed but was limited to administrative and financial systems and could be implemented incrementally.

A 'Management Audit Report' was conducted in tandem with the IG's review with very similar terms of reference. It also reviewed the 1981 Agreement, examined the financial arrangements and assessed the management structure in terms of the Academy's overall objectives. The Audit Report also recommended that the Vice Chief of the Defence Force renegotiate the 1981 Agreement with UNSW and that Defence develop a tool to identify the actual cost of educating an Academy cadet to enable alternative approaches to officer development to be compared and evaluated. Once again, Defence agreed on the need to revise the 1981 Agreement but no substantive progress on a replacement document was made. The task did not seem sufficiently urgent to demand immediate action.

Shortly after the Department of Defence decided that the Academy was running well, it was the turn of politicians to make their appraisal. Although many senior uniformed and civilian officers in Defence supported the Academy and believed it needed more time to prove its worth, others were not convinced and lobbied the Government for another review. In June 1993 the Government was persuaded to support an inquiry into 'the provision of academic studies and professional military education' at all of the ADF's educational institutions, including the single-Service new-entry colleges and staff colleges, the Joint Services Staff College and the Academy. The inference was that each institution could be made more effective or efficient. The inquiry was led by the chairman of the Defence Sub-Committee, Roger Price (Labor Member for Chifley), and the deputy chairman, Ian Sinclair (National Party Member for New England). Sinclair had been the Coalition's Defence Minister for ten months in the last Fraser Government. The sub-committee included the former Labor Minister for Defence Science and Personnel, David Simmons; the Liberal Member for Groom and former RAN Commodore, Bill Taylor; and Bob Halverson, who had served for 25 years in the RAF and RAAF. Taylor and Halverson were the most active members of the sub-committee. Another member, Rod Atkinson, had served in the Australian Army from 1968 to 1975, including an operational deployment to South Vietnam. Price and Sinclair did most of the work and generated most of the ideas, distilling their thinking on a whiteboard in Parliament House during evening meetings.[17]

Wing Commander Ken Given served as the Committee's uniformed advisor and made a significant contribution.[18] Given had been with the Engineer Cadet Squadron (ECS) from 1977 to 1979 and had first-hand experience of the RAAF's association with RMIT. At that time some 150 RAAF cadets were studying for degrees and diplomas at RMIT. They were accommodated at RAAF Base Frognall, located in the Melbourne suburb of Camberwell, and conveyed daily by bus to the RMIT campus for lectures. On completion of their studies, which allowed for formal recognition of their qualifications by the Institution of Engineers, the cadets were sent to the Officer Training School at Point Cook. The ECS scheme started in 1962 and ended in 1985 when the remaining cadets were transferred to the Defence Academy. Given believed the arrangements at both RMIT and Melbourne University (which provided academic support to the RAAF Academy at Point Cook) well served the RAAF's interests and provided excellent value for money. RMIT's engineering faculty had certainly gained an impressive reputation. Notably, Given was never handed any 'instructions' by Defence on the line that he was to advance or the approach he was to take on any policy or practical matter considered by the Committee. Both Price and Sinclair relied on his experience and judgment. He described his own role as 'the "whisperer", prompting questions, interpreting answers and suggesting supplementary inquiries'. He was not involved in setting the Committee's terms of reference but he participated in all private discussions and was involved in all 'white board' sessions, where possibilities were tested and proposals were canvassed.[19]

As the Committee was asked to inquire into the conduct of peacekeeping missions at the same time – Australia had sent peacekeepers to Somalia in 1993 and Rwanda in 1994 – it decided to defer the officer education inquiry until May 1994. This made sense on a number of grounds. There was a strong view within the Department (shared by Tony Ayers) that the inquiry was still premature. The first fully Academy-educated and trained students had only graduated in 1988. They were still junior officers. The main presenting issue was, however, the absence of a contract between the Commonwealth and the University. It was not good public administration to have obligations that were neither defined nor regulated by a contract. The 1981 Agreement was also lacking review provisions and a sunset clause that could allow either party to terminate its operation. Given believed that the military staff posted to the Academy lacked the necessary business skills that the 1981 Agreement required; understandably, they focused on providing well-prepared officers for their single Services, not on financial management. Indeed, the military staff tended to rely too heavily on the goodwill of the Defence Department to cover their funding shortfalls, while its relationship with UNSW was perhaps too casual and too comfortable. Some thought the University had become complacent, presuming too often that Defence would acquiesce to its requests. This was no more than a general impression lacking a firm evidentiary base. But it was clearly sufficient to generate enough suspicion to warrant a closer look.

The mood among University staff towards the inquiry was mixed. Some were resentful; others believed it was an opportunity to expose politicians to the challenges of officer education.

The University, especially the University College staff, saw itself as Defence's 'friend' and considered the inquiry an unjustified intrusion by those who had failed to appreciate the extent of UNSW's commitment to the University College and to officer education. Some academics feared the Academy might be closed and the buildings used to house the three single-Service staff colleges and the Joint Services Staff College located at Weston Creek, an outcome that had been suggested to the Committee. As the Academy had yet to reach its first decade, the Government could claim the 'experiment' had failed, that there was no merit in persisting and that a new approach was needed. As the Academy was seen as a Coalition initiative, a Labor Government could portray such a decision as a regrettable but prudent response to a serious political mistake on the part of the previous Prime Minister, Malcolm Fraser. Although the Government's thinking was not heading in this direction, the University staff had reason to be anxious.

The committee chair, Roger Price, had been Parliamentary Secretary to Prime Minister Bob Hawke. He became Parliamentary Secretary for Defence with specific responsibility for Defence Industry Policy and Exports when Paul Keating became Prime Minister in late 1991. Price was surprised to learn that the new Defence Minister, Senator Robert Ray, 'actually wanted me to do something, i.e., develop a new policy' while the Chief of the Defence Force, General Peter Gration, 'took the view that I should learn as much as I could about the sharp end of the business so that Industry Policy should not become an end in itself [...] It was my happiest time in parliamentary service'.[20] With regard to officer education, Price has stressed that Keating, Ray and Senator John Faulkner (the Minister for Defence Science and Personnel) did not hint at any preferred conclusions or recommendations and, 'if they did, I would not have been amenable. The Committee had a history of being bipartisan'.

Over the ensuing six months, the committee members visited each of the educational institutions listed in its terms of reference, held public hearings and received submissions from individuals and organisations. In relation to the Academy, Professor Niland and Drs Hugh Smith and Anthony Bergin from the UNSW Australian Defence Studies Centre provided detailed written submissions. The Rector and his deputy (Professor John Richards) both gave personal evidence at public hearings together with Dr Bob Hall and James Wood. The University's principal spokesman was the Deputy Vice-Chancellor (Academic), Professor Tony Wicken. Other staff had views but did not make submissions of their own.

The committee made much of the fact that attrition was substantial and graduation rates at ADFA were 'not consistently higher than graduation rates at other universities' although the graduation rate for mature age undergraduates was 100 per cent. The Academy graduation rates were 51.6 per cent in 1988; 51.9 per cent in 1989; 50.4 per cent in 1990; 55.4 per cent in 1991; 50.2 per cent in 1992; 61 per cent in 1993; 75 per cent in 1994. There had been substantial improvement since 1992 but the success rate was far from acceptable. There was also attention to the cost of operating the Academy per student per annum which was $106 087 – a considerable amount of money in 1994 terms. The cost of producing an Academy graduate was slightly more than twice the cost of producing a graduate at the main campus at Kensington but less than the estimate of three times as much when the Academy was first being planned.

The University explained that some costs had declined, such as expenditure on Aeronautical Engineering, which was now wholly provided by UNSW rather than in concert with RMIT and the University of Sydney. The committee noted, with a sentiment approaching irony, that this represented a cost saving of $244 000 each year in a budget of just under $100 million (the combined cost of the education and training programs). The University also stressed the value of its research programs, which had attracted more than $2 million in grants and brought immediate benefits to Defence.

Notwithstanding Labor's long commitment to the Academy concept, it remained a '1980s decision based on a post-Vietnam environment of some antipathy towards the military. The environment has changed considerably since that time'. It seemed that neither the Academy nor the University College had many friends among the Committee members and UNSW staff feared the worst when the report, entitled 'Officer Education: the Military After Next' but widely known as the 'Price Report', was finally released on 23 October 1995.[21] Its conclusions and recommendations were fully supported by the Coalition members of the committee, with the sole exception of Bill Taylor whose opposition was resolved during the drafting process.

Price's tabling speech in Parliament produced the perfect media grab: he referred to the Academy as a 'military nunnery'. This colourful and controversial phrase was bound to bring the Committee Chairman into conflict with the Department of Defence, who clearly wanted to retain the Academy in its existing form, and the University, which saw it as an outreach within the national capital. The report began by observing that:

> Whilst there was some resistance to the establishment of ADFA, the Committee considers that it has to date served the ADF well. [ADFA] has supplied the ADF with a guaranteed stream of university educated officers who have performed well in their limited time in the Services.

It went on to note that the 'principles of jointery which underpin ADFA have been most effective' because '[t]he opportunity which

ADFA provides for the mixing of officer cadets from each of the Services at the first stage of their careers is extremely valuable'. This observation was not sufficient to deflect those who advocated change.

The Report claimed that the 'Military After Next' required an officer corps that 'the sheltered environment of ADFA may not be consistently capable of producing'. In essence, cadets at the Academy were too removed from the community they were going to serve: 'At a relatively early age potential officers are removed from all community influences (family, civilian friends and university peers) and placed in a secluded military learning environment'. That the Academy was not located in a genuinely remote stretch of Jervis Bay but located in the midst of the national capital with its population of 300 000 people did not make a difference because the 'students at ADFA have virtually no opportunity to interact and exchange ideas with the non-military community in a sustained and meaningful way'.[22] Its most contested claim was that 'a decision to join the officer corps as a university graduate will prove superior to a decision to join made in Year 12 in the majority of cases'.

There were echoes here of the report produced by Alan Wrigley, *The Defence Force and the Community*, which appeared five years earlier. This report led to the Commercial Support Program (CSP) and eventually the market testing and transfer of around 10 000 uniformed and civilian Defence positions to the private sector. The Price Report believed that future conflicts would be fought nearer to civilian populations and that better acquaintance with community views and expectations was vital given the politically nuanced nature of ADF operations in the foreseeable future. That most recruits were drawn from white Anglo-Celtic backgrounds meant they were not representative of the broader population and were likely to engage in 'group think'. Attending a civilian university would broaden and deepen their appreciation of human society and promote easier interaction with people of different ages, temperaments and outlooks.

The committee's recommendations were straightforward if not stark, especially number 14: 'The Committee recommends that, on balance, the University College undergraduate program at ADFA be terminated'. In its place, recommendation 16 proposed an 'Undergraduate Sponsorship Scheme' take its place. Around 1000 students would be sponsored by Defence each year under the new scheme. Students could choose their own university and pay for the fees via a voucher system. Before, between and after academic semesters they would receive common military training but live at home and study on their own during semesters. After natural attrition, the best or most suitable 300 would be selected for specialist professional training and commissioning in one of the three Services. The idea of 'vouchers' originated with Price and was later supported by Sinclair. Ken Given was never convinced the scheme would attract much support and feared that managing it would be too complicated.

The bleak proposals continued in recommendation 17: 'the 1981 Agreement between the University of New South Wales and the Commonwealth of Australia be renegotiated'; and recommendation 18: 'undergraduate courses at ADFA cease at the end of 1999'. The reputation and standing, the experience and expertise, the loyalty and commitment of UNSW had no bearing on the committee's recommendations. In sum, too much money was being spent on too few people with too little demonstrable return. Because the provision of a generalist basic degree was not the core business of Defence but of the nation's 37 universities, ending undergraduate education at the Academy would be consistent with the principles of the Commercial Support Program and result in substantial cost savings. Price was, according to Given, privately concerned about an overall reduction in standards, given the proposed involvement of small regional universities with small faculties and limited course offerings in educating young officers, which until then had been provided by leading universities. The content and cost of courses would also differ, alongside semester dates, while local attendance requirements would potentially complicate scheduling training in academic breaks.

But could the Commonwealth simply tear up the 1981 Agreement with UNSW? The short answer was 'yes' because the original document did not 'contain specific termination or amendment clauses or require specific performance information from the University of New South Wales'. In fact, there was no mechanism to review the Agreement or to terminate its operation. The Commonwealth might, however, be responsible for 'any costs associated by the University as a result of the termination of the Agreement' although these could be reduced by redeploying academic and administrative staff and by phasing out rather than terminating the undergraduate program, which would also allow currently enrolled students to complete their UNSW degrees. Although postgraduate degrees might be offered at the Academy once the single-Service staff colleges were amalgamated and, together with the reconfigured Joint Services Staff College, shifted to the Academy site, it was not certain that the contract to award such qualifications would be given to UNSW.

Reflecting on the inquiry and its report in 2015, Roger Price remained convinced that undergraduate education was not a 'core Defence undertaking' and noted that ADFA undergraduates were the 'highest paid' in the country.[23] He felt the money committed to undergraduate education should have been transferred to postgraduate education, which he felt was lacking. When he met with senior Defence leaders to cover the main elements of the report, believing that the Chief of the Defence Force (General John Baker) was onside, he quickly concluded that Baker 'wasn't so supportive'. At a later meeting at the Joint Services Staff College, the Academy Commandant (Major General Frank Hickling) attended and made his opposition clear. Maintaining his conviction that a system of scholarships was a more efficient way of educating young officers, Price conceded that the committee may have been 'remiss in not spelling out in more detail the character of ADFA as a postgraduate institution'. He believed that UNSW had served Defence well and that it had a continuing role 'but not as an undergraduate provider'. Price explained that no-one from UNSW had lobbied him personally during the inquiry and that no other university had declared an interest in the Defence contract. On reflection, he feels 'embarrassed' by the term 'military nunnery', knowing that 'the term would be highly offensive'. His aim was to 'convey that the undergraduates were not interacting with their peers or the community. They were cut off, a closed community'.

With similar hindsight, Given noted the absence of enthusiasm within Defence for the report's recommendations, with some senior people at Russell Hill deciding that politicians knew little about either Defence or education and could be conveniently ignored. Others felt, he recalled, that Defence had lost an opportunity to exert greater influence on the shape and substance of the degree programs offered at the Academy. Inasmuch as the legal, medical, accounting and engineering professions had been quite specific about the education those seeking to enter their profession were to receive, the counter-argument was that military operations are so complex that most academic courses will be relevant in some way to understanding their context and explicating their conduct. Nonetheless, there was a view that Defence was far too passive and should have insisted on reshaping degree programs and recasting course content. It was, some suggested, a 'lazy customer'. The University would have resisted such pressure on the grounds that the needs of cadets were best served by the traditional degree structure and that a 'balanced and liberal education' consisted of courses that transcended narrow vocational confines. The University took from the inquiry the message that greater flexibility was needed, as it did become more responsive to Defence's needs in the next few years.

Commentators lined up to defend or deflect the claim that the Academy was a sheltered environment. Because the claim was essentially a matter of perspective, it quickly became an issue of opinion. Educationalists noted that Academy undergraduates were required to complete a more diverse range of courses than their counterparts at a civilian university. As for the contention that Year 12 students could not be relied upon to make up their minds, the report seemed to overlook the fact that the same young people decided on other professions at the same age and those professions were not objecting to them making an early start in their chosen career. Most attention was given to the report's recommendation that the undergraduate program at the Academy be abolished, with an annual saving of $60 million. To ensure the Academy site was not left deserted, the Australian College of Defence and Security Studies, the Joint Services Staff College and an amalgamation of the three single-Service staff colleges would be relocated to the campus as part of a Graduate School of Defence Studies. The report's proposals would make 190 surplus uniformed personnel available for other postings, although a good number of these people would have been needed to manage those officer candidates who were participating in the potentially complex 'voucher' system.

Media commentators of all persuasions were far from convinced that the Committee had presented a strong case for such far-reaching reform. Verona Burgess, writing in the *Canberra Times*, considered Price's description of the Academy as a 'military nunnery' unnecessarily provocative and predicted a 'political, bureaucratic, academic and military bunfight'.[24] Bruce Juddery, writing in the same newspaper, thought that Price's report was essentially about cost saving rather than educational policy, because the purported practical benefits were illusory:

> The educational and vocational arguments of the Price report – treated apart from the broader higher education debate – are tenuous to the point of nonsense. The real argument is about money. And it would be a very long bow to assume that a government of whatever stripe would gamble on the 'civilian' universities' capacity to provide for the 'military after next' that the Price committee demands.[25]

Writing in the *Canberra Times*, Anthony Laver observed that because he was 'a good Labor man, Mr Price seems to dislike any institution which sets its members apart from the general public'.[26]

Hugh Smith and Anthony Bergin identified what they considered the more obvious difficulties associated with implementing the recommendations. They included:

> uncertainty of the recruiting environment in universities and likely fluctuations in the numbers that could be secured; the difficulty of tailoring a curriculum in a civilian university to meet the needs of officer candidates (for example in engineering); the value of attracting teenagers considering a military career into the ADF before they enter civilian universities and look to other options; the absence of a tri-service environment at the outset of an officer's career; and the loss of an opportunity to influence cadet values appropriately.[27]

In responding to the observation that it cost more to produce a graduate at the Academy than elsewhere, the major offset was the higher level of successful degree completions in the minimum time.[28]

Three Academy scholars, James Warn (Economics and Management), Paul Tranter (Geography and Oceanography) and Glenn Fulford (Mathematics and Statistics) produced a response that challenged the report's limited appreciation of 'the value of tertiary education within a military context' by concentrating on the extent to which UNSW programs met Defence needs and expectations.[29] They noted that universities serve as 'agents for social change by developing among their graduates a capacity for critical reflective thinking', and that the liberal education offered at the Academy served to 'balance the specifically vocational nature of the military training'. Such an education would serve to 'evaluate the relevance of military doctrine' and allow its recipients to 'learn from experience in an intellectually rigorous manner'.[30] Whereas vocational training was oriented towards the trainee's next job, a generalist degree has a 'long shelf life, and offers value over the longer term'. But did the degrees offered at the Academy actually achieve that end?

Warn, Tranter and Fulford surveyed ten cohorts of Academy graduates over the previous decade, asking them to rate the extent to which either their Academy experience or completing academic courses 'had developed certain skills, competencies and attributes'. The survey produced some clear results. For the greatest part, graduates believed they had received a quality education and thought their UNSW degrees were held in high regard. An overwhelming majority believed their capacity for critical reflective thinking was enhanced, with the highest score noted among Arts graduates, followed by Science and then Engineering graduates. Those who completed an Honours year indicated 'a significantly greater degree of development'. There was a consensus among graduates that the 'generalist' program they experienced 'facilitated their self development as well as the development of their subsequent capacity for new learning'. The benefit Defence acquired from these outcomes was plain: it imbued the 'organisational decision processes with a depth and breadth of world views that may otherwise be lacking'.[31] In effect, leadership training was assisted by intellectual development, although it would be difficult to prove that these outcomes would not or could not be achieved – or even bettered – by an education delivered at a civilian university.

The more cognitively mature the leader, the more effective they would be in exercising command. Education facilitated and hastened maturity. Indeed, many graduates found it difficult to dissociate academic learning from understanding and absorbing the principles of leadership. At its best, education was a 'change agent' that created an environment conducive to personal growth. Education and training when conducted concurrently promoted the most efficient and effective transfer of learning, rather than a sequential approach. Education informed the outcomes of training; training sharpened the objectives of education. The two greatest impediments to cadets making the most of their Academy experience was the harassment of both junior and female cadets, and the '51 per cent' syndrome. The authors thought progress had been made to eliminate the first and to diminish the second. Their judgments would prove a little optimistic.

The University's formal response to the report was completed in late October 1995. Its tone was low-key, with Harry Heseltine focusing on the need for the University College to embrace greater flexibility. He explained that the University College was committed to delivering a distance education pilot program between 1996 and 1998, and extending the Master of Defence Studies program

Reviews, reforms and restructures, 1986–1995

to Brisbane and Townsville. His foremost disagreement was with claims about community values and the Academy's social isolation. The report focused on the view from 1995 but 'no attempt seems to have been made to project community conditions in 2005. That is to say, the Report seems to be based on a ludicrously thin slice of time'. A brief prepared by the military staff for the Commandant, Major General Hickling, explained that 'the academic staff, providers of the academic program, are educated, independent thinkers, accustomed to voicing their views in a well argued statement' while the 'tone of the academic environment is set by UNSW and encourages debate on all topics of interest to the students, including community values'. The brief continued: 'the recommendations appear not to have any supporting evidence or studies. Where evidence does exist, the recommendations sometimes appear contradictory [...] the cost savings claimed in the Report cannot be realised; annual savings may not eventuate'.[32] Hickling wrote privately to Sir Edward Woodward, the chair of the Academy Council, on 2 November 1995: 'I have had word from CDF that he would be comfortable with a personal letter from you to the Minister. I have been told to refrain from comment in the media'.[33]

Woodward wrote to Senator Ray on 7 November 1995, worried about media fallout from the Price Report. He claimed the 'cloud of uncertainty' was made worse by the coming election and described a level of resentment among graduating cadets that they were not adequately equipped to lead the 'Military After Next'. He told the Minister that the main recommendations were 'a leap in the dark which would prove unworkable in practice' and that he was 'worried about the short term damage that can be done by the very existence of the report and the way in which it has been promoted in the media'. He urged Ray to 'contain the potential damage to a fine institution' that speculation generated by the report was creating in the community. He received a reply from the Minister on 11 December 1995. Senator Ray noted his concerns but said the Report's challenge to the 'continued validity of undertaking undergraduate training' at the Academy has been 'welcomed by academics and the ADF, and the ensuing debate will prove, I believe, beneficial to all parties'. This was a very positive interpretation of the prevailing mood in the wake of the Price Report. It was an interpretation that Woodward did not share. The Vice-Chancellor, John Niland, also wrote to the Minister pointing to the longstanding and 'special relationship' between the University and Defence and highlighting the positive findings of the 1992 Inspector General's Review. Niland said the report was misguided and short on evidence. He then addressed the report's claims about cadet insularity and did not hold back:

> It was hard to see this as anything but a total nonsense which has been exacerbated by emotional and irresponsible headline grabbing comments about a 'military nunnery'. The short run damage that single comment may have done to recruitment, both staff and students, has still to be realised.

He concluded that 'there may well be a cost premium for producing ADFA graduates but I believe, as does Defence, that the cost is totally justified (and in any event probably exaggerated in the Report)'. He closed by advising the Minister that 'the 1981 Agreement could not be unilaterally discharged'. It was a provocative statement and intended to be so. The University would not be quietly despatched. Senator Ray's reply, dated 29 December 1995, was brief. He told Niland that he had 'no plans to change the structure' and that he 'values highly the special relationship'. All universities are, to some degree, isolated from the community. Hence, the occasional accusation that their members inhabit 'ivory towers'. Some do. All Niland needed to explain was that the Academy was no more isolated than a civilian university from the host population.

The electoral defeat of the Keating Labor Government at the March 1996 poll sealed the Report's fate. Well before the election, the Shadow Minister for Defence, Senator Jocelyn Newman, indicated the Academy's programs would continue in their current form if the Coalition won the election. Senator Rod Kemp tabled the new Coalition Government's response to the Price Report in the Senate on 17 June 1996. It concluded that 'the report lacks substantive and sufficiently detailed reasoning to justify' the termination of undergraduate programs at the Academy. Doing so, the Government said, would be to abandon a system in which UNSW had made a 'substantial investment' that was 'known to produce a high quality product'. There was no evidence of any deficiency in the service provided by the University and, noting concerns about cost, transferring Defence student places to the Unified National System would cost $10 million. The Government felt the voucher system proposed by the Committee 'represents a significant gamble [...] without offering compensating tangible benefits'. But it recognised that the 1981 Agreement needed reviewing to make it consistent with public sector management arrangements, including mechanisms for reviewing financial arrangements and linking performance criteria to present expenditure and future funding. The Government was also adamant that the Academy was a subordinate area of activity within Defence and did not require its own legislation.

Criticisms of the Government's response focused on the cost of the Academy. Senator Stephen Conroy asked Senator Newman (who was appointed Minister for Social Security rather than Defence Minister by Prime Minister John Howard) whether the Government had any plans to reduce funding to UNSW for its service at the Academy, given that all other universities had been warned of Commonwealth funding cuts in the coming year. She denied there were any such plans, but her answer 'does not mean that the operation and funding levels of the Academy are immune' from the need to be cost-effective.[34] She also assured Senator Conroy that Defence resources would not be used to cross-subsidise UNSW. In the House of Representatives, Ian Sinclair declared his disappointment with the Government's response despite being a member of the Coalition parties. He claimed that the Academy was being segregated from every other educational provider in the country and was avoiding public scrutiny. He continued to believe the Academy was not a cost-effective means of providing education for Service officers and that the education they received was not making a demonstrable difference to their professional performance.

The Report was never likely, however, to achieve significant reform with so much institutional resistance if not plain inertia preventing radical change of the kind it was proposing. There is no evidence that the ADF expended much effort on considering the Report's recommendations in relation to the Academy. Other recommendations, particularly those concerning the education and training of middle-ranking officers and the future of the staff colleges, received a more positive reaction. In relation to the Academy, the CDF and the Secretary had already decided they would oppose any wholesale change to the existing arrangements for the education of new-entry officers. Their attitude stemmed, in part, from a belief that the Academy had not been in existence for a sufficient time for there to be sound judgments about the 'experiment'. There was certainly a strong view that new-entry officers ought to be educated together rather than separately and that allowing cadets to choose their own university was a mistake. Not all universities were alike. Courses and standards varied considerably. As UNSW was one of the nation's leading universities, allowing cadets to select a second- or third-tier university would lead to a collective reduction in quality and this could not be contemplated. But the Report lent further weight to the need for a comprehensive tri-Service policy on education and training, and a review of the 1981 Agreement.

The University generally regarded the Price Report as unnecessary and unhelpful. Not much of positive practical benefit was derived from the exercise. The Committee had not explored the underlying reasons for the high attrition rate or the poorer than expected performance of undergraduates. The University's leadership felt it was the target of misplaced criticism and was prepared to defend vigorously the performance of its staff and the quality of its courses. It was also willing to make terminating the undergraduate program a difficult process and an expensive proposition for the Government. In fact, the University contemplated making the 'divorce settlement' so costly that the Government would have second thoughts. This might have appeared vengeful to some within Defence but it reflected the University's firm commitment to educating Service officers and the value that it attached to the experience and expertise that had been accumulated over the previous three decades.

The Report had revealed the extent to which the Academy was the focus of cultural angst and organisational frustration within Defence. Critics appeared to allege that all of the ADF's woes emanated from the Academy or could be traced to a misguided education and training policy. The Academy was apparently over-funded, under-scrutinised, badly managed and out of touch with the 'real world'. Money spent on an expensive education for youths who had displayed no commitment to their Service could have been used to acquire new equipment or purchase additional resources. Those who resisted the transition to greater 'jointery' claimed the Academy was the venue for implementing 'politically correct' social change that was blunting the operational effectiveness and professional identity of the three Services. The University was accused, sometimes openly, of opportunistically exploiting Defence's desire to see the Academy succeed by asking for more money than was needed. There were also murmurings within Defence about the competence of some members of the academic faculty and the quality of its courses. This reached some of the University's staff and produced resentment. While some staff may have been average rather than outstanding performers, many had given more to the Academy and their students than was required by their employment contracts. It was only the personal, unwavering commitment of the Vice-Chancellor, the Rector, the Chief of the Defence Force and the Secretary of the Department of Defence – all strong personalities – that prevented the Academy from being either reformed out of existence or from suffering an institutional implosion. But another threat to the Academy's future was looming. Its existence undermined the viability of the institution and every effort to decrease attrition, increase retention and raise standards. It was an enemy from within and had nothing to do with the UNSW–Defence relationship.

11

Controversy and consolidation, 1996–1998

February 1996 marked the tenth anniversary of the arrival of the first intake of Academy undergraduates. It was a useful time to take stock of progress and to identify persisting problems. The greatest change from 1986 had been the virtual explosion of postgraduate student numbers. There were 137 full-time, 343 part-time and 22 distance postgraduate students. Of the full-time students, 47 were uniformed personnel, 90 were civilians, nine were overseas military officers and 29 were international civilian students. Of the part-time students, 138 were military personnel and 205 were civilians. This growth was neither expected nor made the subject of planning when the Academy opened. There was plainly an unfulfilled demand for continuing higher education within both the ADF and the Canberra community. The Postgraduate Student Committee (PSC) noted that:

> ADFA's postgraduate students say that the relatively
> small size of the place fosters greater familiarity with both
> academic and general staff, and ensures a greater share in
> resources, than their counterparts have at other universities.
> Conversely, most postgraduates mention the isolation
> that occurs at ADFA, and the difficulties of establishing a
> network with their peers. ADFA postgraduates admit that

having fewer student voices means that any student action needs to be more highly organised and concentrated than at the larger universities. Moreover, since ADFA postgraduates do not pay student fees, they have no representation in the Student Guild although the President of the PSC does represent ADFA on the Postgraduate Student Board of UNSW.[1]

During the tenth anniversary year the University College introduced a pilot program for the Master of Defence Studies course, to be undertaken by distance education, to a target cohort of 20 Defence-sponsored students in Sydney, Brisbane and Townsville. There were 90 enquiries and 40 applications, resulting in 29 offers of enrolment being made. The mode of delivery was video conferencing, an email discussion board and a two-day weekend school in both Sydney and Brisbane, together with the distribution of an extensive reading 'brick'.

In November 1996 the School of Aerospace and Mechanical Engineering conducted a seminar to review the competencies required of engineering graduates for the Navy, Army and Air Force. The aim was to ensure that the University College produced the engineers the ADF wanted, and the focus ranged from undergraduate courses to the need for broader or different postgraduate courses. The seminar heard from the Institution of Engineers on its recent review of engineering education in Australia. This gathering led to a review of courses in 1997 and the development of new programs at both the undergraduate and postgraduate levels. The College was trying to anticipate demand and to demonstrate that Defence's engineering needs could be met in more creatively structured courses. In effect, the University was at the vanguard of reforming the profession and helping Defence to see different

Controversy and consolidation, 1996–1998

ways of meeting its technical needs. The University was trying to be proactive in preparing students for a rapidly changing workplace.

By this time Harry Heseltine had decided to retire. The occasion allowed his successor in the Department of English, Professor Bruce Bennett, to comment on the close working arrangements between academic and uniformed staff that had been achieved over the preceding five years:

> The Academy's diarchic arrangement, of government by the Commandant and Rector as equal partners, sets the tone for the whole Academy and the Rector's role has been crucial in it. Throughout his five years as Rector, Harry Heseltine has been a forceful and persuasive advocate of the shared vision of the Academy as a major Australian institution of learning which provides a balanced and liberal education in a military environment.[2]

Believing Alan Gilbert was the most likely successor to Geoff Wilson until Gilbert took a job at the UNSW main campus, Heseltine had described himself as the 'accidental Rector' and as 'Minister for Internal Affairs'. He bore the brunt of opposition and impatience with the University's role at the Academy and was obliged to divert a good deal of his time to the Inspector General's Report and the Price Report, both of which he considered 'potentially quite hostile'.

The vacancy was quickly filled by another internal appointment. John Richards, Professor of Electrical Engineering and Deputy Rector, was appointed to the top job in July 1996. Like his predecessor, Professor Richards had an extensive knowledge of the College's evolution.[3] He was educated at Normanhurst Boys High School in Sydney and was an active member of the School Cadets. A graduate of UNSW with First Class Honours and a PhD in Electrical Engineering, he had worked at James Cook University in Townsville before returning to the Kensington campus in the Centre for Remote Sensing. He came to the University College in June 1987 as Professor of Electrical Engineering and Head of Department. His first memories of the Academy were of new equipment, well-funded research and the valuable contributions made by the Visiting Military Fellows. From December 1991 he had served as Deputy Rector and worked very productively with Heseltine. But the new Rector began at a difficult time for both the tertiary education sector and Defence.

Not long before John Richards became Rector, Air Vice-Marshal Gary Beck was appointed Academy Commandant. As was the case with three of his predecessors and despite earlier affirmations that the Academy command would not be a pre-retirement position, this was likely to be Beck's final posting. Beck had been involved in Air Force education and training for a number of years. Unlike his predecessors, he espoused a very different view of the relationship between Defence and the University. The first Commandant, Rear Admiral Peter Sinclair, had attended to myriad practical issues ranging from financial controls to the Academy workforce. His involvement in academic life was largely restricted to considering exam results, although he did attend occasional seminars and postgraduate classes. The second commandant, Major General Peter Day, appeared to the academic staff to be less 'hands on' than his predecessor. He made much of the University's public standing and referred often to his pride at participating in a very worthy collaboration with UNSW. He would attend seminars in the Engineering schools when the topics related to his own professional Engineering interests, encouraging both staff and students. The third commandant, Air Vice-Marshal Richard Bomball, was more involved in cadets' academic experience, focusing his attention on learning environments and dealing with those aspects of cadet culture that worked against positive results in both the education and training spheres. The fourth commandant, Rear Admiral Gerry Carwardine, tried to raise the Academy's profile in the national capital by highlighting Defence's connection with UNSW, a university that was rising substantially in national and international rankings. His successor, Major General Frank Hickling, was in the post for just thirteen months (becoming the Land Commander in 1996 and the Chief of Army in 1998) and could not have been expected to bring about cultural change. It was with the arrival of the sixth commandant that a very different interpretation of the UNSW–Defence diarchy was presented.

Although there was no formal contract in place (arrangements were still regulated by the 1981 Agreement), Beck believed that the academic staff were 'contractors' and that the Academy was a military establishment, not a university campus. The University's staff were sojourners. Referring to the academics this way naturally created resentment as many believed they were actually engaged in a partnership or, at the least, a collaboration to which they had given a great deal personally and professionally. There had also been a change in the profile of the Academy Council. Big issues of principle had mostly been resolved within the first decade of operations. Patterns had been established; processes had been agreed. The Service Chiefs had asserted and then affirmed their expectations of objectives and outcomes. The Council now dealt with more routine matters and was less activist and interventionist. Deputy Chiefs of Service now attended most of the meetings. For his part, Richards used these opportunities to highlight the University's performance in both teaching and research, and its capacity to 'value add' as a leading university with a global reputation. This was a wise decision as Defence was again signalling an interest in the financial aspects of the Academy's operations.

The election of the Howard Coalition Government on 2 March 1996 had marked a major turning point for both the Defence and Higher Education sectors. Announcing his intention to deal with the $9 billion budget 'black hole' left by the outgoing Keating Government, the new Prime Minister pledged to spare Defence from an overall contraction in Commonwealth outlays even as his government reduced spending in the Education portfolio by several billion dollars. Defence was nonetheless under pressure to find money for new projects from existing allocations. The Minister for Defence, Ian McLachlan, established the 'Defence Efficiency Review' on 15 October 1996.[4] This included assessing all activities conducted within Defence to eliminate duplicated programs, seeking to reduce costs by commercialising support services and reflecting 'modern business practices' while enhancing combat capabilities to 'produce the most efficient and effective Defence Force possible within current budgetary restraints'. The review team consisted of Sir Malcolm McIntosh (Chief Executive of CSIRO), Dr Richard Brabin-Smith (Chief Defence Scientist), Ian Burgess (Chairman of AMP), Andrew Michelmore (Executive General Manager, Western Mining Corporation), John Stone (former Treasury Secretary and Queensland Nationals senator), and Vice Admiral Robert Walls (Vice Chief of the Defence Force). Professor Wolfgang Kasper was the only UNSW staff member associated with the review team, serving as an external advisor.

The review team noted that Defence had endured substantial funding cuts since the early 1990s, with a reduction in uniformed and civilian staff of 16 200 (17 per cent of the workforce) and efficiency gains of nearly $500 million within existing programs, with another $600 million of functions made the subject of commercial contracts. The team noted that much had been achieved 'but this is not the time to rest on laurels' because management could be further reformed and closer links with industry were possible, while even greater efficiencies needed to be explored.[5] The team felt that Australian strategic needs were changing but that the ADF had not adapted well to new contingencies in terms of its organisation and capabilities. While the Government naturally wanted to ensure it was optimising its investment in Defence, there was evidently a sense among uniformed and civilian staff that reform was possible, given the team received most of its submissions (255 in all) from serving officers in the ADF and the Department.

The team decided to examine 'Education and Training' because it was integral to the 'quality of the ADF and its support' and because 'it is a major consumer of resources directly in the cost of facilities and staff and indirectly in the time taken by people who would otherwise be engaged in operations'.[6] Notably, the review report claimed that training was built on the foundations provided by education, which was also important to 'the wider relationships of the Defence Force with the community and in strategic assessment'.[7] After concluding that 'all basic non-military training should be merged across the three Services and the APS, contracted out to recognised civil institutions, and then topped-up on the job', the review team said it was 'similarly concerned' with duplication in the 'education infrastructure':

> ADFA is a magnificent, but expensive, facility in its buildings, equipment and staff. We believe it should be used more extensively and favour making it more available to all members of the ADF and to Defence and other civilians to undertake mature age, full-time and part-time studies. Defence should clarify its requirements from ADFA. Other

officer training institutions have developed and been sustained in relative isolation. We believe that there are substantial benefits to be obtained by merging officer education as a single process, and for there to be greater emphasis on joint officer education. We have been unable in the time available to make specific proposals, but believe the potential for savings and a better product is substantial.[8]

When a rumour spread that senior Defence people would welcome a recommendation that academic programs be opened to competitive tender, Alan Capp of the Training and Development Branch in the Department of Defence informed the Departmental Secretary, Tony Ayers, of the gossip. Ayers instructed Capp to confer with the two officers heading the DER Secretariat, Brigadier Peter Dunn and Mr Patrick Hannan, to have the rumour scotched. Ayers was vehemently against open tendering for the undergraduate program at the Academy, believing that UNSW was playing a crucial role in the cultural change he wanted to see furthered across the ADF. The review did recommend, however, that responsibility for the policy development work preceding the implementation of a much-needed 'joint education system' should be given to the Commandant of ADFA 'in consultation with the Rector, University College, UNSW'.[9]

After the Defence Efficiency Review had pointed to the need for a dramatic overhaul of Australian Defence, the Defence Reform Program (DRP) would implement the Review's main recommendations. After the turbulence of the Price Review the previous year, many University staff were anxious about all Defence reviews and their consequences for the Academy. John Richards tried to set their minds at ease by explaining that, in contrast to the Price Report, the Academy was:

extraordinarily well treated by the DER, and its
recommendations exhibit particular confidence in our ability
to adapt to the changes that will come upon us as the DRP
progressively takes effect. There will be many challenges for
us on the University College side. It is difficult to forecast
what all of them will be but we should be prepared to
embrace a different mix of undergraduate and postgraduate
students, a teaching program that might consist of a blending
of traditional with distance education, and the need for the
College, as a component of UNSW, to enter into partnerships,
franchises and other commercial arrangements with
educational suppliers, to offer a relevant and high quality
educational program for the benefit of the whole Defence
organisation.[10]

Although the University could not dictate to Defence how it ought to restructure itself or where UNSW felt it could economise in areas of expenditure beyond education, Richards identified a special focus for the University College which would prove its worth. He felt it was:

providing an opportunity for our cadets and other students
to obtain a detailed understanding of the affairs of Asia
and the Pacific. Of course we already have an Asia Pacific
Studies program, but it lacks one central element – and that
is language. I have long thought it unusual that a Defence
academy could provide an undergraduate university
education without the ready opportunity for students to
include language in the curriculum.

WIDENING MINDS

He announced that Bahasa Indonesia would become part of the curriculum and be the first element in a larger language program. He would also add weight to the move towards distance education and promote more flexible approaches to study in line with Defence's expectations. As the Academy entered its second decade, he noted that:

> [the] new joint Education and Training Executive within HQADF has already put in place an effective consultative structure to ensure that the Academy, as one component of the post commissioning education and training continuum for Defence, is fully and properly briefed on Defence needs. We are now in a new era, unlike the past, in which we will significantly interact with Defence in terms of their education and training needs [...] we will need to do things differently to secure the benefits that the DRP promises.

While Richards and Beck recognised their view of the relationship between the University and Defence differed substantially, on leaving the Academy and preparing for retirement Beck claimed his views had been 'misrepresented':

> I am the strongest advocate in the ADF for the continuation of the UNSW–Defence relationship but I want to see changes in our business arrangements so that we can attract and be funded for more Defence business. As a strong advocate for a more independent University College, my beliefs seem to have been misinterpreted as a wish to split the Academy. Rather, I sought better definition of our relationship rather than rely on custom and practice, some of which defies explanation. I was interested also in better specification by Defence of our

> educational requirements and renegotiation of the treatment of expenditure and revenue in order to evidence 'value for money' as required of all Defence program managers. I achieved none of these things but I have raised awareness once again of deficiencies in the 1981 Agreement.[11]

He went on:

> the ADFA model is dependent on the academic independence of the College. Determining how this independence is maintained in the face of meeting Defence's educational requirements is central to our relationship. I am not attracted to the soft and illusory argument that the Academy is a partnership or some other blend of good intentions. We need something more substantial given our separate employers and the current level of Defence expenditure. The military and academic arenas – Sparta and Athens – are very different cultures. Mixing the cultures weakens both and confuses outcomes. Our graduates must be comfortable in either environment, not some blend of both.

Having declared his confidence that 'the UNSW quality assurance processes within the University College are sound', he pointed to the 'variability and tenuous nature of the Academy's military environment given nearly 100 per cent turnover of Defence staff every two to three years. Clearly, corporate memory is eroded and reliance is placed on a repetitive and highly documented process requiring little reflective thinking about the nature of our task'. As he passed into retirement, Beck was to be the last 'two-star' Academy commandant. The Academy was to become a component of the Australian Defence College (ADC), The Academy commandant would in future be an officer of 'one-star' rank, subordinate to the ADC Commander – an officer of 'two-star' rank.

As chair of the Academy Council, Sir Edward Woodward was deeply troubled by Beck's approach to the UNSW–Defence relationship, although he said nothing publicly. In a private letter, Woodward described Beck as a 'mover and shaker' who sought to change the UNSW–Defence relationship' by lifting:

> the authority and status of the Commandant above that of the Rector and [making] the College more directly accountable, through him, to the Defence Department for its use of Departmental funds. In pursuing his goals he tried to institute a number of changes which any Rector could only have found galling and, in some cases, offensive.[12]

John Richards, determined to keep his differences with Beck private, revealed some of his difficulties at the end of his brief term as Rector.

> First and perhaps the most frustrating matter facing the managers of the University College is that decisions taken at the level of this Council and at most senior levels of Defence and the Services seem often not to lead to parallel commitment further down the chain of command.

His most serious concern was 'the emerging treatment of the College as a simple "deliverer" of educational services, rather than as part of the family that could assist in advice to Defence and the Services on educational matters'.[13] In his address at the December 1997 Graduation Parade, the Governor-General, Sir William Deane, touched on the 'inevitable' tensions between delivering a military training and providing a liberal education. He contended that the roles were 'complementary' while accepting that the task of 'striking the proper balance [...] is not always an easy one'. From his knowledge of the Academy and its graduates, he felt 'an appropriate balance' had indeed been achieved.

Much of the positive progress being made by both the ADF and the University at the Academy was being undermined, however, by the steady stream of 'bastardisation' allegations attracting media attention. The term 'bastardisation' was imprecise and meant different things to different people. The conduct associated with 'bastardisation' had certainly changed between the 1960s and the 1980s. Bastardisation had been synonymous with senior classes imposing on junior classes additional duties, humiliating tasks and pointless exercises. It sometimes involved behaviour that amounted to common assault. By the 1980s, this kind of behaviour was referred to at Duntroon as 'hazing' or 'bishing'. The ostensible aim of these practices was to have new-entry cadets submit to institutional authority, demonstrate respect for seniority and secure compliance to disciplinary codes. The practices were usually petty and were always resented by those subject to them. The academics hoped, and the senior military staff believed, that the systematic physical and emotional abuse of the most junior cadets had ended when the Academy was opened. It was difficult, if not impossible, to prevent isolated instances of abuse. But it was possible to create the conditions in which systematic abuse was unlikely to gain ground. Many academics believed the conditions would persist while senior cadets disciplined junior cadets with few restraints on their power.

As Commandant, General Peter Day noted the substantial difference between 'misconduct by teenaged boys trying to assert their authority over others', 'bullying of juniors', 'harassment' and 'serious offences [...] that amounted to crime'. Instances of physical and sexual assault at the Academy were not deemed cases of bastardisation. When reported, they were referred to the ACT Police and the individuals involved made subject to administrative discharge if they were convicted. The issue in 1997 was the frequency of such reports. Questions were being asked beyond Defence about the culture within which young officers were being trained and whether it was too distant from the mainstream of Australian society and from the culture of the 'real' ADF. Was the Academy's culture responsible for instances of physical intimidation and sexual violence and was the University implicated in any way?

The Price Report had claimed that Academy cadets were socially isolated and morally immature, with Roger Price referring to the institution as a 'military nunnery' in a press conference. One writer to the *Canberra Times* took a lead from Price's remark: 'ADFA cadets in public (off-duty) have displayed the social under-development only a military nunnery could perpetuate. Many of these male cadets displayed sexism, racist intolerance, open hostility to any non-conformist, and homophobia'.[14] Another correspondent noted that the Academy had no more young men with disagreeable attitudes than the undifferentiated community or other universities.[15] Although the Committee did not interview a single cadet during its review of officer education, the Price Report argued that:

> The insular and institutionalised nature of officer development militates against proper integration and interaction with members of the community. Apart from the provision of lectures by academics, students at ADFA have virtually no opportunity to interact and exchange ideas with the non-military community in a sustained and meaningful way.[16]

There was nothing new in these claims. An editorial in the *Canberra Times* published on 19 June 1974 entitled 'Officers and Students' was adamant that 'the military calling must find its fulfilment within a more democratic framework: soldiers are also citizens and it is in their interest and in that of the community that they feel themselves to be an integral part of the wider national society'. The *Canberra Times* was persisting with the same message four years later:

> There is much to be said for widening the intellectual horizons and the field of social experience of officers in training. The danger of perpetuating or of entrenching a military caste that lives by other social, ethical and legal standards than the rest of the community is ever present.[17]

Within a month of their arrival at the Academy in 1986, female cadets were allegedly targeted by a khaki 'faction' from RMC who disliked the presence of women in the Academy and had pledged to see them gone by 1988.[18] Comparisons were quickly drawn between Duntroon in 1969 and the culture taking hold at the Academy. Three months later a report appeared in the *Canberra Times* of cadets allegedly smoking marijuana, whereas investigations into similar allegations made against students at ANU were never publicly reported. Positive stories about the Academy did not seem to be news. In May 1992 a number of newspapers ran an article on harassment and abuse at the Academy, leading to questions being asked in parliament of the Minister for Defence, Senator Robert Ray. It had already become apparent that public expectations of Academy cadets far exceeded those of other undergraduates studying in the national capital. The court-martial in December 1992 of five cadets accused of inflicting physical abuse – 'woofering' – on one of their colleagues served to confirm community fears that not all was well. Subsequent court proceedings involving cadets and sexual misconduct did not help. After the release of the Price Report in late 1995, a succession of cadets were involved in legal proceedings relating to offences against property and people. These reports gave the collective impression that the Academy's culture was predatory and violent. At least publicly, no one was asking about the University's role in maintaining a safe study environment.

In October 1997, the Minister for Defence Science and Personnel, Bronwyn Bishop, commissioned a review into cadet conduct at the Academy after a series of highly damaging newspaper reports. The most damning appeared in the 14 October 1997 edition of the *Bulletin* and was headed 'Rape, Loot and Pillage'. The need for the inquiry was deeply embarrassing for the Government and a severe indictment of the ADF. Bronwyn Grey, the Director of the Defence Equity Organisation, would lead the inquiry. There were 11 military members and one academic. The inquiry secretariat received formal submissions from many UNSW staff, including Professor John Niland, Associate Professor Hugh Smith, Dr Graeme Cheeseman and Dr C Parker. The review team conferred with University College staff, including Professor John Richards, Associate Professor John Baird, the Reverend Dr Paul McGavin, Mr Laurie Olive, Dr Bill Maley, Dr James Warn, Dr Peter Looker, Ms Kaliope Vassilopoulos, Mrs Joan McPherson, Mrs Jan Martin, Miss Lyn Christie and at the Kensington campus Associate Professor Jane Morrison, Ms Angela Burroughs, Ms Nina Shatifan and Mr Ken Grime.

The release of the 'Report of the review into policies and practices to deal with sexual harassment and sexual offences at the Australian Defence Force Academy' in June 1998 prompted calls from within and beyond the Academy to overhaul its culture. The 'Grey Review', as it was known, concluded in devastating detail that discrimination, bullying and intimidation at the Academy was 'pervasive and public'. The Report noted that the Academy's integration of military and academic components made it different from most other comparable institutions and that 'the tension between the two worlds, which at its best produces a healthy dialogue, remains alive [...] in ways it tends not to elsewhere'. The problem was a lack of balance between military and academic values, with the former regarded as superior, at least among the cadets. The review team noted that academic life traditionally encourages:

> free speech, tolerance and diversity, argument and discussion, dissent, refusal to accept authority as the best way to settle issues, a pleasure in finding new ways to do things, autonomy, flexibility of approach and preference; and excellence of individual performance.

While 'some of these values will always be in tension with aspects of military life [...] at present, however, this appropriate balance does not exist at the Defence Academy'. The dominance of military demands on cadets and midshipmen and the way that undergraduate culture had developed over the years meant that:

> cadets who speak freely in class may be censured by fellow cadets; cadets who try to develop independent or divergent ideas may be treated with derision by their fellows; cadets faced with a choice between committing time and effort to either military or academic demands routinely choose the former; the military routinely directs certain cadets towards some courses (generally science-based) and away from other courses (generally arts-based); cadets, forced to choose between leaving academic classes early or being late for PT,

routinely opt for the former; and, military involvement in the selection of Honours thesis topics may preclude cadets from conducting research into areas in which they are interested but are considered to hold no immediate value to the military.

By this time the Minister for Defence, Ian McLachlan, had decreed that all midshipmen and officer cadets studying at the Academy were to undertake a course in strategic studies and management, both subjects being deemed essential requirements for a career in the ADF. The decree – it was not a request – was controversial among academic staff, who resented political interference in the curriculum quite apart from the difficulty associated with timetabling and delivering compulsory courses. The University College agreed to offer a number of 'General Studies' subjects as part of the undergraduate program. Given the University was being criticised in some quarters for not being sufficiently responsive to Defence requests, augmenting the undergraduate program was seen as a more productive response than resisting what was effectively interference in what had been the University's province.

Noting the University's desire to be more responsive to Defence's needs, the Grey review team recommended a more academic and less anecdotal approach to teaching leadership. The academic demands of the cadet experience needed closer attention, with greater protection afforded to University courses in the weekly timetable. Despite the best efforts of uniformed and university staff, the creation of an 'adult learning environment' had not countered the tendency towards loutish behaviour or encouraged a more mature approach to academic study. The review team noted that although the formal lecture attendance requirement was 80 per cent, the requirement was not enforced and most cadets aimed for 50 per cent attendance.

Controversy and consolidation, 1996–1998

The review team was unequivocal in its recommendation:

> Irrespective of the University College's requirements [...] it should be acknowledged that the Defence Academy is not a civilian institution and, unlike their civilian counterparts, cadets receive a relatively generous payment for their services and may rightly be thought to have extra responsibilities. Consequently, the ADF should require cadets to attend all lectures and tutorials applicable to their degree and that absence from them should be viewed as absence from duty. While this may create tensions between the academic and military components of the Defence Academy over issues such as timetabling, with goodwill these tensions can be resolved.

The review team did not believe there was anything inherently flawed in the structure of the Academy or with its dual focus on education and training. It did recommend a review of the selection, training, appointment and use of Equity Advisers, and pointed to the need for a consistent approach to handling complaints of harassment among uniformed and civilian advisers. Was this part of a wider malaise? In a written submission to the Review Team, a member of the University College staff, Dr Graeme Cheeseman, remarked:

> there are very few female academics at the academy, no women's studies in the curricula and, in spite of the efforts of some female staff members, very few, if any subjects that deal directly with issues of gender or which emphasise feminist thought or practice. This is particularly the case in the areas of history and politics – where we might expect such perspectives to be offered – and represents a significant difference between the University College and most other Australian universities.

To accommodate an enhanced (meaning enlarged) academic curriculum, the review team noted that the number of periods allocated to Common Military Training (CMT) had gradually expanded and could be reduced to lighten the overall burden and allow new material to be introduced.

The Grey Review was widely distributed and scrutinised in detail. Within the ADF, a number of very senior officers were pointing to the Army cadets as the malignant influence and were pushing to have the accommodation blocks segregated by Service. Others painted an even bleaker picture: the culture could not be reformed, the Academy needed to close and perhaps reopen under a different set of operating parameters. Most academics did not think this was necessary as the educational programs were slowly producing better results. The problem, some asserted, resided in inadequate uniformed leadership and the persistence of procedures that plainly failed to counter bad behaviour. The people of Canberra seemed largely indifferent to the Academy's fate. As former UNSW student Chris Alder astutely observes: 'the focus of a steadily increasing amount of disturbing reports until 1998 had exhausted the pool of local sympathy that had offset criticism of the institution in previous years'. When coupled with continuing disdain for the Academy among those ADF members who had trained at the single-Service colleges before 1986, it was apparent the Academy had few friends. While it was easy to blame the press, and the inexhaustible interest of the *Canberra Times* in every instance of misconduct, there was enough evidence for a dispassionate observer to believe the Academy needed a fresh start. A few influential UNSW people at Kensington took the same view.

Not long after the Grey Review was initiated and following discussions within the UNSW Academic Board and the University Council, the Vice-Chancellor announced the University would make its own submission to the review as well as commissioning a report touching on matters 'clearly in the province of the University and the University College' and recommending changes to extant policies or the need for new initiatives to deal with matters that had given rise to the Grey Review. The University had not been pressured to assess the situation itself. It reflected concern that the voices of academic staff in Canberra would not be heard during the Grey Review, given an underlying fear among academics that the Grey Review would not get to the root of the problems or propose adequate solutions. Associate Professor John Carmody, a medical researcher and a staff-elected member of the University Council at Kensington, expressed his concern about reports of cadet behaviour and their potential effect on the University's reputation. He wanted reassurance that the University was not tolerating what ought to be condemned. Other academic staff believed that easier access to the University's procedures would help cadets subjected to unacceptable behaviour, and preserve the Academy learning environment when victims lacked confidence in Defence's guidelines. University counsellors may have been more trusted to provide support than those provided by the ADF.

The 'University Steering Group' was headed by Associate Professor Jane Morrison, the Pro Vice-Chancellor (Development). She was assisted by Ms Nina Shatifan, Director of the UNSW Equity and Diversity Unit, Mrs Joan McPherson, University College Equity Officer, and Mr Laurie Olive, University College Director of Finance, Personnel and Planning. The Steering Committee would conduct its inquiry separately from the Grey review. Indeed, Professor Morrison had no contact with Ms Grey, nor did the two teams ever confer.[19] The focus of the UNSW Group was 'valuing and achieving academic excellence' and determining the extent to which the University's policies 'provides an environment free of harassment and provides support for the implementation of equity policies, requirements and opportunities'.[20] After noting that the differences between 'the military and academic cultures relates to the degree of regimentation and regulation for which each of the cultures strives', the Group concluded that both 'share a strong commitment to intellectual flexibility and creativity in their graduates'. There was an assumption, then, that both the University and Defence wanted to rid the Academy of harassment and discrimination, and an assumption that 'the curriculum is the key mechanism for delivering a balanced and liberal education'. But the Steering Group was 'unclear [...] about what particular actions or approaches were being pursued at University College to ensure that the University met its obligations'.

The Group was critical of the range of academic courses and the mix of electives, believing the range was too narrow and the mix too rigid. It proposed 'a review of all curricula in meeting the obligations of providing a balanced and liberal education'. The Group recommended that the College's Teaching and Learning Committee 'undertake a review of curriculum for inclusivity' and suggested that a School might volunteer to host a 'pilot project on inclusive curriculum'. Other Schools should be required to 'identify inclusive curriculum and gender inclusive teaching as key activity areas' over the ensuing 12 months. The Group also recommended a review of the General Education Program electives to ensure they were sufficiently broad to embody the College's commitment to equity and diversity. The Minister's requirement for strategic studies and management to be included in the undergraduate program was unlikely to broaden the curriculum in this respect. In its more general observations, the Steering Group was troubled that 'some academics have yet to recognise and accept that omission of references to women and gender issues within a discipline can have the negative impact of rendering women's experiences immaterial or irrelevant'. This was particularly concerning within the ADF's own culture, which still struggled with integrating women into its ranks.[21] Therefore, the University College needed to focus more attention on 'attracting and retaining women academic staff because of the barriers presented by some male-dominated disciplines, as well as the military environment in which the College is located'.

The Steering Group felt that most academic staff were ill prepared for handling 'at risk students' irrespective of whether the risk was personal or academic. It also noted that the University's staff was not involved in developing or delivering the 'Unacceptable Behaviour Program'. Troubling to the Group was the low profile of the University's grievance procedures and support structures. The University College was urged to show leadership in encouraging and supporting women's networks in both academic disciplines and professional development, and to increase contact between the Kensington campus and the Canberra campus to assist staff networking and to raise the profile of the particular challenges being faced at each location. Conceding that the University College was smaller and more specialised, it was impractical to apply the targets and measures employed at Kensington. The alternative was evaluating the individual performance of the Rector, academic managers and general staff in relation to equity goals. The Group recommended that the Rector be required to report on how its recommendations were implemented and that plans containing goals, strategies and targets for equity and inclusion be developed for individual schools.

In reflecting on her report more than 15 years later, Professor Morrison was not persuaded that Defence was willing 'to make a serious effort to change its culture'. She felt the refusal to allow the University to contribute to behaviour programs was 'an important failing'. Conversely, the impression she gained from serving on the Academy Council was:

> that most senior representatives of the military were very progressive, and easily embraced social change. The middle ranks seemed to be groups with more entrenched views, and they were the people who were in a position to resist cultural change in relation to equity and discrimination.[22]

The Steering Group's Report was completed in June 1998 (but not publicly circulated until October 1998). The Vice-Chancellor accepted the report and its recommendations in their entirety. The Rector's Advisory Committee was not of one mind on Morrison's definition of 'inclusivity' but agreed that the University College could improve its performance in a number of areas. A local steering committee was established to implement the agreed recommendations and to report to Kensington on progress. Resistance to the Group's recommendations among the University College staff was based largely on the perception that the Steering Group did not adequately understand the dynamics of the UNSW–Defence relationship. The mood was further darkened when a contract to deliver the equity and diversity program recommended by the Grey Review was awarded to the Canberra Institute of Technology (CIT) ahead of the UNSW Equity Unit. CIT had previously tendered successfully for the English Communication Program at the Academy.

As the University dealt with the Grey Review and the Morrison Report, the new Academy commandant, Commodore Brian Adams, had already made substantial progress on a cultural reform program. He had arrived at the Academy in January 1998 as the first one-star ranked officer to hold the appointment. He was newly promoted from the rank of captain and 'hand picked' by the Vice

Controversy and consolidation, 1996–1998

Chief of the Defence Force, Vice Admiral Chris Barrie, for the job. [Barrie was soon to become the Chief of the Defence Force.] Unlike several of his successors who were not university graduates, Adams had completed a bachelors and a masters degree as a part-time student and as a distance education candidate. Adams had been briefed on the need for reform and settled quickly on his priorities. During an initial brief, Vice Admiral Barrie mentioned that closing the Academy was an option if impediments to reforming its culture were insurmountable. Adams had no well-developed views on the 1981 Agreement with UNSW, other than noting the document lacked detail and relied on flows of goodwill on both sides. He also observed that uniformed and academic staff were performing activities beyond the scope of the Agreement. At his first meeting with the Rector's Advisory Committee, Adams explained that although he was 'interested in the academic environment of the Academy, it was not a focus of his attention' as the University College had consistently demonstrated that it could 'produce the results that Defence required'. He thought the issues raised by the Grey Review were not new; they had simply been restated. He stressed the need for academic staff to work with uniformed people to achieve cultural change, including an overall improvement in academic standards and highlighting the importance of an academic education to a successful career in uniform.

Adams visited the Kensington campus during Orientation Week in February 1998 and was struck by the sheer scale of the University. It was plainly a very large and growing enterprise. He met the Vice-Chancellor and members of the University Council. He knew they were concerned about the Academy's cultural problems and their potential to affect the University's academic and community standing. He felt the Council was neither overly pessimistic nor excessively alarmed by some of the media reports. Adams impressed the Council with his grasp of the issues and determination to achieve change. Its members generally accepted his assurances that cultural reform was foremost on his agenda. Also notable was the absence of any ill-will over the Vice-Chancellor's failed attempt to resume control of the University Regiment site adjacent to the main campus several months earlier.

The Regimental training depot was located on campus land subject to an agreement signed in 1956 between the Commonwealth and the Army for a 99-year lease. There was an annual fee of $2. Professor Niland wrote to the Chief of the Defence Force, General John Baker, on 28 July 1997 about the land and its future. He explained: 'We are having difficulty finding suitable space for expansion of core facilities, and, I must say, if the Training Depot were not there the site would be ideal for the purpose'.[23] He suggested relocating the depot to another Defence site in southern Sydney with the offer of 'appropriate compensation to the Department of Defence in respect of the building and other improvements'. A briefing note by Brigadier Phillip Parsonage, the Regiment's former Commanding Officer, noted that:

> UNSW and UNSWR suffered little or no drama over National Service and the Vietnam commitment and the University commenced its far reaching association with Duntroon, then ADFA, [this was] impossible at the time for say Sydney University with its strong anti-Army/Vietnam attitude.

The Brigadier questioned whether Niland really wanted to secure the land for 'core facilities'. Commercial development of the site was more likely. He suggested that Defence should respond by asking for a new site located within the confines of the campus. Parsonage was adamant: if 'the Army leaves it will never be allowed back'. General Baker replied to Niland in September 1997:

> The Regiment is unique in Australian military history having been formed as a result of a ground swell of popular support, both within the Army and the University in the late 1940s. Importantly, the location of the unit within the University's precinct has been identified as a key factor in melding the military ethos with other aspects of university life.[24]

He went on to explain the regiment needed the University population as a recruiting base and felt that the 'relocation of the Regiment to a depot remote from the University could have a detrimental effect on the unit'. General Baker was open to the Regiment being located elsewhere *on the campus* and even asked for alternative sites. None were forthcoming. Making campus land available would defeat the purpose of Niland's original approach to Defence. The matter was not raised when Adams met with the University Council and was quickly forgotten.

In his opening address to cadets in early 1998, Adams had pledged to 'work with the Rector to achieve the Academy's aim of providing you with a liberal and balanced education in the military'. Although the University was not implicated in any of the issues that had led to the misconduct identified in the Grey Review, the academic staff were integral to creating an environment in which positive behaviour was affirmed and rewarded. This environment would make it more difficult for rogue elements to exert a malignant influence on either the academic or military components of the Academy experience. Adams was conscious that some of the academic staff tended to give new uniformed leaders a few months to reveal their intentions before deciding to either co-operate with or to obstruct the Commandant's plans. According to Adams, he encountered academics who regarded the Academy as their possession, resenting any changes effected by the uniformed staff of which they disapproved. A small group preferred to ignore the protocols necessitated by the Academy being a military base and 'did their own thing'; another small group wanted to be accorded military marks of respect by junior uniformed staff; and yet another were unable to see the conflict between the University's ideals and the ADF's core values.

Adams was also concerned that academics made public statements that suggested they were speaking on behalf of the Academy and were counselling cadets with problems that should have been referred immediately to Defence authorities. When taken with a number of hybrid military practices, he felt the Academy had 'slowly drawn away from the Services, a development which simply compounded the problems of the military environment'.[25] In view of the extent of the reforms he needed to institute, Adams wanted to remove the University as much as possible from the military environment. Similarly, he did not allow the Academy Council to be active in the reform process. He reported what he had done and what he planned to do. There was little time for consultation because Vice Admiral Barrie gave him 'just six months to prepare for a new military environment to begin with the 1999 year one entry in January'. Substantial change was required and it needed to happen without delay for the sake of the Academy and whatever remained of its severely tarnished reputation.

The University naturally wanted to be part of reviving the Academy's community standing. Many staff were thoroughly dedicated to the Academy and its objectives and devoted to the cadets and their learning. Those with previous military experience brought with them expertise that enhanced the appeal of their courses. Clearly, individual attitudes loomed large in determining whether the relationship between the University and Defence was positive and productive or negative and destructive. Although some of the academics may have wanted to see Adams moved on quickly because they did not like his methods, he gained a loyal following from those who appreciated his emphasis on education and the importance of gaining a good degree. He was also more imaginative than some of his predecessors in proposing the development of new degree programs to respond to, or even anticipate, emerging vocational needs within the ADF. But sensitivities remained across the military–academic divide.

In an attempt to ensure the Academy did not attract any unnecessary negative attention, Adams wrote to the Rector on 16 September 1998 concerning public comment by academics in print and electronic media. He accepted that 'staff of the University College often write to national and international publications that reflect their personal opinions and judgements on a wide variety of matters' but observed that in adding 'a signature block that uses the ADFA title [...] there is often a perception that they are commenting in an official capacity on behalf of the Defence Academy or the Department'. He did not believe any academic had sought to mislead the public as to the status of their views but feared for the Academy's reputation while it was 'under continual scrutiny for all its actions'. The joint branding exercise with co-located symbols for the ADF and the University completed a few years earlier had already been abandoned.

The Rector, Professor Richards, accepted the point being made and added that 'the University's preferred position is that no university affiliation (and therefore no Defence Force Academy affiliation) be attributed to a correspondent when expressing personal opinions'. There was an obvious tension here: when was an academic expressing a personal opinion and when were they exercising a professional judgment as a public intellectual? Would the forum determine whether the affiliation could be cited? For instance, was it acceptable for an academic to disclose their affiliation with UNSW when publishing a scholarly monograph but unacceptable when offering comment on ABC radio? Indeed, there were occasions when the University wanted to be associated publicly with a person and a policy position or in relation to monographs and journals to be cited as part of its research output. This episode was indicative of the strains that existed within the Academy at that time. Adams' position had definitely exposed the drawbacks of having the ADFA and UNSW logos on the same letterhead, and the logic for the practice being abandoned.

Within two years, however, progress had been made in transforming the Academy's culture. There was a renewal of confidence across the establishment. The cultural reform agenda would still take many years to complete. Programs that address attitudes are always gradual in the influence they exert. But many of the immediate problems identified by the Grey Review had been addressed because Adams began to implement change in January 1998, well before the Review's release in June 1998. Notably, Adams' initiatives included closer interaction with the Canberra community, a community that continued to struggle with the difference between RMC and ADFA. There were many Canberrans with no idea of where the Academy was located or that UNSW was the educational provider. Chris Alder contended that criticisms of the Academy focused on the 'social and cultural isolation of its cadets'. He thought this was principally a function of the 'failure of ADFA effectively to involve and promote itself within the Canberra community'.[26] He argued that the Academy 'to its detriment adopted a reactive rather than proactive approach to public relations' and this failed to deter any media interest 'in ADFA beyond sensational accounts of incidents of harassment and discrimination involving cadets'.[27] There were reasons why cadets tended to be separate or distinct, rather than alienated or isolated, from the local community. The cadets had similar interests, shared similar experiences, sought similar affirmation, and so on. They were not unlike other professional groups in the national capital who socialise together after business hours beyond the workplace because their sense of 'vocation' has drawn them to other like-minded people. The distance between the Academy and the Canberra community appeared to be a permanent fixture.

Adams was selected for promotion to Rear Admiral and would leave the Academy at the end of 1999. His rapid elevation suggested Vice Admiral Barrie was very impressed by his work at the Academy. He addressed the United Services Institute on 3 November 1999 and reflected on the previous two years. He stressed that tertiary education was vital to recruiting; being part of a profession required an education; and tertiary education 'can contribute significantly to producing in officers a better understanding of human beings'. In this context, Adams quoted Gandhi's warning of 'knowledge without character'. He thought that education ought to be delivered when it was vocationally needed but chided Defence as an organisation for not knowing what it wanted in terms of officer education, why it wanted educated people and how it intended to use them. He also shielded the University from criticism, noting that Defence provided the environment in which tertiary education was delivered, reminding his hearers that the University did not control the military environment. But would the changes Adams introduced endure? There were signs they would.

An article published in the *Australian Defence Force Journal* in February 2001 by Squadron Leader John Leonard presented the findings of a survey he had conducted among Academy cadets in 1989 and again in 1998.[28] Using the Jungian concepts of transformation and transcendence, Leonard observed a shift from gratuitous and even illegal applications of authority among cadets that instilled fear and frustration in the junior class during 1989 to 'a leadership relationship based far more on mutual respect' with an emphasis on 'rewards and incentives' in 1998. There was greater respect for individuality and a readiness to accept personal autonomy in 1998, accompanied by recognition that women were equal to their male colleagues and entitled to be free from harassment. Those who refused to participate in the emerging culture were deemed to be dispensable. Notably, the new mood reflected an operating paradigm for the ADF that valued personnel 'who can use their brains, can deal with a diversity of people and cultures, who can tolerate ambiguity, take initiative and ask questions, even to the point of questioning authority'. Leonard believed that the Academy culture that Adams had fostered had even challenged 'conformity and orthodoxy'.

With responses to the Grey Review being implemented and the cultural reform program underway helped by the involvement of academic and administrative staff from the UNSW main campus at Kensington, Defence Minister John Moore ordered yet another review. Concerns about the administrative and financial basis on which the Academy was established and had operated over the previous twelve years remained unaddressed. The review could not have come at a less convenient time for the University. In the same month (October 1998) Professor John Richards announced he had accepted an invitation to become the Deputy Vice-Chancellor of the Australian National University. There would be yet another period of instability for the University College and a further requirement for UNSW to demonstrate the value of its relationship with Defence.

12

Commercial contracts and institutional partnerships, 1999–2004

By the end of the millennium, the Defence Academy was the most closely scrutinised government establishment in the country. It had been the subject of 12 separate reviews since 1990. There was a general weariness on both sides of the diarchy, with successive inquiries into its organisation and operation even before the Grey Review had been published and its principal recommendations implemented in mid-1998. It had also been an unsettled time within the University College, with Professor John Richards concluding his term as Rector on 2 October 1998. At Kensington, the Vice-Chancellor, Professor John Niland, moved quickly to fill the vacancy, believing any delay would hinder the University's contribution to the Academy reform process. An interview panel was assembled that included the Chief of the Defence Force, Admiral Chris Barrie; the Secretary of the Department of Defence, Paul Barratt; the Academy Commandant, Commodore Brian Adams; the chair of the Academy Council, Sir Edward Woodward; and Niland. For their part, the Defence officials wanted a Rector who would help rather than hinder the Academy reform agenda.

Four candidates for the post were interviewed on 30 October 1998. The Acting Rector and future ANU Vice-Chancellor,

Professor Ian Young, had already indicated that he would be leaving the University College for a new position as Pro Vice-Chancellor (International) at Adelaide University. He would not be a candidate. When none of the initial four candidates was considered suitable for the post because they appeared to either misunderstand the Academy's unique mandate or wanted to pursue discordant agendas, Niland asked the recently retired Deputy Vice-Chancellor, Tony Wicken, to be 'on hand' and to assist in Canberra during a likely protracted interregnum. When Ian Young departed for Adelaide, Professor Peter Hall assumed the role of Acting Rector and another canvass for suitable candidates commenced.

Notwithstanding the interregnum and general weariness with reviews, the College was required to be part of the UNSW PREP (Performance, Reporting, Evaluation and Planning) feedback process in the first half of 1999. The Vice-Chancellor was personally involved in the exercise and met with the senior staff in March to discuss 'implementation of the recommendations of the Morrison Report, cooperative interaction with the military arm of ADFA, recruitment of civilian postgraduate and research students, development of distance mode postgraduate courses, Defence's Statement of Requirements [and] the 1998 Staff Opinion Survey'. There was nothing controversial and little that was new in the Statement of Requirement other than the expectation that 'all courses will be delivered at the standard required of a Group of Eight [Go8] University'. This was the first time the 'Go8' had been mentioned in any document produced by Defence. It showed that Defence was conscious of UNSW's national standing, which was continuing to rise. In contrast to an editorial the *Sydney Morning Herald* had published on 25 February 1950 that doubted the wisdom of establishing the new University of Technology and raised doubts as to whether it deserved to be called a university at all, that newspaper's editorial

Commercial contracts and institutional partnerships, 1999–2004

of 3 April 1996 remarked that 'Sydney University as an academic entity probably has to yield to UNSW as the best university in Australia'.[1] Much had changed.

In assessing the performance of the Canberra campus, Niland shared his concern that research within the University College had declined in both 'published output and success rate of ARC Large Grant applications'. He wanted the trend addressed through research mentoring schemes and by possibly appointing an Associate Dean (Research). On the positive side, the College was ranked second only to the Australian Graduate School of Management (AGSM) at Kensington for both overall student satisfaction and good teaching. Niland was conscious, too, of the broader challenges:

> I am well aware of current unease among University College staff engendered by a perception that the general nature of the relationship between Defence and the University has shifted from one of equal partnership to one of purchaser-provider.[2]

He attributed this change of mood to reforms promoted by the Grey Report and the departure of the Rector and Deputy Rector during a time of turmoil. Niland was not, however, overly concerned:

> I do not sense a desire on Defence's part to sever the relationship with the University but I do sense that Defence is currently thinking through its educational and training needs [...] I believe the College should be pro-active in offering various models of change to Defence rather than being passive in the process.

Niland noted that the relationship with Defence was the only area of specific concern identified in the 1998 Staff Opinion Survey. In sum, he said, 'I think the College is travelling well enough overall, especially in a rather difficult time [...] The University College is an integral and important component of UNSW and, as Vice Chancellor, I am determined that it remain so'.

By June 1999, Niland had persuaded the Chair of the UNSW Academic Board, Professor Robert King, to accept the Rectorship – but not without considerable effort. King had no previous involvement with the University College and had not applied for the post. He was acquainted with its senior leadership and familiar with its degree programs from chairing the Academic Board. But he had never read the 1981 Agreement nor was he aware of its flaws. He had not followed the progress of the Grey Review nor the reform agenda the inquiry had prompted, while his personal research interests had not brought him into direct contact with Defence people. He believed, however, that the Canberra campus hosted many gifted people who may not have realised the snobbery that existed against the Academy among some sections of the University's leadership at Kensington. The University College was considered to be a good educational provider and offered high-quality teaching but the general impression at Kensington was that its research capacity and output needed to improve. There were some Kensington academics who resolutely opposed maintaining any link with Defence on political grounds and some who thought the relationship was contrary to the spirit of the University. In other words, the standard objections from the 1960s had persisted among the next generation. These objections were not sufficient in substance or standing to influence the relationship. Niland had a very high regard for the work being done in the national capital and was never among the Academy's detractors.

Robert King was a most unlikely candidate to be Rector. In the first instance, he was (and remains) the only 'external' appointment

Commercial contracts and institutional partnerships, 1999–2004

to the post. His predecessors as Rector and, before that, as Dean of the Faculty of Military Studies, had all been internal appointees or were previously long-term staff members. King had been raised in a pacifist tradition and exercised an entitlement to have his national service call-up postponed until he completed his university studies. By then, Australian involvement in the Vietnam War had ended and national service was terminated in June 1973. When first approached by Niland, King was not interested in the position. His main focus was the Kensington campus. He did not want to relocate to Canberra, in part because his wife Rosemary enjoyed her work in the Sydney area. Niland was, however, adamant that he should consider the position. The Vice-Chancellor did not want to risk another fruitless interview process, especially after several searches had failed to identify a suitable candidate. King was an accomplished administrator with a network of well-placed contacts. He was respected as an academic researcher and was familiar with UNSW's internal procedures. He was eventually persuaded by Niland to take the offer seriously and travelled to Canberra to meet Chris Barrie, Paul Barratt and members of the University College Advisory Council. King recalled that the conversation with Admiral Barrie was frank and forthright:

> 'Have you had anything to do with Defence?' I said 'No' because I really hadn't. He said: 'Good, because whoever runs that place has got to be independent of me and I've got to be independent of them'. He more or less assumed that I was going to take the job. 'I don't want you interfering with Defence matters and I will give you a commitment not to interfere with University matters.' I said, 'I would like you to know that we are going to continue to teach a liberal and balanced degree', and he said yes, that was my business. He

said, 'I'll give you a commitment that when you are in a semester the university studies take priority, and when it is out of semester the students are ours'. 'Fine', I said.

Barrie also pledged his personal support, a commitment that mattered a great deal in the post–Grey Review period.

Having decided to accept Niland's invitation, King made his first visit to the Academy as Rector Designate on 16 July 1999. He attended a meeting of the Rector's Advisory Committee and used the opportunity to discuss the Federal Government's 'Green Paper' on higher education. He outlined the response to the policy document being generated within Kensington. King formally commenced as Rector the following month. He later appointed two Deputy Rectors, one supervising research and the other education. Professor Charles Newton (Education) and Associate Professor Susan Lever (Research) filled the new positions.

As King was preparing to relocate to Canberra, Defence Minister John Moore was persuaded by Defence departmental staff to revise the 1981 Agreement. The terms of the revision were being considered but had not been confirmed by the end of 1999. In May 2000, the University was advised that a new agreement would be submitted to the Service Chiefs for their approval on 14 June, but that 'it would be some months before a comprehensive document would be placed before the University for negotiation'. The University was keen to begin work on a replacement text, especially on the financial arrangements. In early September 2001 the Rector was advised that the latest round of enterprise bargaining would see academic salaries increase by 2 per cent in March 2002 and 3.5 per cent in December 2002. General staff salaries would increase by 3 per cent in December 2001 and 2.5 per cent in December 2002. The advice from Laurie Olive, the College's Director of Finance,

Commercial contracts and institutional partnerships, 1999–2004

Personnel and Planning, was blunt: 'recent increases in the budget from Defence have not kept pace with costs and the decline in the value of the Australian dollar has exacerbated problems. It was expected that some of these issues would be addressed in the renegotiation of the UNSW–Commonwealth Agreement but the process has been delayed'.[3] Over the previous five years, the cost of books and equipment acquired internationally had led to cost over-runs, while internet charges had risen 300 per cent in five years. By the end of September 2001, the Rector confided to his Advisory Committee that the agreement renegotiation had taken much longer than he predicted or anyone expected.

The delay was partly the result of John Moore inviting Professor Ian Zimmer, the Executive Dean of the Faculty of Business, Economics and Law at the University of Queensland, to lead a two-phase review of the Academy. In Phase I, Zimmer would examine and assess the undergraduate programs offered at the Academy and analyse existing and emerging performance benchmarks for the tertiary education sector that would allow him to compare and contrast it with other institutions. In effect, Defence wanted to know from comparisons with other Go8 universities whether it was getting a fair deal from UNSW. In Phase II, he would review the provision of postgraduate education for uniformed officers throughout the country. Zimmer was to ensure that 'these educational programs are being properly managed and targeted, and Defence is receiving value for money from its considerable outlays in these important areas'.

This review was never intended to be a 'grass roots' exercise. The academic or general staff would not be surveyed or interviewed. This was a management-level review that would produce institutional judgments and make organisational recommendations. Zimmer was, at least, familiar with UNSW, having previously completed a PhD dissertation at Kensington on accounting information in corporate lending. It is not known whether the Minister knew of Zimmer's UNSW connection at the time of his appointment, although Zimmer had not been associated with UNSW for many years. There was certainly never any suggestion that Zimmer was biased towards or against UNSW.

Zimmer decided to conduct the postgraduate inquiry first and submitted his report to the Minister on 8 December 2000. After John Moore had resigned the portfolio in January 2001, the Academy review was delayed until the new Defence Minister, Peter Reith, found time to meet with Zimmer in March 2001. The new Minister agreed to expanded terms of reference including a 'cost benefit analysis of the educational service delivered through ADFA and benchmark it against other tertiary service providers'. Zimmer was also to ensure that the education provided at the Academy met the needs of the Services and its students out to 2020. He was to explore the benefits of having civilian students at the Academy and prospects for expanding distance education. Zimmer was asked to report by July 2001, giving him little more than three months to complete the task.

Three things prompted this particular review. The first was the manifest inadequacies of the 1981 Agreement. These had been identified in the Price Report and had still not been addressed. The University now believed that continuing the 1981 Agreement was not in its interests and, with Defence, welcomed this aspect of the review. The second prompt was continuing criticism of the cost of the service provided by UNSW. There was no evidence that another provider could deliver the programs any more cheaply and none had demonstrated or even suggested that they could. Comparing the cost of a civilian undergraduate to a cadet educated at the Academy led inevitably to claims that UNSW was exploiting Defence's goodwill. The third prompt was the desire (also identified in the Price Report) to secure external academic accreditation for the programs being offered at the Service Staff Colleges. Clearly, Zimmer had much to consider in three months and needed help. Before starting, he sought and secured the assistance of Professor Bruce McKern, Chief Executive Officer of the Mount Eliza Business School. In effect, McKern would offer a second opinion if needed and remove any suggestion that Zimmer's report might have unfairly produced a dividend for the University of Queensland where he then worked.[4]

To gain some background insights into Defence generally and the Academy specifically, Zimmer was encouraged to speak with Ken Given as someone familiar with the issues he would need to address. Given had been promoted to Group Captain following his posting to the Joint Standing Committee but had retired from the RAAF during the previous week. What had started as a short conversation between the two men lasted four hours and led to Zimmer requesting that Given be appointed to assist the two-man (Zimmer and McKern) review committee. The final draft of the report would actually be written by Given under Zimmer's direction and with his careful editing. The review team was subsequently augmented with the inclusion of Colonel Clive Badelow, a recently retired Ordnance Corps officer who had previously been Director General of the Defence Corporate Support Program, and Barbara Pepper, a former Army officer with expertise in contract management and logistics, from the Department of Defence.[5] Having observed that the only inquiry whose recommendations had been fully implemented was the Grey Review, Zimmer implored Defence to make his inquiry the final examination of the Academy for the foreseeable future (he suggested at least for the next five years) and that no further reviews be conducted until his recommendations were addressed and absorbed into a new set of administrative arrangements.

Zimmer soon realised that the prevailing view in Defence was that UNSW had 'done well' from the Agreement, while the prevailing view in the University was that Defence was getting more than it was paying for. He was yet to make up his own mind, having no previous involvement with Defence or the Academy Agreement. He also assured the UNSW Vice-Chancellor, the Rector and senior faculty in Canberra that he was not subject to any 'guidance' from the Minister, other than to confirm that Defence was receiving value for money. Zimmer soothed many anxieties with his non-adversarial and empathetic approach. Over the coming months both academics and general staff found they could talk openly and easily with both Zimmer and McKern, who understood how universities were funded and operated.

Zimmer thought the 1981 Agreement was 'outdated, inadequate and divisive'.[6] He noted with a sense of surprise that there are 'no specific arrangements for funding, and the agreement contains no "sunset" or review clause. It is completely "open-ended"'. Although the Agreement had been amended slightly in 1985 to deal with an expanded membership for the Academy Council and a 'Common Services Agreement' had been signed in 1993, the whole document needed substantial revision to reflect the fact that the University's relationship with Defence had become more 'one of partnership than purchaser–provider'.[7] This was a curious interpretation of the previous fifteen years. The trend appeared, in reality, to have moved in the opposite direction – from a partnership, with no financial arrangements, no 'sunset clause' and a heavy reliance on goodwill, to something resembling a purchaser–provider arrangement.

Plainly, according to Zimmer, the 'current funding statement' was a considerable weakness in the document, if only because it aroused suspicion. The educational market for what Defence required had never been tested and UNSW had been protected from competition. A new agreement would need to provide 'greater transparency of outcomes, reporting accountability, firmer measures of performance outcomes and greater flexibility to accommodate the changing needs' of Defence and the University. He agreed that some of UNSW's costs were neither detailed nor obvious and that an 'open and transparent relationship between Defence and UNSW is needed to foster a better contractual arrangement'.[8] Zimmer thought the drafting by Defence of an 'Academic Statement of Requirement' (ASOR) in September 1998 had provided a template for assessing course costs and was a good start. He also noted that UNSW had neither responded to the draft nor agreed its contents. The Rector explained that the document had been considered but a response was delayed pending the outcomes of Zimmer's own report. Zimmer recommended implementing an ASOR immediately to ensure Defence could 'determine if it is getting value for money in a competitive market place' alongside a Service Level Agreement to settle any disagreements on minor operating costs.[9]

Zimmer was adamant that one element in the 1981 Agreement needed to stay. He thought references to a 'balanced and liberal education' prevented the undergraduate program from becoming too vocational and this was vital to the character of the Academy experience. He noted that while the enrolment of civilian undergraduates had appeal on several fronts, he accepted the need to preserve the character of the existing undergraduate program and to restrict it to uniformed personnel. Significantly, Zimmer did not share Price's concerns (indeed, he thought they were overstated) about the social or intellectual isolation of the undergraduate student body but counselled against stressing the 'uniqueness' of the Academy as a military institution. New-entry cadets were not receiving an *Academy* education, he stressed, they were receiving a *university* education from one of the nation's best higher education providers.

Zimmer was highly complimentary in his appraisal of UNSW's

people and procedures, reporting that Defence was satisfied with delivery of both the undergraduate and postgraduate programs, including the range of courses on offer. Although Academy graduates were still relatively junior in terms of rank, they were proving their worth in administrative, technical and operational settings in ways attributable directly to the University College. Indeed, UNSW had shown much greater flexibility in curriculum matters over the previous four years, leading him, perhaps predictably given his own academic background, to recommend introducing a Bachelor of Business degree. This program would incorporate and consolidate the economics and management subjects that were presently being offered. Indeed, the Department of Economics and Management had been among the fastest growing departments during the first decade of the Academy's existence. There was continuing concern, however, about whether UNSW was granting sufficient recognition for prior learning, with the University College being committed to preserving its standards within UNSW as a whole and the University's standards within the higher education sector. Zimmer understood UNSW's reluctance to be generous in recognising prior learning, by which was meant vocational courses, as a member of the recently formed 'Group of Eight' universities. Zimmer did not push this matter nor did he make a recommendation on the Academy taking civilian undergraduates.[10] While he was impressed by the quality of military instruction at the Academy, he recommended against common or Service-specific vocational training being given academic credit. This would work against the academic integrity of UNSW qualifications in both the short and long term.

Zimmer thought that Defence still did not appreciate the importance of UNSW's reputation in attracting numbers of high-quality potential recruits to the ADF. The prospect of gaining a UNSW degree was a significant factor in the minds of young people thinking about their future and a possible career in uniform. Within fifty years of its founding, UNSW had become a major national institution. In 2000, it had achieved an operating profit of $49.8 million (after abnormal earnings of $8 million). Revenue had increased to $604.8 million and spending to $563 million. In that year the University enrolled 34 000 students and employed more than 5000 staff spread across ten faculties, 12 co-operative research centres, five major medical schools affiliated with 18 hospitals, the Australian Graduate School of Management and the University College in Canberra. Indeed, the entire Australian higher education sector had grown considerably, with total enrolments increasing from 350 000 in 1981 to 690 000 in 1999 despite declining Commonwealth funding over the previous three years. By the year 2000, the Howard Government was providing only half of total funding to the higher education sector. In addition to promoting the enrolment of full-fee paying students to boost income, staff to student ratios increased across the nation to reduce expenditure, from 14.8:1 in 1996 to 18.3:1 in 1999.

In terms of teaching and learning, Zimmer reported that the staff to student ratio at the Academy was 8:1 – half the national average – but this was offset by the very large number of courses being offered by the University College (456 undergraduate courses and 243 postgraduate courses) with each staff member offering an average of 4.3 courses a year. This was more than the national average. To explain: the staff–student ratio was a function of a smaller number of students completing a larger number of specialist courses. Putting it another way, the number of undergraduate students would need to double to restore the staff–student ratio at the Academy to more 'normal' university levels. This could have been done with very few extra staff and facilities. The impediment was that these undergraduates were simply not available in the form of midshipmen or officer cadets, while accepting civilian undergraduates was not even contemplated. Defence wanted both breadth and specialisation in its degrees and this directly influenced class sizes and the staff–student ratio. Fortunately, progress had been made on attrition and retention, and on eradicating the '51 per cent mentality' that had led to cadet underperformance over many years.

Alongside these gains was the steadily increasing number of postgraduate students. They accounted for 32 per cent of the student body in 1999. Defence was nonetheless worried that postgraduate growth might detract from the quantity and quality of undergraduate programs. There was no evidence that this was so but Zimmer reported the fear nonetheless. The University responded by highlighting the rise of the University College as a Defence 'centre of excellence', where uniformed and civilian students (the latter being largely civilian staff of the Defence Department and its related agencies) were undertaking research on subjects that were being explored nowhere else in Australia. Indeed, Zimmer recommended reviving the Visiting Military Fellow scheme at the Academy to deepen the links between the Services and research being undertaken on campus. Clearly, there was a co-ordination issue to be resolved so that Defence was both aware of who was studying at the postgraduate level and the nature of the work they were doing, with the cost of research enrolments, according to Zimmer, to be borne by the Department of Education, Training and Youth Affairs, and not by Defence.

On the matter of the cost-effectiveness, Zimmer noted the inherent difficulty of going beyond mere costs to determine real value. Indeed, he concluded that 'quantifying the benefits of the University College is almost impossible'. Beyond the undergraduate and postgraduate programs, Defence used UNSW academics as a standing pool of expertise that it could and would draw upon from time to time. It also used the University College to build links with Australia's friends and allies, and to form useful networks built on the standing of graduates as UNSW alumni, including officers in regional armed forces such as those of Singapore, Malaysia and the Philippines. While it was difficult to bundle together all of the benefits Defence derived from the University College, Zimmer nevertheless felt that a process was needed for 'evaluating the performance of the contract provider that includes quantifiable benchmarks'.[11]

Given the nature and extent of public criticism, Zimmer recommended that Defence affirm its commitment to the original rationale for the Academy and the importance of undergraduate education being provided in a tri-Service uniformed environment. After considering a range of alternatives to the extant agreement, he came to a firm conclusion that, for the immediate future, 'there is no reason to dissolve the current, long-standing association with UNSW' in favour of another provider.[12] Notably, no other university had approached Zimmer while his review was in progress wanting to bid for Defence undergraduate education. Seeking another provider would, in any event, require a competitive tender process. This would take some years to arrange and then complete. Switching providers would be a highly complicated activity that could not be achieved in less than five years. Given the turbulence of the previous decade, he remarked:

> The Review Team is concerned that the quality of education provided by the University College at the Academy would suffer in a frequently changing, open tender environment, as every time the tender came up for renewal, the provider could hemorrhage academic staff unwilling to take a chance on staying with a provider who may or may not win the contract.[13]

He recommended that Defence remain with UNSW 'for the term of the revised agreement, subject to the satisfactory renegotiation of a new contractual agreement'. Zimmer warned against simply rewriting the 1981 Agreement. It needed to be replaced with a document that included a Statement of Academic Requirement, a Common Services Agreement, and detailed schedules covering organisation and finance. The new agreement needed to be in place by January 2003 and should run, he thought, for 'up to ten years'.

Zimmer did not recommend specific changes to the curriculum. He suggested that Defence needed to be clear and consistent on undergraduate course requirements, with appropriate rationale provided to the University College. Conversely, he suggested that the University College needed to be better informed about the changing needs of the 'profession of arms'. While the parties clearly needed to improve communication, Zimmer's proposals did not deal with resolving disputes between Defence and the University on degree structure and course content. It seems he hoped enhanced mutual understanding would lead to a meeting of minds. Zimmer also hinted at the need for both Defence and the University to better 'sell' the benefits of the Academy and the importance of the philosophy and programs that shaped its operations. Clearly, the three Services were not of one mind on what they wanted from undergraduate education, while those who championed the tri-Service cause were alarmed by possible fresh outbreaks of single-Service tribalism. The Academy needed well-placed friends who could speak authoritatively about the importance of education and enthusiastically about what was, and what could be, achieved in and through the University College. While the Academy had received some bad press, and its standing within the Australian community was tarnished as a result of cadet misconduct, the most pressing need was for good internal 'PR' so that, at the very least, Defence and UNSW members would be persuaded that the Academy was a worthwhile endeavour.

The University was naturally relieved and, in fact, greatly encouraged by Zimmer's report. As Zimmer was a senior academic acquainted with university administration and the funding of higher education, he understood the issues from the University's perspective even as he was perceived to be, and indeed was, sympathetic to Defence's requirements. He persuaded the University on the need for a comprehensive contract. This document would later be referred to as an 'agreement' rather than a 'contract' to assuage some concerns that the University was a mere 'contractor', although all of the agreement's provisions were to be legally enforceable. He flagged the likelihood that Defence would argue that the 'premium' paid to UNSW was too high and that too much money was being extracted by Kensington for shared services with little justification. He made no public comment on this perception within Defence although the University repeatedly explained that it made no money from the agreement. Professor Rupert Myers had stressed this point during his evidence to the Parliamentary Works Committee in 1978. Zimmer acknowledged that no university would enter into a relationship with Defence without the prospect of making some money but thought a conventional fee-for-service arrangement was untenable in this instance. UNSW needed to be entrepreneurial and to use its unique relationship with Defence to secure dividends from external collaborations and partnerships.

Negotiating a new agreement began immediately after the Minister received Zimmer's report. Robert King and Laurie Olive were the key negotiators for UNSW. Olive, the College's Director of Finance, Personnel and Planning, was a former Head of the Department of Geography. He had replaced the College Secretary, Terry Earle, in a much enlarged role.[14] From September

2001, Defence referred to the subsequent negotiations as 'Project Sandstone'.[15] This was rather ironic as UNSW was not a 'sandstone university' like the Universities of Sydney and Melbourne. Colonel Clive Badelow, who had assisted the Zimmer Report, was Defence's chief negotiator. He was a highly experienced military officer with a strong background in training and contracting. The original plan was to apply the Commonwealth Procurement Guidelines to the 1981 Agreement and to have a new agreement in place by the end of 2002. Badelow soon found that the extant 1981 Agreement was practically useless and that an entirely new draft drawing on the best legal minds in both the University and Defence was needed.

The residual view in Defence, despite Zimmer's observations and commentary, was that UNSW had done well financially from the Agreement, that the funding arrangements were too generous and that there was unreasonable resistance to programs and courses being changed. The claim that the University was over-charging on on-costs, such as leave and superannuation, and overstating its overheads was not new. Dennis O'Connor, a civilian in the Army's Programs and Budgets Division, recalled travelling to Kensington in 1978 to discuss financial matters and 'came away happy that UNSW was fair and reasonable in its claims'.[16] It was a very different matter when Defence contended that it was paying for a service that did not meet its needs; that is, the product was too expensive and needed to be downgraded in quality. While the University thought application of the Non-Farm Deflator as the means of determining annual payment adjustments was unfair and inequitable, Defence thought otherwise. The Consumer Price Index was considered as an alternative but rejected because it lacked nuance in this context and would have led to substantial reductions in UNSW's funding.

Although the University quietly conceded that it had been well funded prior to 1995, it appeared that generosity had turned to grasping by 1997, according to Laurie Olive, who noted the University was partly surviving on its success in securing additional external funding to cover core costs. In 1995, the block grant was $33.15 million. The next year it was $34.11 million, an increase of 2.9 per cent. Although Australian Research Council grants and fellowships were down from $2.09 million in 1995 to $1.18 million the following year, the decline was offset by infrastructure research grants. They increased by 300 per cent between 1995 and 1996. The overall income was down, however, by 0.4 per cent. In April 1998, Olive reported that staff costs were increasing by 3 per cent annually for academic staff and 2 per cent for general staff. The financial situation was slowly worsening.

By 2000, the annual block grant amounted to $37.64 million. This represented an increase of 1.5 per cent on the previous year. Additionally, Defence was paying for all university staff long service leave as well as superannuation. The 1981 Agreement was open-ended, with escalating annual payments automatically applied. The University still objected to the application of the Non-Farm Deflator as the basis of annual indexation rather than the Consumer Price Index but representatives of the Chief Finance Officer within Defence were insistent. The effect, according to the University, was an annual reduction of between $500 000 and $1 million in real funding which needed to be absorbed within the University College budget. As 80–90 per cent of the annual budget was devoted to staff costs, the solution would be found in reducing the wages bill. This meant shedding academic and administrative staff to bring the staff–student ratio to something approaching 1:12 against the prevailing national average of 1:18.

Defence believed that the annual indexing was more than generous and suspected that millions of dollars were being 'skimmed off' the contract by UNSW officials at Kensington. This was untrue but the suspicion remained. Nevertheless, Kensington continued to require an annual payment to cover shared services, despite the budget pressures. This amount did not reflect the accumulated cost of the services provided but University administrators in Canberra hoped that some of the payment might be waived in the circumstances. There was to be no such relief. Given the extent of the deficit, it would not have provided more than a minor reprieve in any event. Mistaken perceptions of the calculation of the 'Kensington levy' led to the view among financial managers within Defence that the University was extracting an unreasonably high price for its services. Counter-assertions and the production of refuting evidence would not deter the University's critics.

There was also University staff superannuation to consider. The Commonwealth had been funding superannuation for decades. The University would send a report to Defence each year on its accumulated superannuation liability. Although the figure varied significantly from year to year, the gradually escalating amount did not seem to cause any concern on Defence's part. The same was true of the long service leave liability. It was steadily increasing as well. Defence was noting but not responding to the reported details. Indolence was about to be replaced by attentiveness as the bill swelled.

As an aside, the Zimmer Report produced additional resentment towards UNSW within Defence. The Chief Finance Officer's staff had disagreed entirely with Minister Reith's decision on 26 October 2001 to solely source the contract to UNSW (based on Zimmer's strong recommendation) believing that a comprehensive 'market test' involving multiple tenderers to secure the best price for providing undergraduate and postgraduate education at the Academy was more than justified. Reith's successor as Defence Minister, Senator Robert Hill, reaffirmed the sole source decision on

15 May 2002. Disappointed with this outcome, Defence staff immediately began lobbying for a shortened three-year contract instead of the planned ten-year contract. The Chiefs of Staff Committee eventually agreed to a ten-year contract with a mid-term five-year review. Lobbying by the CFO's staff for a different set of contractual arrangements had been unsuccessful in every instance, hence the resentment.

Most of the contract negotiations centred on the cost rather than the content of the educational services the University would provide. A memo headed 'The ADFA 2002 Contract: Quantifying the Premium that Defence Should Pay' prepared by Ken Given on 30 August 2001 identified the two main elements that should be considered in determining any 'premium'. The notion of a premium was based on comparing the cost of delivering an education at a civilian university in a standard setting and the added cost to UNSW of providing educational services at the Academy with its smaller, specialist student community. Defence would absorb the additional cost as a 'premium'. While Defence assumed there would be a premium, quite reasonably it wanted an assurance of value for money. Zimmer argued that the first element in determining the premium rate was the 'unique nature of the University College as a tertiary organisation' and the second was 'the benefits that Defence receives from this exclusive partnership with a tertiary organisation'. The challenge was determining a quantitative value on a qualitative benefit.

Ken Given argued that because UNSW was not permitted within the 1981 Agreement to enrol full fee–paying undergraduate students and that limits on enrolments and class sizes precluded the University from achieving economies of scale, it was entitled to be paid a premium. In terms of the uniqueness of the University College, however, he felt no premium should be paid for the broad range of programs being offered and the provision of ancillary services. In terms of potential benefits to Defence, he also argued that no premium should be paid for the benefits that Zimmer noted accrued to Defence: the value of the University to ADF recruiting, the improving completion rate of students, the existence of guaranteed places, the achievement of high rates of student satisfaction, the low staff–student ratios, the recruitment of highly specialised staff, the contribution of staff as role models, the availability of professional short courses and the presence of international students. In his judgment, none of these things justified an additional payment to UNSW. Given perceived them as being 'benefits of the on-going partnership, not benefits based on additional funding. Therefore, Defence should not pay a premium for the benefits outlined in the Zimmer Report'.

As the lead UNSW negotiator, Robert King enjoyed the complete confidence of the Vice-Chancellor, who played no substantive part in what would follow – other than to secure the eventual approval of the University Council. This approval could be assumed if the replacement agreement did not cost the Kensington campus any money. After his first year as Rector, Niland commended King for 'rebuilding relationships with Defence personnel particularly aided by the Commandant and his Deputy [...] [I] believe that the University's relationship with Defence has improved greatly during your tenure'.[17] King personally impressed the Service Chiefs with his empathetic grasp of the issues facing the military and his tireless efforts to build a partnership based on trust. He was, however, obliged to counter suspicion among uniformed people. Over the previous decade the University had probably taken Defence's goodwill for granted, assumed its position would never be threatened and believed its detractors would not achieve ascendance. The three Services also differed in their attachment to the Academy.

Commercial contracts and institutional partnerships, 1999–2004

Badelow thought the Air Force was the most supportive of the Academy, although it drew least on what the University had to offer. The Army seemed diffident because only one-third of its officers were drawn from the Academy. It was relying more heavily on lateral recruiting from professional organisations, the Army Reserve and its own 'Other Ranks' for new officers. Its most pressing need was engineers. The Navy did not feel the Academy was critical to its needs but was not interested in abandoning tertiary education altogether.

Both sides were clear about what they sought from the negotiations: Defence wanted to pay less; the University wanted to be paid more. Although cost calculations were central, issues of curriculum content were not completely absent. Badelow recalls the tenor of the initial negotiations:

> My opening position was to demand double-digit savings. Professor King's retort was a firm rebuttal. So, after fifteen minutes, we quickly agreed to something close to 10 per cent. In the end, we settled on $20 million savings over 10 years with a further $10 million negotiated later that year (so that I could meaningfully demonstrate to the Chiefs of Staff Committee value-for-money achievement at approximately 8 per cent plus other academic enhancements).

But drafting an agreement that reflected this consensus was taking months and not weeks. Delayed agreement on achieving the second $10 million in savings threatened to stall the negotiations. The unsuitability of the 1981 Agreement as a basis from which to work, the non-existence of an adequate template upon which to base an initial draft, the absence of a comparable Defence contract to direct the whole exercise, and the sheer complexity of the issues meant that extensive information had to be gathered, detailed questions had to be examined and complicated comparisons had to be made. Both sides worked efficiently and assiduously but producing a comprehensive document that established principles for the benefit of leaders and offered guidance for managers took considerable time. There was sufficient goodwill on both sides. The involvement of lawyers was not a sign of distrust but of the need for highly specialist drafting skills. It was plain that the University and Defence wanted to find a way of working together and were prepared to compromise. The only discordant moment in the negotiations arose from a visit to the Academy by staff from the Department of Finance and Administration, who virtually accused the University of stealing money from Defence. When the University College's operations were compared to the Australian Maritime College based in Launceston, UNSW was found to be a cheaper provider of educational services on a per-student basis. The allegation was baseless and rightly rejected.

Such was the intricacy of the negotiations that the Chiefs of Staff Committee could only approve new 'Heads of Agreement' in December 2001. The complete text was some way from being finalised. The University Council agreed to the 'Heads of Agreement' in April 2002. Although many issues were still unresolved, the Chief of the Defence Force, Admiral Chris Barrie, the Head of Defence Personnel Executive, Rear Admiral Russ Shalders, the Vice-Chancellor and the Rector signed a new Agreement on 25 June 2002. It was little more than a statement of intent given the slow progress of negotiations with the finance elements of Defence. The document was designed to reassure the University that the Commonwealth and Defence's leadership were committed to achieving a value for money outcome with UNSW and would not be seeking

Commercial contracts and institutional partnerships, 1999–2004

an alternative provider. In a Defence Department media release, Admiral Barrie remarked that:

> UNSW has provided an excellent range of undergraduate and postgraduate programs at ADFA [...] this signing ceremony paves the way for a more contemporary, transparent and flexible agreement for the future delivery of educational services [...] the Zimmer Report confirmed what Defence already knew – UNSW has provided Defence with high quality educational services for over three decades.[18]

There were hopes the whole document could be agreed and signed in August 2002. This was ambitious.

In July 2002, Badelow drafted a strategy document for Defence's Chief Finance Officer concerning the second $10 million that the department was seeking. After conferring informally with the Secretary of the Department of Defence, the CFO was informed that the negotiation of the new Agreement is '*not* a savings-driven exercise' but advised that 'the opportunity exists to achieve savings efficiencies without compromising desired educational service delivery outcomes'. The problem for Defence was that its own staff had advised Minister Reith (who left the Defence portfolio at the end of 2001) that 'significant savings' could be gained from negotiating a new agreement, and his successor, Senator Robert Hill, was told 'demonstrable savings' were likely. The Service Chiefs had been told a savings efficiency 'in the order of not less than 10 per cent' was the objective of the negotiation. Expectations had been raised and fulfilling them was proving difficult. The Project Sandstone team concluded that 'to achieve the desired value for money outcome, there are only two negotiable Agreement variables, involving a trade off between cost and quality'. In sum, Defence would need to consider a reduction in the educational service delivered by the University despite assertions that the negotiations were not a savings-driven exercise. The outcome would be dominated by numbers and not words.

The annual cost of the 1981 Agreement at that time was $37.644 million with an adjustment of 2.1 per cent due on 1 July 2002. Defence noted that the General staff to Academic staff ratio within the University College was 3:2, whereas the national average was closer to 1:1. Since 1995–96, the University had reduced its overall staff profile by 12 per cent by shedding 15 Academic staff and 36 General staff. Defence believed there appeared to be 'considerable potential for University College initiated staff rationalisation to reduce costs'. In terms of teaching, the University College average staff–student ratio was 1:8, compared with 1:15 at the Kensington campus and the national average of 1:18. To maintain a 'well-above-average' level of service delivery, Defence proposed an increase in the staff–student ratio at the Academy to 1:10 and 1:12 and a 'University College–led review and rationalisation of the wide variation of undergraduate and postgraduate courses currently available'. The University had resisted this change but Defence believed it could and should press the point to achieve the required saving efficiency. There was little alternative for the University but to consider reducing its staff further.

The University remained clear about its objectives as the details of a new agreement were being decided. It sought a ten-year agreement with five years minimum notice of termination, the freedom to deliver postgraduate education on its own terms, acknowledgment that superannuation and long service leave liabilities would be shared (the superannuation liability was estimated to be $23 million and the long service leave liability to be $6.9 million), Defence would have to accept liability for any staff redundancies necessitated by the new agreement, there would be sufficient funding to meet operating costs and overheads, and an assurance of academic freedom for its faculty. Defence was prepared to accept the University's position on postgraduate education, noting that it represented no additional cost to the Department. It would also accept redundancy liability but only for the first three years of the new agreement. The University wanted a longer 'notice of termination' period but Defence was unbending. Five years was the most to which it would agree. Defence did not initially accept that the University had sustained an effective funding cut of 8 per cent since 1997 and was critical of the University for not taking any action to remedy the situation if it had felt aggrieved. It also rejected the University's claim that a reduction in the overall cost of support services had been achieved in recent years.

Defence continued to press the University on reducing its workforce. It should and could take immediate steps to achieve the required 10 per cent savings efficiency by employing fewer staff. It was disappointed that the University's 'preferred approach seems to be more measured and incremental, emphasising progressive change in individual staff profile as distinct from initiating any downsizing action'. The solution, according to the Project Sandstone team, was for Defence to accept a lower standard of service and for the University to rationalise courses and further reduce staff. These goals would be achieved by merging the 12 academic schools into four, with a commensurate decrease in academic, support and administrative staff. A solution was to hand. The eventual 'efficiency saving' was 8 per cent. The most difficult hurdle was Defence's decision not to fund future UNSW superannuation and long service leave liabilities. These were negotiated separately after detailed parallel actuarial reviews. A one-time payment from Defence for both the superannuation and long service leave liabilities was made and the matter was finally settled. A shift was also made from the Non-Farm Deflator to the Consumer Price Index (despite its imperfections), making a big difference to annual adjustments and serving as a mid-way point to the Enterprise Bargaining Agreement becoming the benchmark for salary variations.

A casualty of the negotiations would be the Australian Defence Studies Centre (ADSC). The Centre was established in 1987 with Dr Hugh Smith as the founding Director. During his fifteen years at RMC (1971–86), Smith had convened a number of conferences and seminars focusing on Defence issues and security questions. Some of the early full-time Visiting Fellows at the Centre included Lieutenant General John Coates and Dr Andrew Ross (both military historians). Senior Lecturer Anthony Bergin became head in 1991 when Smith went on study leave. Bergin, who had joined the academic staff at the Naval College in 1981 and made the transition to the Academy in 1986, was promoted to Associate Professor in 1996. Within a decade, the Centre had become a very active organisation that had generated a reputation for creative and critical engagement with contemporary affairs expressed in annual conferences, quarterly seminars, an energetic publishing program and a program of visiting speakers including the distinguished historians Martin Van Creveld, Andrew Krepinevich and Steven Metz. Its profile had probably surpassed that of the Strategic and Defence Studies Centre (SDSC) at the ANU that had been in existence since 1966. Funds were generated from Centre activities and from sponsorships. A new building to accommodate the Centre was completed in February 1995 and opened by the Chief of the Defence Force at a cost of nearly $1 million.

There were, however, academic detractors on the Canberra campus who believed that the Centre's output was not being measured – as they believed it should have been – in terms of refereed

Commercial contracts and institutional partnerships, 1999–2004

articles in scholarly journals. These critics, mainly from the Department of History who believed their own discipline's potential to contribute was being overlooked, felt there were an excessive number of public policy position papers lacking analysis and depth. Other detractors felt the Centre did not adequately identify itself with UNSW or the Academy. For instance, they contended, its logo did not feature the University's badge. Zimmer was of a different view. He noted that the 'Australian Defence Studies Centre, an important "think tank" on Defence and security issues, is funded from within the block grant' to the University College.[19] In other words, it was a valuable asset that did not involve additional expense for Defence. A little later in his report Zimmer observed that:

> [with] its Defence-related academic expertise, the Defence Academy is a valuable provider of short courses and conferences with a Defence focus in Canberra. The Strategy Group stated that 'ADFA is able to host conferences, and espouse ideas and concepts that are not necessarily directly associated with Defence. Often times Defence is constrained in its ability to speak on issues, while ADFA is a useful conduit to explore matters in a manner that is academic rather than official'.[20]

When the College needed to find 'savings' to fund the 10 per cent operating reduction demanded by Defence, the Centre was considered for possible closure along with a range of other cost-cutting options. ADSC had little involvement in undergraduate teaching and was not a commitment that had featured in either the 1981 Agreement or that appeared in the revised agreement. A committee of representatives from the Arts schools was assembled to consider the Centre's future, although it had recently been the subject of an internal University review headed by the Dean of Law, Professor Leon Trakman, who recommended its continuation. Those from the School of History with the exception of Dr Frank Cain voted to close it; those from the School of Politics voted to retain it. The Rector was advised that the committee's recommendation was closure and the recommendation was accepted. The Centre closed in December 2003. The Centre's funds, approximately $400 000, were then available for redistribution.

There were mixed reactions to the decision within Defence on both the uniformed and civilian sides. The closure was not discussed with Defence before the final decision was made to determine whether sponsorship or support was possible. Outwardly ambivalent, given it was a UNSW entity and one that it did not own, some within Defence could not understand the decision as the Centre was an important venue for engagement with a range of security issues and policy challenges. Its closure left SDSC as the only university entity in the country focusing specifically on national defence and allowed the ANU to re-assert its involvement in contemporary Defence policy. SDSC was recovering from a period in the doldrums when its adversaries in the ANU, who believed its staff should have been located in the Service staff colleges and not the university, successfully managed to reduce its funding and marginalise its influence. As Professor Paul Dibb noted at the launch of its fiftieth anniversary history in 2016, SDSC had a split personality, caught in 'the void between policy-relevant work and "pure academic work"'.[21] The ADSC remit at the Academy was partly revived a decade later with the formation of the Australian Centre for the Study of Armed Conflict and Society (ACSACS), although ADSC's demise continues to arouse strong feelings among those who felt UNSW had turned its back on a vehicle for engaging directly with Defence issues.

Commercial contracts and institutional partnerships, 1999–2004

The final ripple from the Zimmer Report was a major reorganisation of the UNSW academic community. There had been some discussion among members of the Rector's Advisory Committee during early 1999 of reducing the number of academic schools to overcome duplication of support services. When it became apparent that Defence would seek some 'efficiencies' from the University as part of the new agreement in 2001, the Rector's initial approach was to review the funding arrangements for each school 'in a consistent way'. King believed that Chemistry, Physics and Geography were over-funded in terms of student load; other schools such as Computer Science and Economics and Management were under-funded. If transferring teaching load did not address the inequities adequately, 'other interventions would be required'. While there would be separate discussion about the number of schools, King was 'cognisant of the dissipation of energy when staff members become heavily involved in defending their discipline'. He hoped to avoid 'major staff disruption'. After several options were considered, other interventions were required. The number of schools was reduced from 12 to five on 1 July 2003. The School of Humanities and Social Sciences consisted of the old schools of History, Politics, English and Indonesian. The new School of Business subsumed the former school of Economics and Management. Engineering was divided into two new enlarged Schools: Electrical Engineering and Information Technology, and Aeronautical, Civil and Mechanical Engineering. The two engineering schools were later combined into one. The fifth new School was Physical, Environmental and Mathematical Sciences. It incorporated the former schools of Geography and Oceanography, Physics, Chemistry and Mathematics.

There were different views on the wisdom of amalgamating schools. Some of these views were discipline-specific. The chance to work more closely with colleagues in related disciplines was welcomed by some but was seen as irrelevant by others. The benefits in teaching were probably more apparent than in research. In terms of research, creating networks was more important than establishing organic structures. Irrespective of formal arrangements, academics would reach out to scholars of repute and they would establish partnerships on the basis of converging interests. Defence had no view on the efficacy of the school amalgamations. It wanted to see financial savings and if organisational efficiencies meant staff could be reduced without narrowing the range of courses on offer, it would not object.

The Rector attended the Chiefs of Staff Committee (COSC) meeting on 18 September 2002. He reported to his Advisory Committee that COSC had endorsed the principles of the new agreement and the University was 'now able to move forward to plan for the next ten years' although the final agreement was not signed until 12 December 2003. It had been a long and laborious process but both sides concurred the 2003 Agreement (which was actually a contract) was infinitely better than the 1981 Agreement (which was essentially an arrangement). There had been goodwill on both sides. But now the relationship had changed. It was now a clear fee-for-service arrangement and the possibility that the two parties might decide to separate had been made much more explicit. The University Council had also endorsed a significant name change.

Robert King had observed that 'the confusion between ADFA and the ADC [Australian Defence College] or Duntroon is prevalent even in Canberra'. He noted there were 'clear benefits in being recognised correctly. Correct attribution can help raise the profile of the University and the Academy, promote the concept of the University College as a centre of intellectual excellence, and allow [staff] to share and contribute to the prestige of the University'.[22]

He thought references to the 'University College' had meaning within UNSW but might have implied to outsiders that it was 'not quite a university' and when abbreviated to 'UC' led to confusion with the University of Canberra. He proposed the introduction of four new descriptions including 'UNSW at the Australian Defence Force Academy' and 'UNSW@ADFA'. By the end of 2001, the title 'University College' was abolished and no-one seemed to lament its demise. In time, the academic component of the Academy would become better known as 'UNSW Canberra' and signify that the University had actually outgrown the confines of the Academy. It was a remarkable turn-around of fortunes given fears that UNSW might have found itself without a place in Canberra just five years before. It was now time to consolidate the gains made since Zimmer's inquiry was foreshadowed.

13

Stability and sustainment, 2005–2017

With the 2003 Agreement in place and the reorganisation of the 12 schools completed, Robert King returned to Kensington in February 2004 as the Deputy Vice-Chancellor (Academic). He went back to a campus that was enduring protracted organisational turmoil. John Niland had resigned as Vice-Chancellor in 2002 after two five-year terms, having allegedly 'lost the support of the Council' over the previous 18 months.[1] He was succeeded by Professor Rory Hume who, like Niland, had seriously considered applying to enter Duntroon as a teenager.[2] Deciding instead on a career in dentistry, Hume joined the Army Reserve, serving through non-commissioned ranks until being commissioned as a Dental Officer and eventually serving as Commander, Dental Services for the Australian 2nd Division. Hume had more military experience than any of his predecessors and came to the University with personal goodwill and professional empathy towards Defence.

Vice-Chancellor Hume believed the relationship with Defence was 'not only a great service to society but something that strengthened UNSW itself'.[3] Hume thought a 'highly educated, socially aware officer corps [was] much more likely to develop when there is an appropriate balance between university-based tertiary study and military training'. His visits to the Canberra campus were usually for ceremonial events, with the routine business of the

University conducted in conjunction with the Vice-Chancellor's Advisory Committee (VCAC) or over the telephone. Like Niland, Hume did not feel any need to intervene in the College's affairs. The Rector attended VCAC on a weekly basis and briefed the Vice-Chancellor and the other members of the University's senior executive staff on happenings in Canberra.

The energetic Hume initiated a major planning exercise throughout the University before he suddenly resigned in April 2004 after only 21 months in the post. A public statement cited a 'breakdown in his relationship with members of the University Council'. A report in the *Sydney Morning Herald* mentioned no names but attributed the apparent breakdown to Hume's approach to commercialising research, collaborations with industry, his American-style management philosophy and, particularly, his handling of academic misconduct charges against Professor Bruce Hall in the School of Medicine.[4] (Hall's long-running defamation action against the ABC's *Science Show* would not be concluded until 2014.) Professor Mark Wainwright, one of the Deputy Vice-Chancellors, was appointed Acting Vice-Chancellor and was later confirmed in the position. He would serve until a new Vice-Chancellor was found. It could have been Defence's turn to be worried about its partner. Although the University had three Vice-Chancellors in the space of two years in what was the most turbulent period in its 50-year history, Defence was persuaded that the governance issues at Kensington had not, and were not, affecting the education being offered to the cadets.

Professor David Lovell became the Acting Rector on 1 March 2004 ahead of the new Rector, Professor John Baird, arriving on 20 September 2004. John Baird was born and raised in Brisbane and spent five years as a stockman in the Northern Territory and Cape York after matriculation. He moved to Canberra and enrolled at the

ANU, graduating with a BSc (Hons) and a PhD in Physics specialising in hypersonics. He was initially appointed Teaching Fellow in Mechanical Engineering at Duntroon and taught courses in fluid mechanics and thermodynamics. He transitioned to the Academy on its formation. While on sabbatical leave (in partnership with an American wind tunnel manufacturer and a German energy company), Baird was responsible for designing the world's largest free piston shock tunnel facility for the German research organisation DLR at Goettingen in the late 1980s. Promoted to Associate Professor in 1990, he was appointed Head of Mechanical Engineering the following year and later led the introduction of Aeronautical Engineering programs at the Academy. Baird left the Academy to become Professor and Dean of the Faculty of Engineering and Information Technology at the ANU in 2001. Although recruited from the ANU to serve as the fifth Academy Rector, Baird had spent the bulk of his academic career at Duntroon and then at the Academy. He was known by practically everyone at UNSW Canberra.

Baird's principal task was to 'embed' the new Agreement in the academic and administrative culture of the Academy. He stressed that the role of UNSW Canberra was to 'support the educational needs of the ADF' and outlined four priority activities: learning and teaching, academic research, personnel management, and business development.[5] Responding to demands from Defence and the wider community for expanded course offerings had been a looming challenge. The new millennium had marked a considerable shift in attitudes to higher education in Australia and within the ADF. In addition to the steadily increasing number of graduates and the expanding number of programs being developed by universities to cater for the emerging needs of a range of professional groups, the ADF had for some time wanted its courses recognised by Colleges of Technical and Further Education (TAFEs), Institutes of

Technology and universities. External recognition had become a key element in the ADF's recruitment and retention strategy. Recognition of prior learning was a challenge for the established universities, including UNSW, which was keen to ensure that their core business was not disrupted by the creation of new universities as part of a thorough structural overhaul of the higher education sector and the rapid enlargement of the TAFE sector under the Hawke Government (1983–1991).

The Army had first posed the possibility that UNSW degree programs might recognise professional military training in 1967. The University made its position clear: training was training and education was education. This was a rather arbitrary distinction given there were common objectives and shared outcomes. The real issue turned on the standard of military instruction and the expertise of military instructors. But the desire for vocational learning to be acknowledged did not subside. In 1993, UNSW was asked to consider the possibility of the Academy's 'military education' program, principally the psychology and leadership components, being recognised as general studies units within the degree structure. The University still opposed crediting military training completed by undergraduate cadets but was willing to explore delivering some courses, mainly within the domain of Behavioural Science, as part of the degree program if the military staff teaching these courses were qualified for appointment as UNSW adjunct faculty and the course content was at the requisite level. The ADF also approached the University to deliver certain modules at the Australian College of Defence and Strategic Studies (ACDSS) established in 1994 on the site of the former Joint Services Staff College in the Canberra suburb of Weston Creek. The University was keen to submit a 'tender' although the then Rector, Harry Heseltine, wondered whether the University College had the expertise required. As the modules crossed academic disciplines, Heseltine wanted the UNSW tender to be 'unified' and to include possible involvement from the language schools at the main campus, especially if the operations of the ADF Language School were opened to a commercial tender, such as a university, and relocated to Canberra. One in three of the University's ACDSS bids would be successful.

The University was also the preferred tenderer to develop and deliver 'Military Communication Training to Support AMET [Academy Military Education and Training]' in November 2001.[6] The support program included diagnostic literacy testing, a communication skills training package, and competency assessment. The other tenderers were the Canberra Institute of Technology, the Royal Melbourne Institute of Technology and Raytheon. The tender evaluation process concluded that the UNSW bid was the best in terms of detailed assessment and price, both of which were assessed as 'superior'. A key factor was UNSW having 'considerable experience in providing educational services to Defence and ADFA in particular'. The University had separately been involved in teaching at the Army's School of Signals at Watsonia and in delivering the Navy's Principal Warfare Officers' Course.

Throughout the late 1990s, the Army maintained its interest in broader national accreditation to ensure, in the first instance, that completing the RMC initial entry course assured admission into a degree program and, in the second instance, completion of the Command and Staff College course could lead to enrolment in a higher degree. The Directorate of Personnel Plans affirmed the Army's continuing commitment to generalist undergraduate education and flagged a shift towards more vocationally oriented study at the postgraduate level. There was an obvious need for the University to invest more strongly in distance education. Some progress had already been made. During his Rectorship, John Richards had tasked Dr Allan Arnold from the Department of Chemistry with exploring the technical aspects of distance education, conscious that Defence might source courses from another education provider if UNSW was slow or unwilling to expand and diversify its offerings. Unsurprisingly, given manpower pressures during the mid-1990s, there was declining support within the Services for enrolling students in Honours programs. The fourth year of undergraduate study did not seem to fit into the ADF's education continuum and its emphasis on the importance of cadets finishing their degrees in the minimum time and commencing effective uniformed service. Having students complete their Masters degrees at a mid-career point was seen to better fit evolving personnel plans in each of the Services.

Defence was finally thinking creatively about its postgraduate needs and the relationship between postgraduate study and the command and staff courses being offered to middle and senior ranked officers. The convergence of postgraduate programs and staff courses led the University College to review its structures in 1999 and to propose the creation of a 'Graduate School [...] to unify the present postgraduate course structure and raise enrolments in coursework and research based degrees'. If UNSW failed to widen access to its own programs and make study more flexible, the prospect of external university accreditation of staff college courses meant that Defence students would be much less likely to enrol in separate University College programs. They could effectively complete the requirements of the staff college and a university at the same time. On the other hand, UNSW needed to maintain the standard of its offerings (and entry requirements) both as a matter of principle and to attract students who wanted a quality degree.

The answer appeared to be expanding the number of Masters programs to cater for the specialist needs of particular professional groups. For instance, the University of Southern Queensland and the University of Sydney were offering Masters programs in project management and enrolling Defence members as students. A comparable program needed to be considered in Canberra. Given the 'changing Defence climate', the University College explored the possibility of a suite of courses from Certificate to Masters level that covered 'statistics, finance and administration, military technology and human resource management'.[7] The courses would be developed in consultation with the Australian Institute of Project Management, and the Directorate of Acquisition Training and Management within the Department of Defence. The University College would also consider offering professional doctorates based on coursework and research in response to Defence's desire for in-service training and technical expertise to be recognised at the highest level.

As the University tried to anticipate evolving Defence needs, the Department had embarked on a series of dramatic reforms of its own. In response to a recommendation from the 1997 Defence Efficiency Review, the Australian Defence College (ADC) opened on 18 January 1999. The College was headed by an officer of two-star rank and incorporated the Australian College for Defence and Strategic Studies (later renamed the Centre for Defence and Strategic Studies CDSS), the Australian Command and Staff College (ACSC) and the Defence Academy (ADFA), with each element headed by a civilian or uniformed officer of one-star rank equivalent. The ADC essentially merged, expanded and replaced the Joint Services Staff College and the single-Service staff colleges (i.e. it created a new staff college course for more senior ranks in the form of CDSS). It offered its first courses in 1999, drawing on study modules provided by a number of Australian universities, including UNSW. The delivery of these modules was awarded after an open tender process that led UNSW to seek a number of collaborative partners, including defence contractors and suppliers, over the ensuing years.

Although the contracts at Weston Creek were not large and the statement of academic requirement was unclear and liable to change at short notice, it appeared that Defence was acknowledging the need for a holistic approach to learning. An expansive vision for education was beginning to emerge. In his appearance at the Senate References Committee on Employment, Workplace Relations, Small Business and Education on 13 August 2001, one of the few Defence officials with expertise and continuing experience in higher education, Brendan Sargeant, explained that:

> Defence is turning increasingly to the universities for educational services [...] [it] has about 14 partnerships with universities that specifically address Defence requirements for higher education. The most substantial of these arrangements is with the University of New South Wales for the provision of undergraduate degrees for officer entrants to the ADF and postgraduate qualifications for the broader Defence community.[8]

He went on to mention, with the aid of a rather familiar phrase, the importance of 'offering a balanced and liberal education that encourages its students to think as well as to know'. After explaining that he saw education and training as being located on a continuum and not as distinct activities, he stressed the value of:

> access to people who may not be part of the [Defence] community and that we hear what people have to say. We need more voices, more ideas and we have the ability to deepen our understanding of where we are in the world and

where we might be. We are more likely to find those people in universities than anywhere else. One of the aims of our higher education system in Defence is really to open up that world to our people who will be the future decision-makers.[9]

He also outlined Defence's support for establishing a national quality assurance agency that would work with the university sector to mediate standards across the country. Defence was confident its own training standards met VET sector accreditation requirements. The more pressing issue for the Department was defining the vocational outcomes of tertiary education, and it was conducting some rudimentary curriculum mapping exercises at that time. In its written submission to the Committee, Defence mentioned inadequacies in the administration, organisation and resourcing of universities and claimed that 'universities have some way to go in becoming effective business partners'. In response to questioning, Sargeant pointed out to the Senators that:

> universities have quite a different culture from that of a large bureaucratic organisation like Defence. People who work in universities are driven by slightly different values, or in some cases radically different values, and as we try to build partnerships with those people, there is a fairly serious process of dialogue to find out how the different worlds that we have come from can meet.[10]

His remarks were not without a practical dimension.

The ADC Strategic Studies program was then being run by Melbourne University Private and had been since 1999. Melbourne University Private, a profit-making vehicle that sought to bypass government restrictions on the tertiary sector, was Professor Alan

Gilbert's most controversial initiative during his time as Vice-Chancellor of Melbourne University. A CDSS staff member had described the provider in an internal memorandum as 'incompetent, hopeless, bumbling, prone to mismanagement and lacked quality control of its administrative side'.[11] The contract was to end in December 2001 and new tenders would be called. When asked whether the criticism of Melbourne University Private was fair, the College's Commander, Rear Admiral Raydon Gates, avoided giving a direct answer. He said progress had been made rectifying the issues and that the service provided now met expectations. Students, especially international students, also resented receiving a degree marked 'Melbourne University Private Ltd'. It appeared to be a second-class testamur. At that time the University of Canberra had the contract as the educational provider for the Australian Command and Staff College. While Defence might have lamented the cost of working with UNSW, it had never expressed doubts about the quality of the product. The use of alternative educational providers at the ADC and its subordinate units demonstrated that UNSW not only understood Defence and provided competent staff and consistent programs but that other universities were struggling to meet their contractual obligations because they could not, in part, discern Defence's needs. They were also employing staff who were often unfamiliar with the nuances of the uniformed learning environment. UNSW, however, had its own challenges.

While UNSW had developed a range of postgraduate coursework programs across the 12 schools that existed at the time, the relationship between each of these programs was limited and their focus was on providing traditional university postgraduate education in the separate disciplines. The course offerings were quite diverse and there was no clear relationship between these courses and the professional development needs of more senior Defence officers, both military and civilian. The latter, the Rector believed, was important to developing a more comprehensive partnership with Defence. Accordingly, Robert King established a new Graduate Studies Institute in the second half of 2001. Its task was to develop a much more integrated set of postgraduate coursework offerings across the campus while providing the flexibility to tailor particular programs within that umbrella to support Defence's professional development needs. Professor Stewart Woodman, who had previously been responsible for postgraduate coursework at the ANU's Strategic and Defence Studies Centre, was the Inaugural Director. The Institute's aim was to 'bring together the University College's extensive graduate course work offerings into an integrated program well matched to the evolving postgraduate educational and professional development environments [...] particularly as they relate to Defence higher education needs'. In a discussion paper prepared for the Rector and senior staff, Woodman noted that:

> In the post-Zimmer era, postgraduate coursework has become a clearly defined element in the University's teaching responsibilities [...] There are signs that many areas of Defence have a significant interest in professional development, linked to possible higher degree candidature [...] the difficulty is that different areas of Defence are developing their programs independently. A clear picture of the overall Defence requirement is unlikely to emerge, at least for some time.[12]

In the absence of clarity, Woodman suggested that the University needed to 'shape the agenda to ensure that the programs are as coherent and sustainable as possible from a University perspective'. He flagged the need for greater flexibility in the structure of courses.

The multi-disciplinary content needed to be increased alongside a willingness to allow 'professional expectations to shape programs'.

Specific programs with an inter-disciplinary focus were developed in several areas, including the RAAF Engineers' Retention Program and professional development for the Defence Imagery and Geospatial Organisation, while the Institute also supported a range of executive short courses delivered by the various Schools. The Institute was closely involved in the significantly expanded number of online courses, promoting the quality and consistency of their delivery. To strengthen the University's ability to support international students under the Defence Cooperation Program, tailored English language courses were introduced across postgraduate offerings, while the Institute took responsibility for supporting those students at the Academy.

These changes marked a highly significant attitudinal shift since the 1970s. There was closer attention to revamped strategic and operational priorities within Defence and a greater readiness to have academic programs shaped by vocational needs. Over previous decades, academics had been strongly inclined to resist suggestions about possible course content and frequently resented interference by 'outsiders' in their discipline. The request that undergraduate Defence students be obliged to complete units in strategic studies or leadership principles initially met with solid opposition. There were academics who were committed to the pursuit of 'pure knowledge' and unconcerned about how it would be applied. Others believed the University set the requisite educational standards for each academic discipline at the undergraduate level and feared the professions would dilute the cogency and quality of degree programs with vocational demands. The decision in 1996 to work with Defence in promoting language studies, especially the introduction of Bahasa Indonesia, was a turning point. Because Defence was committed to a completely open tender process for programs offered at the ADC, and Defence people were 'shopping around' for the postgraduate degree that best met their vocational needs and professional aspirations, the University had to focus closely on changes within the Defence organisation and, indeed, across government more broadly. By the end of 2004, UNSW Canberra had six specialist contracts in addition to the Academy's Military Communication Program. These contracts were with each of the three Services, the Defence Science and Technology Organisation, the Defence Imagery and Geospatial Organisation, and the Australian Maritime Safety Authority.

The first Triennium Review of the 2003 Academy Agreement was conducted in late May 2005.[13] The Rector, John Baird, reported that the total number of undergraduate students in 2004 was 983, compared to 977 the previous year. The University had based its costing estimates on 1120 undergraduates. This situation revealed another long-running issue for the University. It was obliged to manage the consequences of indifferent performance by Defence recruiting, an organisation that had consistently refused all offers of help from the University. The number of Bachelor of Arts students in 2005 was down by 13 per cent from 2004, although the number of enrolments in the new Bachelor of Business program nearly doubled in the same period. The number of students enrolled in Science degrees was only 82 per cent of the expected figure (referred to as the 'Agreement Base Number') while the number of Engineering students fell to only 71 per cent of the expected figure. This was a decline of 10 per cent on the previous year and reflected a national pattern. Enrolments in the Bachelor of Technology degree nearly doubled expectation, offsetting the decline in Engineering enrolments.

In his submission to the review, the Rector reminded Defence that the University 'has control only over the quality of candidates not the quantity'. The University offered to develop a joint recruitment project with Defence to address the ADF's needs. The response was now more receptive. The most positive news was that the Student Progress Rate (SPR) was 95.3 per cent, well above the UNSW-wide rate of 87 per cent. The usual pattern for universities was a 10–15 per cent failure rate in first year, reducing to practically zero in fourth year. Enormous advances had plainly been made since the early 1990s when the overall 'wastage' rate was approaching and, in some instances, exceeding 50 per cent. Much of the previous wastage was due to first-year dropouts before the first academic term began – an issue successfully tackled by the military side.

For its part, Defence noted that 'ADFA is unique with a separate but coordinated education program [...] measurement must be confined to what can be practicably measured [...] no single metric will provide the means to effectively measure ADFA's performance'. One external measure was comparison 'against similar international academies in the area of attrition rates, and both academic and military achievement'; another was against the performance of officers recruited through other schemes, including those who were recruited with a degree. As part of the Triennium Review process, the University and Defence agreed that accurate and reliable internal measures would be applied in future. They would essentially be relative, with performance reported in terms of movements up or down on a scorecard.

The Agreement also allowed for the offer of 50 fully funded, full-time postgraduate research places. John Baird was adamant that Defence should make the most of these opportunities and hired an additional staff member to promote interest across the Services. The University noted that there were differing interpretations of who should be allocated the places and in what order of priority. It accepted that Defence needed to demonstrate value for money when assessing the research effort its members at the Academy were undertaking. The University also explained that its postgraduate research efforts were an essential element of academic professional fulfilment and a key to the effectiveness of all teaching in terms of its broader reputation as one of the nation's best universities. Of the approximately $3.6m (or approximately 10 per cent of the overall contract value in 2003 dollars) allocated to research in the Agreement, Defence wanted an identifiable amount of money set aside for clearly defined Defence-related research outcomes. In reply, the University stressed that the core of the Agreement was providing a 'balanced and liberal education' and this necessitated non-Defence research. As the majority of the $3.6m research allocation was made up of academic salaries, restricting this funding to Defence-related research would be a major cause of concern to UNSW.

As a compromise, the University accepted that Defence should be able to connect the resources allocated for research with a transparent value-for-money determination. The University agreed to distinguish Defence-related and non-Defence-related research currently being conducted at the Academy. The Rector had earlier explained that 'academic staff are required to carry out research as part of their contract of employment and their area of research cannot be directed by UNSW Canberra'. He pointed out that a great deal of research was conducted in Defence-related areas without any external pressure. Defence wanted its research investment to be reflected in a 100 per cent focus on Defence-related activities whereas the Rector proposed the figure be 60 per cent. Defence remained open to describing 'Defence-related' very broadly in order to satisfy the value-for-money demand.

The Triennium Review also addressed the continuing need for a 'transition out' plan in the event that either the University or Defence decided to terminate the Agreement. This was a major concern for the University given its commitment to staff and students, and its desire to make some long-term infrastructure investment in Canberra. The University was more worried about the contract being terminated than Defence was, although both parties would be affected if this occurred. The University would inevitably suffer reputational damage while Defence would face enormous upheaval in educating its personnel. The absence of an agreed plan reflected the inherent complexity and potentially enormous cost of either party walking away. Two separate scenarios were explored as part of the review. The first involved contracting a new educational service provider other than UNSW. The second involved replacing the Academy with a new structure that did not require a sole-source educational provider. Neither scenario was straightforward. The production of previous draft 'transition out' plans had never ventured beyond broad principles. Both parties realised it would be a very complicated divorce and five years (the length of the longest course undertaken by a newly enrolling undergraduate student) was probably an unrealistic time frame for the separation. John Baird explained that 'five years in the life of the university was not long and any transition to significantly revised or new arrangements would cause considerable disruption. If the provider changes every five years, there will be considerable disruption and little in the way of output information by which to assess the contract'. He pointed out that at the Royal Military College of Defence Science at Shrivenham in the United Kingdom, where Cranfield University was the sole educational provider, 'there is a 22 year contract, including a 10-year transition-out provision'. The University added that 'in the transition period, running two universities would result in significantly higher costs'. Not surprisingly, there was considerable resistance to including a five-year period as the extent of the transition-out period.

Although some issues remained unsettled, the first review of the 2003 Agreement was conducted with considerable 'co-operation and trust' and produced a number of recommendations that included raising priority 'for the longer term facilities requirements identified in the revised ADFA Master Plan'; rolling all UNSW's contracts into a single document; devising a realistic 'Transition Out' plan and generating some metrics for determining value for money. Defence noted that 'the fee levels for international and local fee paying students [elsewhere in Canberra] were comparable' although the University of Sydney and Canberra University were cheaper for international students.

As the Review was completed, UNSW was awarded a contract with the Australian Command and Staff College (ACSC) to provide a core Strategy module, two Strategy electives and four Management electives in the academic program from the beginning of 2006. ACSC wanted UNSW academic staff and not 'Visiting Fellows' to deliver the modules although the University could outsource the associated academic administration. The negotiations were co-operative and positive, with an attractive feature being the articulation of modules into postgraduate programs at UNSW. ACSC-delivered elements would not be given credit. Course members would also be enrolled as part-time students of UNSW and granted full access to the Library.

Fulfilling the ADC's requirements proved to be an arduous exercise. The first year was fraught with difficulty as the University did not do well at interpreting Defence's expectations, not helped by Defence being unsure of what it wanted and the inconsistency of course objectives. Professor Stewart Woodman brought stability and substance to the program. Fundamental to the ACSC program was achieving an effective balance between recognition of the course members' extensive professional development studies and the postgraduate courses delivered by UNSW staff. A rigorous framework to ensure high-quality content and consistent assessment was developed across the University courses. While the pressure to simply recognise the professional modules delivered by military staff at the Defence College was resisted, a review of the content and assessment of those programs led to them being recognised as an elective 'Joint Operations' component within a university degree plan tailored for mid-ranking military officers. Achieving that balance was no small achievement, given the wish of some senior military officers to have the University effectively 'rubber-stamp' Defence's professional programs. Subsequent years featured requests for curriculum changes and amendments to the contract with no additional funds for variations or compensation for the time taken to manage the changes.

When the ACSC contract came up for renewal at the end of 2011, the University was conscious that it needed to avoid being seen as 'double-dipping', accepting funding for postgraduate programs at the expense of its undergraduate funding from Defence. Its concern to fully cost the program left it open to underbidding by other potential providers who were more able to absorb supporting costs elsewhere in their programs. The Australian Defence College signed a ten-year agreement with the Strategic and Defence Studies Centre (SDSC) at the ANU to provide substantial elements of the course from 2012 at an annual cost of $1.78 million. The ANU had indeed undercut UNSW on price. There was a collective sigh of relief in some quarters but disappointment elsewhere in Canberra. While some UNSW academic staff were offended at the imputation that another institution could do better in meeting Defence's needs, others felt it allowed a clearer focus on Academy programs. Academics teaching popular and vocationally relevant courses at the Academy, such as military history and strategic studies, could not be in both places at once. By way of contrast, SDSC staff did not have the same competing undergraduate teaching demands. The contract had not been sufficiently generous to cross-subsidise any other activities at the Academy campus. In any event, the University was managing to find some stable sources of funding beyond Defence.

By 2006, the University's total revenue from all sources in Canberra was $50.6 million, of which $39.7 million was derived from the main Defence agreement. A strategic planning exercise conducted that year aimed to increase the number of more senior academics and their collective research output, enrolments and completions in higher degree research programs and the research funding each academic was able to attract. The Chief of the Defence Force Students (CDFS) Programs were established in 2005 to attract and challenge high achievers at the Academy. These programs were available across all disciplines for new-entry midshipmen and officer cadets with the prescribed ATAR and to second-year students who achieved a high Weighted Average Mark in their first year at the Academy. One group of Academy Engineering CDFS students competed in the 'Engineers Without Borders' design awards and were placed second in the finals. First place was awarded to another UNSW team. The various CDFS programs have reflected a productive alliance between the University and Defence in an attempt to draw academic high achievers into the ADF. It appeared from the experience of the programs' first five years that 4.5 per cent of the student population could complete the program at the requisite level over three years of degree studies.

While undergraduate numbers were static (in 2007 there were 321 new-entry students, down from 330 in 2006; 804 undergraduates, up 31 from the 2006 level with a total enrolment of 992 undergraduates and 1139 postgraduates), there were hopes of a 50 per cent increase in both postgraduate fee-paying students and higher degree research students within three to five years.

Enrolments had increased gradually over the previous three years, the School of Business growing largely at the expense of the School of Humanities and Social Sciences (HASS). But HASS had benefitted financially from the University's contract to deliver part of the educational program at ACSC. Enrolments in Engineering (in all sub-disciplines) and Information Technology remained static. Similarly, although the campus was a consistently higher achiever in learning and teaching in terms of UNSW internal rankings (Canberra was only second by 0.2 per cent to the Faculty of Business in learning and teaching performance indicators), its research performance was lagging behind.

The research areas where UNSW Canberra had become an acknowledged leader included: advanced materials, artificial intelligence and complex systems, astrophysics, autonomous vehicles, control theory, geographical information systems, hypersonics, leadership and management, military history, photonic quantum information technologies, strategic studies, systems engineering and project management. To maintain its leading edge, the University realised it was obliged to compete for staff with defence industries who usually offered better remuneration. But the prospect of conjoint appointments appeared to offer scope to improve the conditions of service that could be provided to newer appointees. The Defence and Security Applications Research Centre was established in 2006 to provide closer and stronger engagement with Defence. The next three years reflected a steady increase in both research funding and publications. UNSW Canberra received $2.8 million in research funding in 2007, increasing to $4.1 million in 2008. By 2010, the annual figure was in excess of $5 million.

The one area of the UNSW–Defence relationship that appeared stagnant was governance. The composition of the very large Academy Council had been the subject of intense interest in the 1980s.

In the five years after the Academy opened, the number and rank of uniformed attendees declined. The Service Chiefs evidently believed they had made their mark on the Academy and felt able to send their deputies. The University saw the Council as a means of influencing thinking within Defence and generating a store of personal goodwill towards its people and programs. The chairmanship of Sir Edward Woodward had nonetheless brought gravitas to the Council. Sir Edward had been a distinguished jurist. Rectors and commandants sought his advice and tried to enlist his support. His contribution to sound working relations across the diarchy was unseen but significant. It was only when Sir Edward decided to retire in 1999 (and was replaced by Justice Ian Callinan of the High Court) that the Council's indebtedness was apparent. Callinan had been the Defence Minister's nomination and, the University was led to believe, the only candidate the Minister considered suitable. The University had no-one else in mind and acquiesced. Despite initial reservations because he was the Minister's nominee, Callinan made many helpful contributions and was a strong supporter of academic excellence at the Academy.

The role of the Academy Council had, however, diminished greatly. The Academy was now an element of the Australian Defence College and the Academy Commandant reported to its Commander. The University had entered into a settled pattern of activity and very little close consultation with Kensington was required. In the absence of any overt disagreements between Defence and the University, the Academy Council was, in one sense, not needed or wanted. The 2003 Agreement provided for a new Consultative Council with a much reduced mandate. Clearly intended to be an advisory group like the Academy Council it replaced, it did not make decisions, confirm appointments or approve procedures. But even as a forum, it was ineffective. Contentious matters tended to be debated elsewhere. It did not meet at all in 2008 and was replaced in 2009 by the ADFA Working Group, which would meet twice each year. Plainly both sides were content with their own governance arrangements while managing the diarchy presented few new issues or compelling problems.

Noting the lengthy negotiations that preceded the 2003 Agreement and mindful that its term was ten years with a review to be conducted every five years, in 2006 both the University and Defence began to ponder the possibility of another new agreement. In his final report as Commandant (February 2006), Commodore James Goldrick noted that the 2003 Agreement had put 'the academic–military relationship on a sound footing […] [and given a] much clearer understanding within UNSW that the Academy's central function is to provide the future leaders of the ADF'. He was adamant that 'unless we maintain a relationship with a Go8 university, such as UNSW, we cannot be sure of a sufficient level of educational quality in ADFA's degree programs to justify the operation'.[14]

The new UNSW Vice-Chancellor, Professor Fred Hilmer, did not take long to recognise the need for further work on the 2003 Agreement. Hilmer arrived at UNSW in early 2006, having previously been the Dean of UNSW's Australian Graduate School of Management (1989–98) and CEO of John Fairfax Holdings (1998–2005). His personal views would have clearly favoured placing any new agreement on a sound financial and organisational basis. Unlike Rory Hume, Hilmer did not have any military experience.[15] Prior to being appointed Vice-Chancellor, he had no contact with Defence. Based on previous conversations with successive Rectors at VCAC, Hilmer believed that the Academy was an asset to the University and never considered the possibility that UNSW might walk away from the Agreement. Coming from a business background and applying very specific resource-based metrics, he believed that the Canberra campus was operating well but felt teaching was too resource-intensive and that its research activity had declined (along with that of all of UNSW). Canberra was then in the bottom half of the University performance table and not the top half. He wanted greater focus and more intensity. Hilmer also felt the University was required to cross-subsidise too much activity in Canberra and that Defence ought to be covering more of the University's overheads. UNSW Canberra could, in his view, become a world leader in a number of fields and attract more PhD students.

Hilmer was concerned about the 2003 Agreement on multiple fronts, most specifically the Defence 'opt out' provisions. He wanted an agreement that was more stable and which allowed for longer-term investment on the University's part. While Defence was drawn to subjecting the Academy Statement of Requirement to market forces, he understood how disruptive tendering would be for both staff and students at the Academy. Indeed, the very existence of a sunset clause in the Agreement was having a deleterious effect on its operations. It had generated its own turbulence by requiring major reviews that may not have been needed if both parties were content with the arrangements and the ADF was satisfied with the University's performance. His own preference was a 15-year agreement. This would take some of the urgency out of strategic planning and allow for a longer-term approach. He also wanted the agreement to have an 'evergreening provision': if there had been no problems with the University's performance over the previous year, the term of the agreement would start again. The University would essentially earn running contract extensions.

There appeared to be an appetite within Defence as well to amend the Agreement further. The Chief of the Defence Force, General David Hurley, formed a warm personal relationship with the Vice-Chancellor and supported reviewing the Agreement. Hilmer could see that Defence wanted to sharpen the connection between research and capability development. Hilmer believed that the first Defence Minister in the Rudd Labor Government, Joel Fitzgibbon, understood the need for the proposed arrangements but that his Prime Minister, whose affiliations and affections were solidly with the ANU, did not. Hilmer's first task was, then, to persuade the Commonwealth to grasp the importance of scheduled reviews of the Agreement happening on time to allow the University some certainty about its future contribution. If the Commonwealth wanted to engage another educational provider or seek tenders, the University needed to know of its plans and to know early. An attitude of indifference to scheduling reviews was not helpful, even if Defence gave every hint that it was satisfied with the service provided by UNSW and intended to extend the agreement.

Negotiations on yet another revised agreement began in October 2007 in the hope that a new document could be signed in April 2008. The University wanted the period without the need for formal renewal extended from five years to ten years; an improved Academic Statement of Requirements; an adjustment of the Commonwealth's liability for UNSW staff redundancies; inclusion of a revised indexation factor (that took into account a number of indices including average weekly earnings and the Consumer Price Index); and renegotiation of the five-year transition to another educational provider 'to something that is practical and implementable'. A revised Agreement was eventually signed in December 2009, extending the relationship with UNSW to 2023.[16] An Annual Agreement Review Working Group was established to provide quality assurance and to have oversight of its steady evolution. Rear Admiral Raydon Gates chaired the first Annual Agreement Review on 11 October 2010. UNSW was judged to have provided 'all agreed services at an appropriate standard or level'.[17] Subsequent annual reviews have produced similar outcomes and precluded any movement within Defence to consider another educational provider.

The positive outcome from the Agreement Review was not unexpected. UNSW was assessed as the nation's top university for teaching and learning, for both performance and improvement according to the Commonwealth's Learning and Teaching Performance Fund, which paid $6.9 million to UNSW in 2008 – the highest payment to any university in the country. The UNSW Canberra Learning and Teaching Group was established in 2010 to bring together all parties involved in enhancing the student experience and optimising the benefits of new and emerging technology. The Learning and Teaching Course Evaluation Questionnaire circulated in 2011 found that 87.95 per cent of UNSW Canberra graduates were satisfied overall with their degree program compared with the national benchmark of 83.56 per cent and the Go8 benchmark of 83.71 per cent. In terms of teaching, 74.04 per cent of UNSW Canberra students believed they had experienced good teaching, compared with the national benchmark of 68.77 per cent and the Go8 figure of 66.57 per cent. On the whole, UNSW Canberra students were less satisfied than Kensington students in first year but more satisfied in third year. This shift in satisfaction might have reflected the general anxieties of being a first-year cadet rather than a new undergraduate. There was also a rapid expansion in the number of professional short courses offered by the University, principally to uniformed people, civilian staff in the Department of Defence and the growing Defence industry sector. In 2008, there were 116 separate courses attracting 1040 participants, with 79 per cent coming from Defence. These courses produced $1.3 million in income for the University and plainly met a pressing need. By 2012, the number of courses had been rationalised but their value exceeded $2 million.

In 2010, John Baird decided to retire. His health had been

variable over the preceding few years and he had fulfilled many of the objectives set for his rectorship, including a revised agreement with Defence. He was made a Member of the Order of Australia (AM) on Australia Day 2012 for 'service to higher education, particularly through the Australian Defence Force Academy, and to the discipline of engineering as an academic and researcher'. Baird was replaced by another experienced UNSW Canberra insider and engineer, Professor Michael Frater, the Head of the School of Engineering and Information Technology (SEIT). The new Rector had completed his electrical engineering and science degree at the University of Sydney, and later obtained a Masters in Higher Education from UNSW and a PhD in systems engineering from the ANU. He was appointed to UNSW Canberra in 1996 and rose to become Presiding Member of the University College Academic Board in 2006 and Head of School in 2007. He maintained research interests in video communications and underwater networks. Unlike his predecessors, Frater had served in the ADF. He was an officer in the Army Reserve (Royal Australian Corps of Signals) from 1996 until 2005, serving as a troop commander, squadron operations officer and second in command of a signals squadron, gaining expertise in military communications systems and electronic warfare. Michael Frater is the youngest academic appointed to the post, at age 45.

One of the first projects to involve the new Rector was developing and gaining approval for the statement of ADFA–UNSW Canberra common attributes. Based on relevant UNSW and Department of Defence documents, this statement identified the capabilities and attributes required of both UNSW graduates and junior officers in the ADF, and broadly set out how they would be developed in the UNSW academic programs and the Academy Military Education and Training (AMET) program. Frater's foremost objective for UNSW Canberra was developing 'globally focused graduates who are rigorous scholars, capable of leadership and professional practice in an international community'. He intended to make the most of UNSW's reputation as only one of five Australian universities to 'make the top 50 in the QS World Universities Ranking in 2011'. UNSW was also the 'first Australian university to be awarded five stars in the new QS Stars rating system which measures performance against international benchmarks'. Things appeared to be going well until cadet misconduct again drew unwanted public attention to the Academy.

What became known in early 2011 as the 'Skype scandal' attracted media interest across Australia and overseas. It undid much of the effective reputational work that had followed the Grey Report more than 15 years earlier. Indeed, a number of media commentators connected every cadet scandal since 1969 as the basis for asserting that nothing had really changed in the formation of junior officers over the preceding 40 years. It was a tawdry episode. In March 2011, a male Army cadet had consensual sex with a female Air Force cadet and secretly streamed the encounter via Skype to a computer in another room where a group of cadets was watching. The female cadet later became aware she had been filmed without her consent and reported the matter to Defence investigators. Civil charges were subsequently laid. The Minister for Defence, Stephen Smith, stood down the Academy Commandant, Commodore Bruce Kafer, amid claims he had mismanaged the situation. (Commodore Kafer was reinstated nearly 12 months later and, in a letter obtained by the *Australian*, complained that the Minister had denied him natural justice and tarnished his reputation without cause. Commodore Kafer was later accorded the rare honour of being made a Fellow of the University and was promoted to Rear Admiral.[18])Rear Admiral James Goldrick returned to the Academy as Acting Commandant. In October 2011, the male cadet who participated in the sex act and the cadet to whose computer the act was conveyed were both found guilty of a series of offences in the ACT Supreme Court. They were discharged from the ADF. The five cadets who viewed the incident were also expelled from the Academy. A number of separate inquiries were initiated in the wake of the scandal, including a review into how women were treated at the Academy. Elizabeth Broderick, the Commonwealth Sex Discrimination Commissioner, conducted the review. Defence accepted 25 of the 31 recommendations contained in her report.

In the Academy's 'Annual Status Report' for 2011, the reinstated Commandant thought that 'despite the upheaval caused by the Skype incident and its aftermath, ADFA's military and academic staffs continued to deliver high quality education and training programs'. The number of cadets counselled for unsatisfactory performance was actually lower than previously: 13 per cent in 2011 compared with 17 per cent in 2009. The resignation–discharge rate was almost unchanged from the year before. Therefore, Kafer claimed: 'these figures indicate that the Skype incident had no significant effect on the attrition rate, whilst the negative publicity in the aftermath possibly spurred the cadet body to higher achievements in their education and training'.

The University had not been mentioned in media reporting of the scandal and had made no official comment. Because it was a disciplinary matter for Defence, the University was without a formal role in the subsequent inquiries. On 3 November 2011, Dr Kathryn Spurling, an Honorary Visiting Fellow at UNSW Canberra, participated in a segment broadcast by ABC TV's *Lateline* program which focused on the continuing need for cultural reform at the Academy.[19] Spurling told interviewer Steve Cannane: 'I've had so many young women coming to me with the most terrible abuse cases [...] it wasn't improving, so I'm really delighted that [the Broderick] inquiry has taken place this year, but it's been at a cost to a lot of young women whom we've lost over the last ten years'. Cannane then stated: 'But Kathryn Spurling won't be around to see if the treatment of women changes. After ten years of teaching at UNSW at ADFA, she recently had her contract terminated. She says it's because she's been outspoken about the abuse of women'. The report ended with Spurling commenting that her experience 'says something about anyone who speaks out'. The program's transcript ended with a clarification: '*Lateline* incorrectly stated that Dr Kathryn Spurling had a teaching role at UNSW at ADFA. Her role as Visiting Fellow was to give guest lectures and publish articles on women in the military'. Another ABC News website that featured an interview with Spurling carried an 'Editor's Note': 'UNSW denies Dr Spurling's Visiting Fellowship was terminated due to her public comments. It says her contract was due to expire at that time'.[20]

While UNSW might have been criticised for its apparent 'silence' and censured for a lack of moral courage, it was not in a position to contribute helpfully. Although there were some unwritten dimensions in the division of responsibility between the Rector and the Commandant, the 'Skype scandal' was clearly a matter of cadet discipline. Defence had assumed legal responsibility for the incident and did not want either a second opinion or rival public commentary. The University deplored the conduct of these cadets. That was a given. There was a risk, however, that ill-timed remarks from academics containing more heat than light might have excluded the University from any reform process if UNSW appeared to be either too close or too removed from Defence. With hindsight, allowing UNSW to contribute to a joint statement with Defence might have answered any criticism that the University was indifferent to what had occurred or that it did not fully support further cultural reform.

A statement of this kind would have assured the parents of children considering entry to the Academy that the University was committed to providing a safe study environment.

Such a statement would also have embodied the University's concern about the scandal's effect on academic performance. This was difficult to determine given the existence of so many variables influencing results. For instance, did the 'Skype scandal' deter higher-performing students, particularly women, from joining the Academy? If so, would the absence of these women be apparent in the overall academic performance of the cadet cohort that joined the Academy in 2012? It would be easy and convenient to conclude that the consequences of the Skype incident had no bearing beyond those who were immediately involved and personally affected. A study of first-year Physics enrolments covering the period before and after the 'Skype scandal' revealed that the number of women enrolling in the subject halved in 2012 and there was an appreciable drop in the results of the women who did enrol (in comparison to their male counterparts).[21] In the absence of other variables, it is possible that the scandal may have had a bearing on the academic calibre of the 2012 entry. Put simply: the brightest women decided on another career. The full impact of the whole incident is unknown as the University has yet to undertake a comprehensive review of academic results to draw its own conclusions.

The conduct of the cadets involved in the 'Skype scandal' pointed to a broader problem: that sexual abuse and harassment seem to be endemic in Australian universities. Indeed, the problem had become so acute in higher education institutions that the Australian Human Rights Commission launched a project in December 2016 to examine the incidence and handling of sexual assault and harassment within the nation's universities.[22] Of course, the Government and the public are entitled to expect higher standards at the Academy on several grounds. First, cadets are partly selected on character grounds, whereas most other universities offer places based only on academic results. Second, cadets are not only under discipline at the Academy, they are closely supervised by serving officers with strong powers of punishment. Again, this element is absent on other campuses. Third, abuse among cadets can create career-long conflicts and antagonisms, as happened with bastardisation at Duntroon. Most other university students enter widely dispersed occupations and locations after graduation. For Defence's leaders, the 'Skype scandal' affirmed the continuing need for vigilance. Reputational gains and cultural advances made with one student cohort at the Academy could be lost within the next.

The steadily changing student community was also influencing the character of the Academy. After 1986 there were progressively more women, more postgraduates and more civilians on campus. Each group in its own way has slowly and imperceptibly shaped the campus mood. The biggest difference between the Kensington and Canberra campuses, other than size, was the complete absence of civilian undergraduates. In response to external advocacy, their exclusion was reconsidered in late 2015. The idea was not new. In the 1970s, the University had suggested that a small number of civilians studying at the new Academy would diversify the learning environment and possibly enrich the cadet experience. Defence rejected the idea on the basis that the presence of civilians would weaken the military ethos of the Academy and adversely affect cadet discipline. However, five civilian students had been allowed to study 'miscellaneous subjects' at RMC as part of their UNSW degree programs – principally engineering – because there was no other engineering faculty in Canberra. They were admitted 'inadvertently' and in 'good faith' but they would be allowed to continue 'as a service to the continuing education of professional engineers in the ACT'.[23]

The possible admission of civilian undergraduate students was raised again in March 1984, this time by the Chairman of the Canberra Division of the Institution of Engineers, Mr SD Hardy. He pressed the Minister for Defence, Gordon Scholes, for his help in addressing the national shortage of engineers.[24] Scholes replied that Defence's facilities were fully stretched and that admitting civilians would work against the 'cohesive nature of the total cadet body'. Later the following year, the Academy Commandant, Rear Admiral Peter Sinclair, noted that a 'reduction in Air Force requirements and the transfer of aeronautical engineering students to RMIT for the final two years of their course leaves us with some spare capacity'. He explained to Defence Headquarters that the admission of a small number of civilians 'could have some advantages for the Academy, the Department of Defence, and the Canberra community in general'.[25] A decision was deferred.

A few months later the Canberra College of Advanced Education (CCAE) announced it would offer the first two years of an engineering degree and was looking for a partner institution.[26] The Minister for Education, Senator Susan Ryan, wrote to her colleague the Defence Minister, Kim Beazley, in July 1986 pushing for the admission of successful CCAE students into UNSW programs and proposing an inquiry into engineering education in the Australian Capital Territory.[27] The first civilian students would seek enrolment in early 1988. The reaction of Sinclair and Wilson was initially negative, although they thought that 'goodwill would be generated by accepting them [...] and the ground rules could be settled'. The Academy could train a pool of civilian engineers from which it might recruit without any public expense for their education. But the initiative changed direction in 1989 with the formation of a loose alliance between ANU, CCAE (soon to become the University of Canberra), the CIT and the University College. It was called the ACT School of Engineering. Although the School existed only briefly, it did, with the UNSW delegate in the chair, broker a co-operative model for civilian undergraduate engineering courses in the ACT delivered by the University of Canberra and the ANU.[28] Nearly thirty years later, an announcement was made that UNSW Canberra would accept its first civilian undergraduates – up to 40 engineering students – in 2016. The initiative was designed to boost 'study options' in Canberra and enhance the learning environment for Defence students 'by bringing a new group of high-performing students into the classroom. Industry is also highly supportive of the announcement'.[29]

The inclusion of civilian undergraduates signified the highly fluid nature of the relationship between UNSW and Defence over the past five decades. The initial arrangement to educate undergraduates at Jervis Bay and Duntroon had morphed into an agreement to provide undergraduate education to all three Services in addition to a suite of postgraduate courses and research programs that met the continuing education needs of uniformed personnel of all ranks as well as Defence civilians. In the 50th year of enrolling students in Canberra, there were 978 undergraduate enrolments: 189 enrolled in Arts degrees, 237 in Business and Information Technology degrees, 148 in Science degrees, 48 in Technology degrees, and 356 in Engineering degrees; there were 330 research degree students including 277 doctoral students; and 1493 postgraduate enrolments, of which the largest courses were the Masters of Business and the Masters of Project Management, making a total of 2861 students, with another 60 students taking courses that were not leading to degrees.

The UNSW Canberra School of Engineering and Information Technology (SEIT) exemplified the evolving co-operation between the University and Defence. In addition to hosting the Capability

Systems Centre and the Australian Centre for Cyber Security, the School hosts a biannual External Advisory Forum to obtain feedback from military and civilian representatives. It also supports the 'Defence Civilian Undergraduate Sponsorship' scheme sponsored by the Defence Capability and Sustainment Group and is developing new Honours courses and dual degree combinations of Engineering and Science to meet emerging needs and future aspirations.

Plainly, the University had become much more than a mere provider of educational services to Defence. It had helped the Services to reposition academic study within their own learning continuums, included vocation-specific courses in the undergraduate program, introduced Business degrees, and expanded the range of Masters coursework degrees and made them available by distance.[30] Suggested innovations were seldom rejected out of hand. The University was also playing an active role in shaping Defence's approach to research and development and influencing the workforce it was assembling to conduct its core business. That the relationship between UNSW and Defence would span 50 years was not a prospect anyone had entertained when the initial agreements were signed with the Navy and the Army in 1967.

14

Observations and conclusions

Fifty years ago the attitude of most Australian universities to national security and the armed forces was driven by ideological commitments that had more to do with political advocacy than academic integrity. Disagreement with Western involvement in the Vietnam War and opposition to Australian support for the Saigon Government appeared to preclude a considered appreciation of why uniformed officers needed a tertiary education, and prevented many academics from discerning how they could influence their nation's diplomacy. The Vice-Chancellor of UNSW, Professor Sir Philip Baxter, and his senior leadership team saw an opportunity in 1967 to shape national life through educating the armed forces at a moment of escalating regional instability and deepening domestic upheaval.

Baxter could see no reason why the University should not offer its help to the Navy and the Army. After all, the Air Force had entered into a relationship with Melbourne University in 1951 – another time of global conflict and social unrest, with the Korean War and the Communist Party Dissolution referendum. This relationship had established a notable precedent. While the idea of establishing a tri-Service academy had been openly canvassed in 1959 as the best long-term option for viable officer education, the Navy and the Army desperately needed an interim solution. When Baxter offered the assistance of UNSW in 1966, he no doubt expected some reciprocity of goodwill from the Commonwealth Government, although he never made this prospect a condition of helping. Baxter was simply committed to advancing what he believed was the common good. As a scientific researcher and technological innovator, he was intellectually fearless and practically tireless. As an educational administrator he assumed an authority that none of his successors would dare claim for themselves. Following Baxter's decision to assist the Services – a decision that involved no soul-searching, protracted consideration, broad consultation or accurate appraisal of what would be involved – the relationship between UNSW and Defence has been close but complicated.

What began as a selfless offer of assistance that was never intended to extend beyond a decade has continued and broadened into a major institutional collaboration. For 50 years UNSW has been the foremost provider of tertiary education to Australia's uniformed men and women. Has the UNSW–Defence relationship been a success? In a word, yes. While success can be measured in many ways, that the relationship has endured for half a century reflects a strong belief within both UNSW and Defence that the outcomes have exceeded expectation. The academic results of Defence students have steadily increased alongside retention rates. It took several decades to achieve these outcomes at the Academy but the much-improved results and higher retention reversed what had become a source of deep concern at the single-Service colleges. Certainly, for the Navy and the Air Force, the affiliation models were unsustainable. They were costly and the 'wastage' was simply too great. Beyond the education of new-entry officers, there has also been a very rapid expansion in the number of postgraduate students. These students have included serving uniformed personnel and civilians examining a range of defence and security questions. The latter have provided the critical mass necessary for a robust scholarly community whose conversations have greatly benefitted the ADF. The Defence relationship has significantly expanded the UNSW student base, given the University an opportunity to influence thinking within a substantial Commonwealth entity, created an enlarged platform for technological research and scholarly inquiry, and afforded the University a presence in the national capital. The Academy has earned itself a reputation for academic excellence in many fields and is seen by Defence as a rich resource that enhances capability.

In sum, the University provides a high-quality academic education within a military environment. Could this objective be achieved by another means? Possibly. But the Government has decided that the Academy model best meets the present and future needs of the ADF. It has affirmed its preference for this model consistently over the past three decades and expressed confidence in UNSW as the optimal partner. Indeed, the UNSW connection is considered a central pillar of the Academy model and its foremost strength. The possibility of obtaining a UNSW degree from the Academy is attractive to both prospective undergraduates and postgraduates. Although overlooked by some Academy critics, the quality of the UNSW Canberra degree program plays an important role in new-entry recruiting and mid-career development.

Each year more than 650 Service officers graduate with UNSW degrees in addition to an average of 100 Defence civilians. These qualifications have fitted both uniformed and civilian personnel for effective service and provided men and women with the basis for fulfilling and productive careers. UNSW graduates are considered well educated and properly prepared to engage with the issues and challenges that confront not only the Defence community but the Australian nation as well. They have contributed to policy and practice across many realms and have attributed their achievements in part to the education they received from the University. The Academy's graduates have embraced the importance of taking a tri-Service perspective on defence questions because they have been persuaded that each Service makes a complementary contribution to Australian security. They have also become more aware of the ability of civilians to interpret problems and propose solutions, and their capacity to support ADF operations in multiple domains. Despite the fears and objections of the doubters and detractors, the relationship has neither damaged the University's scholarly integrity nor harmed the ADF's disciplined culture.

Defence has, of course, wanted more from the University than degrees for its young officers. It has sought teachers and researchers who would personally embody the possibilities of a balanced and liberal education to assist in promoting a cultural shift that esteems higher education and the value of academic learning. Defence wanted the benefits of being part of a community – the academic community – and to be party to scholarly conversations in order to reshape the corporate mood and defeat anti-intellectualism. In contrast to 50 years ago, when fewer than 3 per cent of uniformed officers possessed tertiary qualifications, the proportion will soon exceed 50 per cent, with 25 per cent possessing a higher degree. The complexion of the officer community has, therefore, changed dramatically. This change can be attributed to UNSW academics and administrators and the strong and lasting relationships they have formed with uniformed staff and students. In contrast to the number of Service personnel who have served at the Academy as trainers, the average stay of a tenured academic is, on average, five times longer. The educators rather than the trainers essentially carry the story of the Academy and largely preserve its corporate memory. Most UNSW faculty have been personally and professionally devoted to their academic discipline and to the cadets. For them, working within the University at the Academy is more than a job. Teaching and research embody their sense of vocation and express their identity. The longevity and devotion of the University's staff has helped Defence access experience and expertise that might otherwise have remained untapped and unknown. Despite occasional jibes about Left-leaning political sentiment and preoccupation with obscure subjects, UNSW academics are well regarded and widely respected for their commitment to the learning needs of their Defence students. The success of the UNSW–Defence relationship owes much to the UNSW staff.

The ability of the RMC Faculty of Military Studies and later the Defence Academy to attract quality staff is reflected in the subsequent appointment of four Canberra faculty as university Vice-Chancellors: Geoff Wilson to Central Queensland and then Deakin, Alan Gilbert to Tasmania, Melbourne and then Manchester, Bruce Thom to New England and Ian Young to Swinburne and then the ANU. This is a remarkable achievement for a relatively small academic community, suggesting that the experience of working with Defence has enhanced the skills and enlarged the capacities of those who gained senior leadership experience in Canberra. The influence of UNSW academics has been reflected in a number of other measures, including the award of seven Rhodes Scholarships to ADF students since 1967. Notably, six have been Army officers, with the seventh from the Air Force; two in the period of the Faculty of Military Studies (1977 and 1985) and five in the University College period (2001, 2002, 2003, 2009 and 2013).

The relationship between the two institutions has naturally meant different things to the University and Defence. UNSW exists to serve the educational needs of the nation, whereas Defence's primary role is undergirding national security. For UNSW, working with Defence has not been crucial to the University's overall

Observations and conclusions

evolution but it has been the most durable element in its community outreach and industry engagement. The relationship has advanced some general areas of University expertise, added weight to some programs and provided valuable organisational synergies. The large number of UNSW alumni extends the University's influence at home and abroad, and highlights its ability to be an institutional partner that is flexible and able to compromise. UNSW has managed its responsibilities without major conflict or controversy. The University has alerted Defence to looming threats and offered solutions to persistent problems. But what of criticisms?

At different times since 1967, Defence officials have complained that the University has engaged in over-charging, that it has been indifferent to reasonable requests and elitist in its attitude to modifying the curriculum, that it has failed to respect Service personnel policies in relation to cadet conduct and has routinely taken Defence's goodwill for granted. Some of these complaints are justified. Conversely, Defence has been able to 'set and forget' many educational policy settings, knowing that the University has been attentive to changes within Defence and has generated its own expertise and experience in working closely with ADF personnel. The University has given Defence cost-effective research and development resources and been an advocate for Defence needs in a range of academic and technical forums.

Notably, neither side has sought an elusive and unachievable sense of unity. The University and Defence have enjoyed a good rapport over five decades because the essential differences in the mindset of the two parties have been recognised and accepted even when they are trying to achieve rather different things with the undergraduates/cadets they seek to shape. Attempts to consolidate the two institutions and their programs are unwelcome to both because they would involve unacceptable compromises with core values. This is one of the few cases where 'creative tension' has actually worked. The Academy experience in the United States, where the academic and military 'missions' are identical, has been rejected in Australia and for good reason. The Academy 'model' is also considered the most effective by Australia's allies, who envy the nuanced relationship that exists between a leading university and the uniformed community. It is, some contend, world's best practice. Not that everyone agrees. The former Treasury head and Nationals Senator, John Stone, expressed astonishment at:

> [some of] the views being publicly advanced by some members of the ADFA teaching staff. I am the last person to wish to suppress ideas, but I have found it hard to see how these staff members could, while holding these views, continue in good conscience purport to educate young men and women setting out to serve their country in the honourable profession of arms. To the extent that these staff members were or are provided by UNSW, I would suggest that the University has been in breach of its contract – if not in a legal sense, then certainly in a moral one.[1]

He went on to ask why the ADF needs to 'enter into a contract of this kind with any university. Does West Point or the US Naval Academy at Annapolis do so? I think not'. He suggested it would be better for ADFA to 'employ academics (I am thinking here of historians but the point is more general in application) of its own choosing by offering teaching contracts and itself selecting among the applicants'. One could reasonably ask on what basis these applicants would be selected and whether Defence has the competence to make the best academic appointments.

There is a real danger in either the University or Defence

recruiting scholars who possess views that the military 'approves' or 'endorses' rather than scholars who are thought to be the leaders in their respective fields of research. The potential for ideological and pedagogical bias in terms of what and how material is taught is very real. Hugh Smith notes that:

> the successful operation of the Faculty of Military Studies (FMS) at Duntroon showed how genuine university status could be secured for the education of future officers and how difficulties could be overcome. Transferring this experience to a tri-Service institution was a relatively simple matter from an academic point of view. The success of the FMS 'experiment' [at RMC Duntroon] also demonstrated what might have to be given up if an autonomous body was created, even if it conferred its own degrees and even if it was called a university.[2]

Whatever the views of academics, the overall aim is to produce officers capable of free thought and independent action. In this respect, the Academy model has been effective. ADF officers are known for their creativity and ingenuity as well as for being soundly educated in the liberal tradition. UNSW Canberra has encouraged a strong commitment to the 'contest of ideas', evidenced by the vigour with which defence policy and national strategy is debated in the various professional journals to which uniformed officers contribute. Former ADF officers and UNSW graduates frequently engage in public debate and have shown their intellectual skills are equal to the task of analysing a range of complex problems. In the spirit of promoting intellectual inquiry at the Academy, the University has sponsored speakers and supported symposia reflecting the full spectrum of political belief and philosophical conviction in contemporary Australia. By way of illustration, the former Labor Defence Minister, Kim Beazley, delivered the first University College annual lecture in 1992. He was followed by a number of national figures including the journalist Paul Kelly, former Liberal Prime Minister Malcolm Fraser, former National Party leader and Deputy Prime Minister Tim Fischer, former Communist Party member and ABC broadcaster Phillip Adams, one-time Greens senate candidate Professor Clive Hamilton, former Australian Democrats leaders Janine Haines and Natasha Stott Despoja, former Federal Sex Discrimination Commissioner (and future Governor General) Quentin Bryce and well-known social commentator Dr Hugh Mackay. The notion that the Academy staff are predominantly left-wing and the students are predominantly right-wing would be rejected by anyone with a first-hand experience of the lecture theatres and the seminar rooms.

The relationship between the University and Defence has nevertheless endured real stresses and strains. From the University's perspective, the principal difficulty has been Defence's inability to be clear and consistent about the connections between cadet education and officer employment. How does education prepare a young officer for the requirements of their Service in a resource-constrained environment? A number of answers have been offered over the past 50 years, some constructive, others destructive. It has been said that training is a necessity and education is a luxury; the former can be mapped alongside duty statements while the latter is indirectly applicable and only then in a very generalised way. In the context of imprecise outcomes, it is prudent to maximise training and minimise education. Conversely, it has been said that training relates to vocation while education serves to build character; as officers must lead and leadership draws directly on character, the need for education is practically limitless. As training relates to performing

Observations and conclusions

specific tasks and education focuses on understanding the 'big picture', assessing possibilities and exploring potentialities, an uneducated officer class will lack initiative and insight.

Although the character of ADF service has changed substantially over the past 50 years and the demands on ADF officers have evolved considerably, the nature of leadership and the expectations of command have remained constant. While it could be argued that the *role* of education may have changed, the counter argument is that the *place* of education has not. In the context of continuous change, it is still possible to maintain a vision for education. Such a vision could have provided continuity to the UNSW–Defence relationship and consistency to the education that was delivered. Both the University and Defence bear responsibility for the inability to articulate a clear vision for education. By way of mitigation, the University would claim that it was often difficult to know who spoke for Defence about education. When addressing Defence as a collective entity in the 1960s on policy matters, the University dealt mainly with the College commanding officers at Jervis Bay and Duntroon and designated staff officers in the Single-service headquarters. Between 1970 and 1974, it dealt primarily with the Chiefs of Staff Committee and the Departmental Secretary, Sir Arthur Tange, on matters of principle (there were liaison committees to do with practical issues at RANC and RMC). From 1975 to 1984, it dealt usually with the Interim Academy Council on policy matters and with the Chief Project Officer on practical matters. From 1984 to 1998, liaison was conducted via the Academy Advisory Council, the Academy Commandant (who was also the Chief Project Officer until 1987) and with the Head of Joint Education and Training within Defence. Most recently, the University has dealt with the Academy Commandant and the Commander of the Australian Defence College. Along the way there has been a series of ADF liaison committees to ensure the undergraduate program has met the workplace needs of professional groups and the requirements of external accreditation bodies. It has always been difficult for Defence to come to a clear and consistent view on education policy for the entire ADF because the three Services are unable to come to a clear and consistent view of their own needs. In trying to speak with a single voice, civilian and uniformed tri-Service representatives have tried to approximate what the Services think they want while trying to harmonise competing occupational requirements and contrasting learning continuums.

Research has been just as difficult to co-ordinate. There needs to be a coincidence of an interested academic researcher and a related Defence need for effective collaboration to occur. Occasionally, academics have shaped their research in line with present and future Defence needs. One of the restraints on collaborations developing has been the need for academics to contribute to their own disciplinary community as a vehicle for ensuring their currency with new thinking. Academics also have duties and responsibilities that transcend the confines of the Academy and the needs of Defence. The Army accepted this reality in the early 1970s with the formation of the Faculty of Military Studies. Academics are obliged to undertake theoretical or pure research that is sometimes without an obvious practical or immediate application to ensure they remain at the leading edge of their discipline and continue to attract the respect of their colleagues. Very often it is personal contact between academics and their former students that allows Defence to make the most of the University's research capacity.

Despite education being its core business, the University might have done more to promote wide-ranging discussion within Defence about the importance of academic education. Recognition of the place of education and the value of tertiary qualifications within the

Defence community was usually assumed. This was mistaken. The need for a discussion about the role of education was probably least pressing in relation to Engineering courses and greatest in relation to Arts courses. Within Engineering, the educational requirements are clear. The products of an Engineering program are professional engineers. The course is closely associated with vocational outcomes and external accreditation. Within Science, the educational requirements are slightly less clear. The graduate is acquainted with the scientific method but in many instances completing an Honours or Masters degree is essential for employment in a discipline like Physics, Chemistry, Geology or Mathematics. Within Business, the educational requirements are precise for some specialisations and less so for others. It is possible to complete a Business degree and gain employment in a vocation or admission to a professional body – depending upon the vocation and the profession.

Humanities are different because the educational requirements are unclear for those seeking a career other than within the University. Humanities disciplines like History, Politics, Language, Literature, Ethics and International Relations are nonetheless important because they provide a context for all human endeavour. Whereas the other disciplines focus on 'what' and 'how', the humanities disciplines ask 'why'. The humanities are concerned with values, virtues and principles. These disciplines ask questions relating to right and wrong, and the point and purpose of human striving. At the Academy, the undergraduate program contains mandatory humanities courses for all students, focusing on Strategic Studies, Law and Ethics. This subject mix reflects the requirement for the undergraduate education experience to be 'balanced and liberal'. The University has struggled to explain the meaning of this phrase and why a 'balanced and liberal' education is a laudable goal.

Although it remains an apt description of the education UNSW has pledged to provide, the phrase 'balanced and liberal' may have allowed the University to avoid critical self-reflection. The University has sometimes used the phrase like a talisman to ward off attacks from uniformed critics who are unwilling to risk appearing ignorant by asking for a definition. The University could not, however, explain why this phrase was apparently the best summary and the strongest argument for what it had to offer. In fact, it would struggle to demonstrate that it was actually teaching young officers to think and would have difficulty articulating a compelling account of leadership. The University claimed to be a 'thought leader' when it was sometimes merely a generator of ideas – some unworthy of serious consideration. The most serious indictment of the University's leadership has been its failure to explain the phrase 'balanced and liberal' in consultation with Defence in the context of emerging challenges to what constitutes the public interest and the common good. Either the University does not want Defence involvement in discussing its mandate or perhaps does not believe that Service officers are sufficiently acquainted with the charter of universities to make a constructive contribution to why the higher education sector even exists. This is mistaken on both fronts. The University is, however, entitled to claim some mitigation.

Constantly changing institutional leadership within Defence has played a large part in prolonging the struggle to develop a shared and substantial vision for education. In terms of political leadership, Malcolm Fraser has been the only Minister of Defence who has held the portfolio over the past 50 years to take a close interest in officer education, and there have been 21 Defence Ministers since 1967. Fraser has been the only parliamentarian to provide firm direction on how that education could and would be delivered. Every few years since 1967 there has been a new Chief of the Defence Force and new Service chiefs, a new uniformed director-general of

Observations and conclusions

Service education, and a new Academy Commandant. Among uniformed leaders, General Sir John Wilton stands out as the foremost senior officer with a clear vision of, and a consistent commitment to, officer education and its broader strategic and social purposes. After the establishment of the Academy was approved in 1974, Admiral Sir Anthony Synnot, the Chief of Defence Force Staff from 1979 to 1982, was probably the foremost practical advocate of tri-Service education and the needs of the new institution. In terms of civilian leaders, Sir Arthur Tange, grasped firmly the pressing need for better-educated officers in the higher echelons of Defence. Intelligent people did not threaten Tange, whether they wore a uniform or a suit. Because Tange and Synnot were both very gifted leaders in their respective sides of the Defence diarchy, they were able to keep the Academy vision alive and attractive. Curiously, the fulfilment of this particular objective – to help senior Service officers be better policy advisors – has never been validated. There is yet to be a detailed study of whether and in what ways a university education has enhanced their performance.

Beyond Fraser, Wilton and Tange, the majority of senior leaders in Defence have mostly managed what they have inherited rather than exercising leadership to create new possibilities and better educational opportunities. Fortunately, the foresight and influence of Fraser, Wilton and Tange was coupled with the intellectual determination of Sir Philip Baxter and the negotiating skills of Sir Rupert Myers to provide a sustainable platform for effective officer education after 1967. Since Myers' retirement in 1981, few senior officers in the University have glimpsed the potential and the possibilities for collaboration that were readily apparent to Myers and his colleagues, especially Al Willis and Rex Vowels. For the greatest part, the task of setting a vision has rested with the Academy Rector and each has attempted to articulate their sense of what 'balanced and liberal' means, and what it might mean in the years ahead, as the University invested in a future it could not predict.

While it might be countered that those who led UNSW during its first 30 years (1949–1979) had the freedom to ponder big ideas and to focus on principles because they were creating something new, they had to work against the closed single-Service college arrangements then in place. They were also were obliged to counter those who said plans for a close relationship between UNSW and Defence would diminish the University's academic standing and undermine Defence's disciplined culture. From 1965 to 1981, there were people within both UNSW and Defence working actively to discredit the evolving relationship; those who felt that tertiary education for junior officers was either an indulgence or a waste; those who believed that attention to vocational training and practical experience was more important and more pressing than providing young officers with academic post-nominals; those who argued that free and independent scholarly inquiry could not be conducted within the regulated confines of a military establishment; and those who disdained the profession of arms because they thought its practitioners were anti-intellectual, uncivilised or both. That the relationship withstood all these pressures and went on to greater things suggests the fundamental rightness and robustness of the arrangements that developed.

Against all of these considerable pressures, Fraser, Wilton, Tange, Baxter and Myers were determined to fashion a new system for educating young officers. They could not see the future but they knew enough of recent experience and sufficient of emerging needs to know they had to be innovative. Baxter wanted UNSW to expand well beyond its origins, which resided in technological know-how. He wanted to show that UNSW was focused on practical problems and could deliver attractive solutions. He also wanted to demonstrate that a university could be entrepreneurial even when local vested interest and narrow political belief sought to assert itself in a manner that was indifferent to worthwhile intellectual objectives or educational outcomes. Fraser in particular realised that the intellectual culture of the Services was lagging behind the civilian community, that failing to offer a tertiary education would affect recruitment and retention, and that Service officers were constrained in their capacity to offer advice on policy without a tertiary education.

In terms of maintaining the relationship, the quality of Defence's interactions with the University has depended substantially on the attitude of officers appointed to the Academy, to staff positions within Defence and to the single-Service headquarters. Some of these officers had no experience of universities, were without an undergraduate degree and did not have any sense of the legislative framework or the economic constraints in which Australia's universities must operate. At times, the University may have had excessive regard for the opinions of these officers and given their views a standing they did not deserve. At other times, the University has shown little regard for the opinions of these officers and has stalled, obfuscated and waited for them to be re-posted. The relationship has occasionally been imperilled by the attitudes and actions of a small minority of uniformed officers who have seen the University as a mere provider of educational services rather than as a partner in promoting professional development. As a more general observation, there has also been the persistent problem of assumed knowledge.

Most Australians have views about learning and its mediation because they have attended a school and been a student at some point in their lives. Unlike some areas of public life where people feel less able to assert expertise or even familiarity, such as economics, agriculture and medicine, every Australian has an opinion about education: why it is important and how it is delivered. But each person's experience of learning is slightly different. Individuals have their own learning strategies. Because learning is highly personalised, it substantially influences attitudes to education. The delivery of education is also liable to subjective assessments on who, what, how, why, when and where individuals should be educated, quite apart from idiosyncratic estimates of the amount of education needed for someone to contribute to society and to fulfil a vocation. These unexamined opinions often flow into general judgments about the benefits of education.

On the Academy's 20th anniversary in 2006, Lieutenant Colonel Chris Field suggested the academic program could be bolstered in two ways.[3] His first suggestion was 'the award of scholarships to staff at Top 50 Australian companies who would study at ADFA, obtain a superior degree, establish friendships with cadets and gain an insight into Defence requirements and challenges'. The second was a change to the basic degree structure to allow graduation within two years rather than three. His reason? Officers would be produced more quickly and return of service obligations would be shortened, making uniformed service more attractive for young Australians. Both ideas arose from Field's own experiences of what has worked and what could be improved. His instrumentalist approach to education has its own merit. But it is, however, just one view among many. There are those within his own Service who would prefer to see civilians excluded from the Academy at all levels and those within the University who believe that degree programs are already too compressed. The insistence on a common academic standard at the Academy has avoided the vast variations in the quality of degrees offered at Australian universities that would have resulted from the voucher system proposed by the Price Committee.

Observations and conclusions

Chris Field's views illustrate my point: there are many competing and contrasting ideas and insights, agendas and aspirations when it comes to education, and many are valid. The existence of these views has kept the UNSW–Defence relationship very fluid. Whether the University ought to have given a firmer lead on its thinking and a clearer indication of its position on issues such as program content and course duration remains a moot point. Offering advice is not, of itself, intrusive. Defence has always been at liberty to make its own decisions. But as an institution that desires to lead thought more than to follow it, the University's experience and expertise if offered regularly and vigorously might have helped Defence to narrow the range of options under review by excluding those unworthy of serious consideration.

The existence of so much supposition and so many suggestions on what could or should be done in the field of education would not matter in most situations. Indeed, the existence of many interested voices would be welcome. But the willingness of well-meaning individuals within Defence to propose significant change in the relationship with UNSW based on little more than their own limited experience and personal intuition is worrying. Some very senior ADF officers without a university degree have made questionable decisions about education policy, with profound effects. There have been Academy commandants with no experience in the dynamics of learning exercising oversight of an institution devoted to the intellectual development of young minds. It is the likely intrusion of personal preferences and prejudices in the absence of professional expertise that is alarming. While senior uniformed and civilian Defence personnel are routinely required to make judgments about matters in which they could not claim expertise, it is the mistaken presumption of insight and intuition with respect to the principles and practice of education that makes designing and delivering academic programs and courses more susceptible to whims and fads.

The continuing challenge for Defence is to achieve some consistency in the officer posting cycle, end the short-term nature of postings to the Academy, nurture a cohort of officers who have gained some long-term familiarity with providing and delivering higher education and training, secure a collective uniformed view on education and its relationship with training, and ensure that the ADF's operational tempo leaves room for people to be educated. Academy postings for uniformed personnel ought to be 3–4 years rather than 1–2 years for corporate wisdom to accumulate. The urgent need for trained operators should not inhibit long-term investment in nurturing well-educated leaders. This is a serious matter with wide-ranging consequences. Professor David Lovell, the long-serving former head of Humanities and Social Sciences at UNSW Canberra, noted that the Department of Defence was the Commonwealth government department 'most concerned with the professional education of its members' and is 'perhaps the largest single education and training organisation in Australia'.[4] After more than a quarter of a century at the Academy, he identified four personal frustrations in working with Defence. They were:

> A very high turnover of Defence staff at all levels within Defence education institutions, which is disruptive in itself and may leave non-Defence staff as the custodians of their corporate knowledge; a determination among many senior staff to initiate conspicuous change and thus add to their case for promotion, whether or not the change is sensible or thought through; a hostility to any courses of study not deemed to be 'relevant', where relevance is conceived instrumentally; [and], a conflation and fundamental misunderstanding of the differences between education and training.

Observations and conclusions

While military commanders were required to operate against the backdrop of rapidly changing tactics and emerging weapons systems, Lovell conceded that education was an imprecise activity that yielded general rather than specific benefits: 'learning is not linear [...] it is influenced by a number of factors and in unpredictable ways and its results are genuinely novel'.[5] While many within Defence have embraced the managerial dictum – if you can't measure it you can't manage it – courses in humanities and social sciences in particular 'provide a basis for, and a love of, inquisitiveness and even serendipity, so that new and unexpected connections can be explored'.[6] As a senior UNSW Canberra academic, he sought:

> [a more] sophisticated understanding of the educational task as being more about developing knowledge and insight than about ticking boxes in templates [...] the adoption of a consistent approach that does not regard formal education processes [...] as subjects to be added or deleted at the whim of perceived immediate 'needs' [...] and courage to appreciate that education is not something to be 'done' to others but requires that learners be given latitude to educate themselves.[7]

In contrast to the diversity of arrangements between military establishments and academic educators that exists across Western nations, he reflected that 'ADFA is testament to the ability of Defence to allow the latitude required for the foundations of a genuine education'. Notwithstanding this latitude, there have been too many commandants and too much staff turnover on the military side for Defence to achieve the substantial benefits that come with continuity of leadership and for the University to enjoy a consistency of approach to difficult issues.

What of the formal agreements that have undergirded the relationship between the University and Defence? In sum, these agreements have been problematic on two grounds. First, they have reflected the inability and the unwillingness of Defence to commit to a long-term relationship with the University. Second, they embody the sheer difficulty of encompassing every aspect of a relationship in a set of words that are acceptable to lawyers and accountants. The agreements giving expression to the relationship have suffered from what they don't say rather than from what they do say – which is the bare minimum. The core problem is the absence of a shared vision of education that can be cited to resist or, at least, to deflect the preferences and prejudices of a particular uniformed officer, civilian official or curious politician. In his thoughtful critique of the 1981 Agreement, Major Klaus Felsche noted the absence of a conflict resolution mechanism:

> ADFA is a uniquely Australian institution, formed by the Australian political and military environment and contrasting sharply in structure with its non-Australian counterparts. No other nation has attempted to establish a military academy without a mechanism for governing the whole institution internally, and without a formal mechanism to resolve conflicts between different components of that institution at the highest levels within the institution.

His depiction of the Academy overlooks, however, the intention of the drafters of the 1981 Agreement to create a diarchy – something that plainly troubled him. Indeed, many Service officers are concerned about the all-encompassing military–civilian diarchy within the Department of Defence itself, with the Chief of the Defence Force heading the uniformed side and the Secretary heading the civilian side. Many Defence people (uniformed and civilian) have an unstated preference for a unitary organisation, believing that it is always neater, tidier and administratively less troublesome. But means cannot prevail over ends, especially in education. The distinct features of the military and the academy could never be combined without one being damaged. The two cultures must co-exist and accept the possibility of misunderstanding and disagreement from time to time. It cannot be any other way if the relationship is to survive and if the parties are to thrive. This is neither an admission of defeat nor a surrender to tribalism. Co-existence need not be damaging to either party. A report on officer education produced by the Australian Strategic Policy Institute (ASPI) in August 2012 commented favourably on the very arrangements that Klaus Felsche criticised two decades earlier:

> One reason for the success of the arrangement is the separation between academic and military organisations. In contrast to United States cadet academies and staff colleges, there's no attempt to integrate academic staff into a single military hierarchy or place them under the control of the Department of Defense [...] the apparent anomaly of having two chief executives on campus – the Rector of University College and the Commandant of ADFA – has been overcome by a clear division of labour between the two organisations and by a shared preference for settling problems on campus rather than taking them up with the UNSW or Defence hierarchy. In sum, the relationship between Defence and UNSW has become institutionalized. Both sides have come to understand each other and learned how to work together.[8]

Clearly, none of the written agreements between UNSW and Defence have been perfect. Indeed, all have been flawed, especially the original 1981 Agreement, whose shortcomings were apparent within twelve months. But these agreements have been made to work by people on both sides of the diarchy establishing professional relationships that have transcended the gaps in the formally prescribed arrangements. The importance of these relationships was recognised by Geoff Wilson and Harry Heseltine, who worked closely and well with Peter Sinclair, Peter Day, Richard Bomball and Gerry Carwardine. Had their relationships been difficult, the Academy as a whole would have suffered. It is a tribute to them that they (and their staffs) managed to find ways through or around administrative and operational problems. The need to focus their attention on these challenges may have diverted their efforts from dealing with the deteriorating cadet culture that had become a matter of serious political angst and public concern by 1997. When Brian Adams arrived as Commandant in 1998, the future of the Academy rested not on nimble administration across the diarchy but on forceful intervention from Defence. Adams deserves credit for saving the Academy from reputational oblivion. His methods were not always appreciated by academics who did not recognise the need to reconfigure the Academy culture from the ground up. Adams helped preserve the reputation of UNSW by never implying that the University or its staff were complicit in the problems Defence was attempting to address.

The many reviews conducted into the Academy during its first two decades touched on a range of important matters but yielded very little in terms of the Academy's founding philosophy and the existence of the military–academic diarchy. These reviews were probably unavoidable given the extent of public investment and the promised benefits – many of which took some time to deliver.

Observations and conclusions

Where they might have been more direct and forthright was in proposing a holistic vision of education and a partnership with UNSW that transcended the cadet experience. The Academy is, of course, more than a venue for undergraduate education and the UNSW–Defence relationship goes well beyond teaching activities in Canberra. For Defence, the Academy was part of a significant cultural change process designed to recast single-Service loyalties and further the cause of Service 'jointery'. For the higher education sector, the UNSW–Defence relationship is the longest 'industry collaboration' in Australian history. The relationship demonstrates that a university can partner with a substantial public entity and not be compromised. Perhaps surprisingly, the University has not felt the need to conduct its own thoroughgoing review of the Defence relationship other than to initiate an inquiry to complement the Grey Review, largely for the purpose of reassuring itself. The University has never been indifferent to events at the Academy but has never seriously considered walking away either. Why?

First, the Academy was a local initiative and what happened in Canberra did not usually involve the rest of the University. Second, the budget was self-contained and the arrangements were strictly 'not for profit'. Money could not be siphoned off and sent to Kensington. If the Canberra campus operated within the budget, it was not a drain on the University. Third, change in Canberra was largely internal and mostly management driven. Enrolment procedures, approval and modification of courses and subjects, staff promotions and discipline could largely be handled by Canberra, with exceptional matters referred to Kensington for confirmation or concurrence if required. Fourth, innovations in terms of new degrees and course structures were evolutionary and not revolutionary. Kensington's procedures for creating and changing programs and courses largely met Canberra's needs.

The bystander mentality towards the Academy among some officials at the main campus was rather odd. The University College exploited the UNSW corporate 'brand'. The College was obliged to maintain the University's standards and to safeguard its reputation. The performance of Canberra academics and administrators could have dragged down UNSW's national and international ranking. The various Duntroon and Academy controversies and scandals could have diminished the standing of UNSW within the wider community. Ensuring standards were preserved and the University's reputation was protected has always been a critical task for the Rector. Because the arrangements appeared to be working, there has been little appetite for a wholesale review or drastic reconsideration of the way ahead. The leaders at Kensington and the staff in Canberra were more or less content despite the problems.

The change of name from the 'University College' to 'UNSW Canberra' was highly symbolic in two ways. First, it acknowledged that the University's presence in Canberra transcended the Academy. Even within Defence, the University had extensive links as a research partner and consultant. Second, the standing of the University's staff and the standard of its teaching, research and administration had to be every bit as good as at the main campus. There could be no sense that the Canberra campus was 'UNSW-lite'. Everything in Canberra had to stand the test of whether it compared favourably with Kensington. And it has.

The Academy campus in Canberra also offered UNSW a national platform. Its strategic position has, however, never been fully exploited. The Academy's location could have been utilised by Kensington to enhance the University's reputation internationally through consistently engaging with the diplomatic community. The specific research interests in Canberra and the University's special access to Defence funding could have broadened and deepened work being done at the main campus. Canberra could have promoted UNSW programs to non-Sydney-based postgraduate students. But few academics at Kensington have taken an active interest in what happens in Canberra and even fewer have wanted to explore its potential. Canberra has been allowed to operate as a semi-autonomous franchise of the main campus. UNSW is yet to develop a whole-of-university strategy for the national capital. Of course, the main campus is vast, the student population is huge and the need for structural change has often been urgent and compelling. It would be easy to lament lost opportunities. Fortunately for the University, the possibilities in Canberra remain and the potential for a greatly expanded UNSW presence in the national capital is undiminished.

What will the future bring? In the next 50 years, the relationship between the University and Defence is likely to feature less uncertainty and reflect greater stability. There will inevitably be some misunderstanding between the two parties given the complexity of the interactions. But this will be offset by a strong sense of empathy. The University hopes that Defence will commit to a long-term agreement, perhaps in excess of 25 years, which will avoid the substantial costs that have been associated with maintaining the relationship over the past 50 years as a non-permanent, renewal agreement. If a long-term agreement can be negotiated, both parties are more likely to invest additional resources in shared activities, affording greater consistency. Education will have a greater profile within Defence as it does across the undifferentiated civilian community. The increased number of undergraduates will create further demand for mid-career educational opportunities at the postgraduate level, while an expanded pool of postgraduates will allow Defence to make an even greater contribution to the delivery of undergraduate and postgraduate programs by officers posted to the Academy as Visiting Military Fellows – if, of course, the ADF gives this program sufficient priority. Consequently, developing a postgraduate policy ought to become an urgent necessity. Defence has insisted upon a single source for undergraduate education but does not have a comparable view on postgraduate education, merely that those completing staff college courses receive a qualification that is externally accredited. This inconsistency is something of an oddity given what is at stake.

The absence of a postgraduate policy highlights the continuing need for Defence and the University to articulate vocational needs, to align education and training with the intellectual complexities of providing for Australia's defence and security, while maintaining a commitment to the 'balanced and liberal' education that empowers informed discussion and undergirds intelligent debate. Clearly, the UNSW–Defence relationship is an effective one that has stood the test of time. The Academy system is working well. But it can be made more effective and efficient through evolving teaching and learning strategies that make the most of diverse media, and lecturers and supervisors with personal experience and professional discipline. The converse challenge is to avoid making education a slave to vocational imperatives or a servant of the job market. That is the task of training. Education is focused on an open-ended quest for knowledge in which the journey and the destination are equally important. There are also challenges in dealing with plagiarism, as the military encourages teamwork and almost the entire undergraduate student body (with the exception of the small number of civilian engineering undergraduates) is accommodated on campus. UNSW Canberra has led the way in both the efficiency and quality of student assessment, and in curriculum mapping, while Academy academics are confronted every year with the products of eight different State educational bodies.[9] The UNSW–Defence relationship has essentially embedded higher education and lifelong

Observations and conclusions

learning within the ADF and in the Department. There are pockets of anti-intellectualism and people proud to dismiss pursuits of the mind. Those disinterested in teaching and learning are, however, few and far between and their influence is negligible and declining. Although more policy work should have been done to clarify objectives and outcomes in the 1970s and 1980s, a willingness to invest financial and human resources shows that Defence is clearly committed to lifelong learning.

For the University, there are two main threats to its relationship with Defence. One is external; the other is internal. The external threat is the possibility that Defence will insist on opening the Academy Agreement to tender, resulting in Defence seeking another partner institution. Should the ANU win the tender as the only real competitor, the Academy campus would in all likelihood become a virtual ghost town. Under the pretext of economy and avoiding duplication in teaching (the only way that the ANU could underbid UNSW and extract a financial dividend from the contract), the uniformed cohort of undergraduates would be obliged to join with civilian counterparts in classes at the ANU's Acton campus, which would not have the distinctive defence and security focus that accompanies teaching at the Academy campus. Eventually, there would be little sense of Defence students being part of a cohort; they would be dispersed among the large undergraduate population at Acton and be indistinguishable from its other students. In a sense, it would be a virtual return to the pre-1986 period.

The internal threat resides in the possible emergence of divergent mindsets within the University itself. UNSW is a very large institution. Its Kensington campus is home to over 55 000 students and staff. Its annual budget runs to nearly $2 billion. The Canberra campus is one of nine faculties. The University does not rely on the Canberra campus for its future wellbeing or financial viability. In many respects, the Canberra campus is not crucial to fulfilling the University's current corporate plan: Strategy 2025. To ensure its survival and to safeguard its integrity, UNSW Canberra has developed a life of its own apart from Defence. Its outlook is now much bigger than its Agreement with the Commonwealth and although its prime 'customer' is the ADF, it has the capacity to conduct teaching and research beyond the uniformed community. These are positive developments for Defence, which ought to welcome the evolution of UNSW Canberra as an educational partner that is neither inward-looking nor in slow decline. The rapid expansion in postgraduate numbers since 1997 has produced many benefits, including a large alumni of Service officers located throughout the region. Neither the University nor Defence has really exploited this network of professional goodwill and, in some cases, a sense of personal indebtedness. Conversely, the growing cohort of UNSW graduates from the Academy is likely to increase awareness of defence and security issues within the entire University, which has, to date, tended to leave these subject areas to the Canberra academics. The main campus has absolved itself of responsibility for teaching and research on armed conflict and operational deployments because these areas had been effectively 'delegated' to Canberra. This attitude must change. The main campus needs to embrace activities and academics in Canberra to ensure the experience and expertise available at the Academy is not overlooked.

To conclude where we began: the Australian people have high hopes for what education can achieve. These hopes have extended to funding a quality education for the majority of the nation's uniformed leaders. But the expectation of producing an educated officer class has tended to focus on pragmatics rather than principles. In other words, the content and delivery of education has been considered ahead of why education is important and how it will benefit Defence. As the majority of Service officers (and other ranks too) now have university degrees and many in the rising generation have postgraduate qualifications, the investment of time and money needs to be substantiated.

The 50-year relationship between UNSW and Defence has produced many significant achievements in teaching, learning and research. UNSW Canberra can be proud of the contribution its staff has made to the personal and professional development of Australia's defence leaders and the standing of the University within and beyond the nation. The relationship with Defence has been made to work despite myriad distractions, disturbances and disruptions. The core of the relationship – the provision of a balanced and liberal education – has produced at least two generations of officers who can advise the national Government confidently and creatively on defence and security issues. The Government continues to rely on the ADF's expertise and experience, knowing that its contributions will be nuanced and insightful. UNSW has been responsible for creating an environment beyond the Service colleges and the Academy in which critical commentary and forthright conclusions are encouraged and esteemed. This environment has influenced the quality of advice the Government has received. It is the type of contribution that Professor Sir Philip Baxter hoped the University might make to national affairs. When the UNSW–Defence relationship began in 1967, Australia was involved in a controversial counterinsurgency campaign in South Vietnam. Half a century later, Australia is involved in a complicated counterinsurgency conflict in Afghanistan. In such a context, the relationship remains of abiding importance to the nation and its interests.

Notes

Preface

1. Ian Pfennigwerth, *ADFA: the First 25 Years of the Australian Defence Force Academy*, Creative Media Unit, UNSW Canberra, December 2012.

Foreword

1. General Sir John Hackett, 'The Education of an Officer', *Journal of the Royal United Service Institute for Defence Studies*, 1961, pp. 32–33.

Introduction

1. In her book *Basser, Philip Baxter and Goldstein: The Kensington Colleges*, UNSW Press, Sydney, 2015, Claire Scobie mentions a number of naval undergraduates who lived in the Kensington Colleges but does not discuss their collective contribution.
2. Patrick O'Farrell, *UNSW: A Portrait*, UNSW Press, Sydney, 1999.
3. Norman F Dixon, *On the Psychology of Military Incompetence*, Pimlico, London, 1976.
4. Dixon, *On the Psychology of Military Incompetence*, p. 19.
5. Dixon, *On the Psychology of Military Incompetence*, p. 159.
6. Dixon, *On the Psychology of Military Incompetence*, p. 161.
7. Dixon, *On the Psychology of Military Incompetence*, p. 162.
8. Dixon, *On the Psychology of Military Incompetence*, p. 20.
9. Alexis de Tocqueville, *Democracy in America*, chapter 22, see <fabiusmaximus.com/2009/08/07/detocqueville/>

1   The universities and service officer education, 1901–1966

1. For a general description of the growth of the tertiary education sector in Australia see Hannah Forsyth, *A History of the Modern Australian University*, NewSouth Publishing, Sydney, 2014. Despite the novelty of both institutions, there is no mention in Forsyth's book of either the Faculty of Military Studies at Duntroon or the Australian Defence Force Academy.
2. Hannah Forsyth, 'Academic Work in Australian Universities in the 1940s and 1950s', *History of Education Review*, vol. 39, no. 1, 2010, pp. 38–50.
3. Applications closed on 30 June 1950 with the final recipients receiving assistance over the ensuing 12 months.
4. Patrick O'Farrell, *UNSW: A Portrait*, p. 32.
5. For an account of Baxter's life see Philip Gissing, 'Sir John Philip Baxter (1905–1989)', *Australian Dictionary of Biography*, vol. 17, Melbourne University Press, 2007.
6. For a general account see Bob Bessant, 'Robert Gordon Menzies and Education

in Australia', in Stephen Murray-Smith (ed.), *Melbourne Studies in Education*, Melbourne University Press, Melbourne, 1977, pp. 163–87.
7   Robert G Menzies, 'Speech on Universities Committee report', CPD (Reps), 28 November 1957, pp. 2694–2702.
8   See Anthony P Gallagher, *Coordinating Australian University Development: A Study of the Australian Universities Commission, 1959–1970*, University of Queensland Press, St Lucia, 1982.
9   The text of the three-volume report entitled 'Tertiary Education in Australia' can be downloaded as PDF files from <www.voced.edu.au/content/ngv%3A53781> and URL <hdl.voced.edu.au/10707/228215>
10 See David J Deasey, *The History of the University of New South Wales Regiment, 1952–1977*, Haldane Publishing, Sydney, 1978, pp. 7–10.
11 The background to the clause was explained in the Launceston newspaper the *Examiner* on 20 March 1900, p. 4.
12 The full text of the *Defence Act* (1903) can be downloaded as PDF C1903A00020 from <www.legislation.gov.au>
13 For general histories of the RAN College see Frank B Eldridge, *A History of the Royal Australian Naval College*, Georgian House, Melbourne, 1949; Ian Cunningham, *Work Hard, Play Hard: the Royal Australian Naval College, 1913–1988*, AGPS, Canberra, 1988; and Peter D Jones, *Australia's Argonauts: the remarkable story of the first class to enter the Royal Australian Naval College*, Echo Books, Canberra, 2016.
14 Reports of exam results, comparisons and wastage are held in the RANC Museum Collection, files A1913–28, 19 June 1928, part 2.1, 'Report on the wastage of the 1913–1928 entry'; A1914–1929, part 10.1, 'General RANC Admin pps staffing, ministerials, POP Exam Results'; A1915–1927, part 6.1, 'Exams and results' (incomplete).
15 The University of Sydney was persuaded to extend recognition in 1949.
16 Secretary of the Navy to Henry Barff (Warden and Registrar, Sydney University) dated 27 June 1921, RANC Historical File A21–24.4.2.
17 'Scheme for RANC – policy submission to Naval Board', file A1922, RANC Historical File, part 1.1.
18 Report to Federal Parliament quoted in Hugh Smith, 'Educating the guardians: the politics of the Australian Defence Force Academy', *Politics*, vol. 19, no. 1, May 1984, p. 26.
19 'Transfer of RANC from Jervis Bay to Flinders Naval Depot', file A1930, part 1.2, RANC Historical File.
20 The 'Special Entry' was terminated in 1950.
21 Annual Report for 1954, Royal Australian Naval College, copy held in the RANC Museum Collection.
22 For a general account of long-term trends in the education of British naval officers see Harry Dickinson, *Educating the Royal Navy: 18th and 19th Century Education for Officers*, Routledge, London, 2008.
23 Naval Board file 44006/121/54, letter dated 20 June 1958.
24 'Murray–Dartmouth Review Committee Report', RANC Historical File A58–60.2.1., paragraph 69.

25 'Murray–Dartmouth Review Committee Report'.
26 'Murray–Dartmouth Review Committee Report'.
27 The 'New Scheme' for junior officer training was outlined by the commanding officer of RANC, Captain Eric Peel, in the *RAN College Magazine*, 1961, pp. 16–17.
28 Gerry Purcell, email to author, 13 August 2014.
29 *RANC Yearbook*, 1964, pp. 12–13.
30 *RANC Yearbook*, 1968, p. 5.
31 When Foster was appointed to the post of Chief of the Australian General Staff at the end of 1915, a suitable replacement could not be found and the university department ceased to exist.
32 For general accounts of the establishment and evolution of the Royal Military College Duntroon see Chris Coulthard-Clark, *Duntroon: The Royal Military College of Australia, 1911–1986*, Allen & Unwin, Sydney, 1986; Darren Moore, *Duntroon: The Royal Military College of Australia 1911–2001*, Royal Military College of Australia, Canberra, 2001 and Steve Hart, *Duntroon: Its Heritage and Sacred Legacy*, Defence Publishing Service, Canberra, 2009.
33 See Chris Coulthard-Clark, 'The case of the wandering college', *Canberra Times* [hereafter *CT*], 18 November 1979, p. 7.
34 MP742/1, item 48/2/227, 'Report on the Royal Military College of Australia', 30 November 1944, known as the 'Vasey Report'.
35 For an assessment of Blamey's role see Jeffrey Grey, *Australian Brass: the career of Lieutenant General Sir Horace Robertson*, Cambridge University Press, Melbourne, 1992, p. 114.
36 'Duntroon under fire', *SMH*, 22 May 1946.
37 Rowell Report, p. 9. Quoted in *SMH*. For further information see both general works on Duntroon, Coultard-Clark and Moore.
38 A copy of this report was included in a folder compiled by Mr Len Hume marked 'Chronology of considerations and events leading to the establishment of the Australian Defence Force Academy', Academy Library Special Collections 85/327(1).
39 'Inter-departmental committee on affiliation of the RAAF Academy and the University of Melbourne', NAA A463, 1961/6144.
40 Roy E Frost, *RAAF College and Academy, 1947–86*, RAAF, Canberra, 1991, p. 45.
41 Quoted in Frost, *RAAF College and Academy*, p. 52.

2 **Firm foundations and the faculty solution, 1964–1967**
1 Details of General Wilton's interactions with the ANU are contained in the Wilton manuscript collection held in the Academy Library, MS 231, BRN: 222709. The following narrative is drawn from these papers. The material in the collection ranges from personal correspondence dated 1940 to press clippings and correspondence generated in 1978–1980 relating to the establishment of the Defence Force Academy.
2 See 'University Facilities in Canberra, Report of a Committee Appointed to Report on the Provision of University Facilities for Residents in Canberra', 23 April 1926, *Parliamentary Papers*, 1926–1928.
3 Major General Charles Finlay (known widely as 'Basil' Finlay), RMC Commandant 1962–68, Oral History Transcript, CN 841/9.

4   Material in this chapter relating to the period 1964–1966 is drawn from CN 269 'Royal Military College Duntroon' – a collection of papers assembled by Professor Rex Vowels. It contains reports, proposals, correspondence, minutes of meetings and examination results.
5   See 'Tertiary Training in Relation to Royal Military College and RAAF Academy, 1960–1963', NAA A703, 674/10/51.
6   For a career outline see Selwyn Cornish, 'Herbert "Joe" Burton, 1900–1983', *Australian Dictionary of Biography*, vol. 17, Melbourne University Press, Melbourne, 2007.
7   Major General CH Finlay, RMC Commandant 1962–68, Oral History Transcript, CN 841/9.
8   For extended treatment's of Clark's life and times see Stephen Holt, *A Short History of Manning Clark*, Allen and Unwin, Sydney, 1999; and Mark McKenna, *An Eye For Eternity: The Life of Manning Clark*, Melbourne University Press, Melbourne, 2011.
9   See 'Report of the Committee on the Future of Tertiary Education in Australia to the Australian Universities Commission', *Parliamentary Papers*, 1964–1965.
10 Material in this chapter is drawn from UNSW Files 16583 CN 1160 (Box 268) 'Agreement UNSW & Minister Defence Dept, Military College Affiliation' and 'Organisation FMS, Agreement UNSW and Minister Defence Department', 1966–1967.
11 Emphasis added. *New South Wales University of Technology Act 1949*, section 29.1.
12 Baxter to Cutler, CN 1160, 28 April 1966, UNSW Archives.
13 UNSW Council Resolution, 66/74.
14 Roy Pugh, 'Minutes of the meeting of a joint committee set up to explore the possibility of an association between the University of New South Wales and the Royal Military College on 12 and 13 May at Duntroon', UNSW archives, CN1160, B268, dated 15 May 1966. See also Fred Ayscough, 'Notes on RMC Duntroon discussions between members of the Professorial Board's Standing Committee on RMC Duntroon and staff of RMC, 12–13 May 1966', CN 1160, B268, UNSW Archives.
15 Jeremiah Hirschhorn to Rex Vowels, letter dated 24 May 1966, CN 1160.B268, UNSW Archives.
16 Report of the Interim Committee, 7 June 1966, CN 269, UNSW Archives.
17 Malcolm Fraser to Rupert Myers, personal letter dated 19 July 1966, CN 269, UNSW Archives.
18 Ayscough, 'Report of the Interim Duntroon Committee of the Professorial Board', UNSW archives CN 1160, B268, 7 June 1966.
19 Professor John Clarke died suddenly on 4 June 1967. Clarke was replaced by two newly appointed Pro Vice-Chancellors – Albert Willis and Rex Vowels. In July 1967, Clarke's portfolio of work was given to Willis who had special responsibility for the service colleges until his retirement. AH Willis Oral History, 18–19 August 1986, CN 841/19(A), UNSW Archives.
20 AH Willis Oral History, 18–19 August 1986, CN 841/19(A), UNSW Archives.
21 Baxter to Vowels, letter dated 22 July 1966, CN 269.B2, UNSW archives.
22 For Arthur Corbett's account of his appointment to RMC and service as President of the Institution of Engineers see oral history interview, CN 841/4, UNSW Archives.

23　Commandant RMC, Supplementary Report to the Interim Committee, 9 September 1966.
24　Sutherland retired in 1967.
25　Richard Walsh (although published anonymously), 'The university student – 67', *Current Affairs Bulletin*, vol. 39, no. 8, 13 March 1967, p. 124.
26　Professor Ian Lowe email to author, 27 April 2015.
27　Dr Bob Hall, email to author, 5 May 2015.
28　Ian Lowe, President, UNSW Students' Union, 'Submission to a sub-committee of the Professorial Board of the University of NSW', UNSW archives CN 269.B2. Lowe later published a reply to Baxter in *Tharunka*, 'The president replies', 14 June 1966, p. 5.
29　Philip Baxter to Ian Lowe, personal letter dated 16 May 1966.
30　Philip Baxter, 'The University of New South Wales and the Royal Military College', *Tharunka*, 14 June 1966, p. 4.
31　Graeme Dunstan, 'Duntroon: A consideration', *Tharunka*, 14 June 1966, p. 8.
32　John Bannon to Baxter, personal letter dated 19 June 1967.
33　*The Making of the Australian National University*, Stephen Foster and Margaret Varghese, ANU Press, Canberra, 2009, p. 276.
34　Jason Andrews and James Connor, 'UNSW and the establishment of the Faculty of Military Studies at the Royal Military College, Duntroon: 1965–68', *History of Education Review*, vol. 44, no. 2, 2015, pp. 153–69, p. 166 quoted.

3　Establishing a tradition, 1967–1972

1　UNSW Council Resolution 67/118, 'Organisation, Military Studies Faculty, Affiliation Agreement UNSW & Minister Defence Department, 1966 to 1967', CN 1160 Box 268. Most of the correspondence and committee minutes from this period are located in CN 1160, UNSW Archives. The items are often duplicated and held in multiple files.
2　Email from Colonel Peter Rose, 9 February 2017.
3　Army Press Release no. 5590 dated 17 December 1968.
4　Copies of the RMC Interim Council minutes are available in the Bridges Memorial Library at RMC Duntroon and in boxes 291, 311 and 312 of CN 1160, 'Interim Council, Faculty of Military Studies, Organisation', UNSW Archives.
5　See Gerry Walsh, oral history transcript, CN 841/18, UNSW Archives.
6　See Alec Hill, oral history transcript, CN 841/10, UNSW Archives.
7　Al Willis, oral history, CH 841/19, UNSW Archives.
8　Lance Barnard, CPD (Reps), 25 September 1969.
9　CPD (Reps), 26 September 1969, Question no. 1963.
10　John W Masland and Laurence I Radway, *Soldiers and Scholars*, Princeton University Press, Princeton, 1957, p. 509.
11　Faculty resolution 70/4, Executive Committee of the Interim Council, 11 June 1970, 'Organisation, Military Studies, Faculties and Board, 1968–1982', CN 1160, Box 159, UNSW Archives.
12　Letter to Professor John Burns, 7 April 1972, 'Standing Committee, Faculty of Military Studies, Correspondence, 1968–1984', CN 1160 Box 70, UNSW Archives.
13　Alan Barcan, *Radical Students: The Old Left at Sydney University*, Melbourne

University Press, Melbourne, 2000, p. 148.
14 Ian Wilson, 'Brian Beddie (1920–1994)', Obituaries Australia <oa.anu.edu.au/obituary/brian-beddie-94>
15 Chandran Kukathas, 'Professor Brian Beddie', Academy of the Social Sciences in Australia <www.assa.edu.au/fellowship/fellow/deceased/100086>

4   **The University as a unifying agent in Defence, 1967–1975**
1   Robert Menzies to Athol Townley, confidential memo, 11 December 1958, copy held by author.
2   For a balanced account of this period see Graham Wright, 'Organisation for the Administration of Australian Defence Policy, 1958–1974', *JRUSI*, October 1974, pp. 128–41.
3   Michael Howard and Cyril English, *Report of the Committee of Inquiry into Service Colleges*, UK Ministry of Defence, London, July 1966.
4   Entry for Sir Leslie Harold Martin, Australian Dictionary of Biography website <adb.anu.edu.au/biography/martin-sir-leslie-harold-14939>
5   Peter Howson, *The Howson Diaries*, Viking Press, Melbourne, 1984, p. 320.
6   'Report of Tertiary Education (Services' Cadet Colleges) Committee on a Proposed Tri-Service Academy, 1970', NAA M1376, p. 1.
7   Philip Ayres, *Malcolm Fraser*, Heinemann, Melbourne, 1987, p. 165.
8   Cabinet submission 299: 'to obtain Cabinet endorsement of a proposal to establish a tri-Service Cadet Academy on the site of RMC Duntroon, as proposed by the Martin Committee'; Ayres, *Malcolm Fraser*, p. 167.
9   Howson, '19 October 1970', *The Howson Diaries*, p. 664.
10 Cabinet Decision no. 727 on Submission no. 299, NAA: A5869, 299, 29 May 1970. The submission contains a copy of the Martin Committee's report. See Gorton's reply to a parliamentary question, CPD (Reps), 30 March 1971, p. 1132.
11 AEE Waller to McMahon, nd; McMahon's notes, 8 June 1970, NAA: A5882, CO 931 and for a summary of Cabinet's reasons, Bunting's letter to Fraser, 16 Oct 70, ibid. Gorton's views were accurately summarised by Hugh Armfield, the *Age*, 16 October 1970; interview Fraser, 19 September 2000.
12 Oral history interview transcript, Professor Sir Leslie Martin, 10 August 1972 CN 24/5 (NLA 11).
13 Peter Edwards, *Arthur Tange: Last of the Mandarins*, Allen & Unwin, Sydney, 2008, p. 16.
14 Sir Arthur Tange, edited by Peter Edwards, *Defence Policy-Making: A Close-up View, 1950–1980*, ANU E-Press, 2008, p. 119.
15 Tange, *Defence Policy-Making*, p. 120.
16 Edwards, *Arthur Tange*, p. 247.
17 Edwards, *Arthur Tange*, p. 247.
18 Dr John Sneddon, Elected member of the Interim Council to the Chair of the Interim Council (Major General Hay), 17 May 1973.
19 Professor Tom Chapman, Chair of the Faculty of Military Studies to the Commandant of RMC (Major General Hay), 7 June 1973.
20 Barnard to Myers, 27 July 1973.
21 Hugh Smith's paper was subsequently published as 'Internal Conflict in an

Independent Papua New Guinea: Problems of Australian Involvement', *Australian Outlook*, vol. 28, no. 2, August 1974, pp. 160–67. Smith was identified as a member of the Faculty of Military Studies. The Department did not express further concern about the seminar or any published proceedings.

22   Myers to Barnard, 7 February 1974.
23   Myers to Hay, Commandant of RMC, 2 May 1974.

5   The Academy commitment, 1974–1976

1   Cabinet decision 2138, 19 March 1974: 'to obtain approval in principle to establish, by legislation, of a tri-Service Academy on a new site in Duntroon area'. Copy held by author.
2   Roger Thompson, President of the RMC Academic Staff Association, to Myers, 8 July 1974.
3   Hugh Smith, 'Training Australia's future officers', *CT*, 15 April 1974 and 'Officer training: the case for university education', *CT*, 16 April 1974.
4   'Officers and Students', *CT*, 19 June 1974.
5   'Organisations, Faculty of Military Studies, Autonomy including Proposed Tri-service Academy', CN 1160, Box 357, Folio 89, Record of informal meeting of FMS on 4 Sept 1974 'to discuss the proposed Australian Defence Force Academy', UNSW Archives.
6   Interview with Len Hume, 14 July 2015.
7   Wilson to Myers, 6 September 1974, Folio 86, CN 1160, UNSW Archives.
8   Hugh Smith, 'Officer Education', *Australian Quarterly*, September 1974, p. 32.
9   'Organisations, Faculty of Military Studies, Autonomy including Proposed Tri-service Academy', CN 1160, Box 330, formerly 27250 Part F now 370024, UNSW Archives.
10 'Organisations, Faculty of Military Studies, Autonomy including Proposed Tri-service Academy', CN 1160, Box 356, UNSW Archives.
11 UNSW Archives, 27250 Part B new number 370020, CN 1160, Box 356.
12 UNSW Archives, 27250 Part C new number 370020, CN 1160, Box 356.
13 Letter to the Editor, *CT*, 12 August 1978.
14 Percy Partridge, *Educating for the Profession of Arms*, Canberra Papers on Strategy and Defence no. 5, ANU Press, Canberra, 1969, pp. 8–9; and Michael Howard, *Captain Professor: The Memoirs of Sir Michael Howard*, Continuum International Publishing Group, 2006, p. 187.
15 The author went by the pseudonym 'Grey Rock', 'The Royal Navy and its Graduates', *Naval Review*, vol. 67, no. 3, July 1979, p. 220.
16 Quoted in Squadron Leader AM Newbould, 'The Graduate Entry Scheme: A Milestone in Cranwell's Development', *Royal Air Force College Journal*, July 1969, p. 29.
17 Cathy Downes, *Special Trust and Confidence: the making of an officer*, Frank Cass, London, 1991, p. 39.
18 Cathy Downes, *Special Trust and Confidence*, p. 134.
19 John Sweetman, 'Another Yellow Brick Road in Pursuit of Educational Fantasy', *British Army Review*, no. 57, December 1977, p. 16.
20 Josie Castle, *University of Wollongong – An Illustrated History, 1951–1991*, University of Wollongong, Wollongong, 1991.

6 The joint educational enterprise, 1970–1980

1. 'Symposium on the Education of Naval Officers held at RANC Jervis Bay, 3 July 1970 in Conjunction with UNSW', printed proceedings, photocopy held by author.
2. Naval Officer Education Symposium, p. 38.
3. Naval Officer Education Symposium, p. 54.
4. Naval Officer Education Symposium, p. 59.
5. Naval Officer Education Symposium, p. 62.
6. Naval Officer Education Symposium, p. 66.
7. Naval Officer Education Symposium, p. 67.
8. 'Survey Questionnaires, 1973–1974', Joint Advisory Committee of UNSW and RANC, series CN 1100, UNSW Archives.
9. Address by CO RANC, Passing Out Parade, 5 July 1979, *RANC Magazine 1978–79*, p. 42.
10. The RAAF–Melbourne University connection has not been the subject of much attention. *A Short History of the University of Melbourne* by Stuart MacIntyre and Richard Selleck (Melbourne University Press, Melbourne, 2003) makes no mention of the RAAF Academy at Point Cook; Stuart MacIntyre and Richard Selleck, *A Short History of Melbourne University*. An earlier work by John Poynter and Carolyn Rasmussen has one paragraph on the RAAF affiliation but limits the commentary to 1961–1963; John Poynter and Carolyn Rasmussen, *A Place Apart – The University of Melbourne: Decades of Change*, Melbourne University Press, Melbourne, 1996, pp. 214–15.
11. Roy Frost, *The RAAF College & Academy, 1947–86*, RAAF internal publication, 1991, p. 17. Frost completed his manuscript in 1987 but it was not published for another four years. RAAF Radschool Association website at: <www.radschool.org.au/Books/RAAF%20College%20and%20Academy.pdf>
12. See the obituary 'Walter Davis Hardy: one of the "fathers" of the RAAF', *Sydney Morning Herald*, 16 April 2015.
13. Frost, *The RAAF College & Academy*, p. 101.
14. See Defence file, AF 92/9285/1 – 'NSW University Squadron'.

7 From autonomy to uncertainty, 1975–1978

1. John Burns oral history transcript, CN 841/2, UNSW Archives.
2. 'Organisations, Internal, FMS, Autonomy 1975–76 inc. Tri-service Academy', CN 1160, Box 356 (new 370021) Part C, UNSW Archives.
3. Myers to Burns, 8 March 1976, 'Organisations, Internal, FMS, Autonomy 1975–76 inc. Tri-service Academy', CN 1160 Box 356 (new 370021) Part C, UNSW Archives.
4. FMS resolution (76/15), 'Organisations, Internal, FMS, Autonomy 1975–76 inc. Tri-service Academy', CN 1160, Box 356 (new 370021) Part C, UNSW Archives.
5. The first draft was circulated on 2 July and amended after a Faculty meeting on 16 July 1976.
6. Item EC 4 DC Executive 29 June 1976, 'Assessment of Student Numbers'.
7. Email from Dr John Sneddon, 11 February 2017.
8. See also 'Duntroon Civilian Students, 1977–78', CN 264, Box 1, UNSW Archives.
9. See also Brian Beddie, oral history transcript, CN 841/1, UNSW Archives.

Notes to pages 208–227

10 The subsequent quoted and cited correspondence is Part D, CN 1160, Box 356, UNSW Archives.
11 Cabinet approval no. 715 dated September 1976, NAA: A12909.
12 Cabinet submission no. 715 dated 22 September 1976 and Cabinet decision no. 1636 dated 12 October 1976, Department of Defence file 'ADFA 18/2 Part 2', folios 23 and 38.
13 Ken Fry, Member for Fraser, Adjournment debate, CPD (Reps), 2 November 1976.
14 *Australian Defence*, 1976 Defence White Paper, AGPS, Canberra, p. 37.
15 Cabinet submission no. 834 dated 10 November 1976 and Cabinet decision 1838 dated 16 November 1976, Department of Defence file 'ADFA 18/2 Part 2', folios 63 and 67.
16 Willis to Myers, 1 September 1976.
17 CN 264, Box 1, UNSW Archives.
18 CN 1160, Box 329, Old file no. 27250, Part J, UNSW Archives.
19 Draft Bill dated 11 March 1977 was considered by the Executive Committee of the Development Council.
20 Cabinet decision no. 3271 dated 12 July 1977, Department of Defence file 'ADFA 18/2 Part 4', folios 32 and 36.
21 Myers to Killen, 26 July 1977. This and the following correspondence is drawn from Box 1, CN 1166, James D Killen, 'Papers referring to the establishment of the Australian Defence Force Academy', and Boxes 1–3, CN 264, 'Personal papers of Pro-Vice-Chancellor Professor Ray Golding', UNSW Archives.
22 Burns to Myers, 1 August 1977.
23 Paper by GV Wilson, 27 July 1977.
24 Burns to Myers, 10 August 1977.
25 File 1985/84 Part 1, 'Committees – Australian Vice Chancellors Committee, 1977–1988', NAA Series A7996 – 'Proposal to establish a University College in a Tri-service Academy'.
26 Myers to Killen, February 1978.
27 Professor Ray Martin to Myers, 14 March 1978.
28 Low to Myers, 16 March 1978.
29 Letter to Myers, 13 March 1978.
30 Letter to Myers, 14 March 1978.
31 Telex to Myers, 11 May 1978.
32 Killen papers, Folder 5, 'Miscellaneous/Correspondence etc. 1977–81', CN 1166, UNSW Archives.
33 Email to the author, 5 October 2016.
34 Francis West, 'Notes on the News', 24 May 1978, ABC Radio National, copy and commentary supplied by Professor West.
35 Noted in Killen to Myers, 28 June 1977, MS 239, ADFA and UNSW Canberra Historical Collection, Box 1, Folder 4.
36 Tange, *Defence Policy-Making*, p. 121.
37 Dunbar to Casey, 8 June 1978, Noel Dunbar papers, MS301, Special Collections, Defence Academy Library.

8 Parliamentary works and political will, 1978–1981

1. Much of the material in this chapter is drawn from the series CN 1166 'Papers referring to the establishment of the Australian Defence Force Academy, 1970–80' held in the UNSW Archives in two boxes divided into ten folders originating from James D Killen, Minister for Defence, 1975–1982. The material is arranged either chronologically or sequentially according to the type of document. This series contains many duplicates of the same documents and does not identify the final draft of any document. It is essentially a collection of papers and includes the proof of evidence presented to the Parliamentary Works Committee. Complementary material (and some duplicate correspondence) is also contained in CN 1160, Box 330 and Box 331, 'Military Studies Faculty, Autonomy, Proposed tri-Service Academy', 1978–1979 and 1978–1982. The files appear to be post-facto accumulations of related documents and are without any folio numbers.
2. For transcripts of evidence provided to the PWC see 'Policy – Proposal to establish a University College in a Tri-Service Academy', NAA series A7996, 'Committees – Public Works Committee', file 1985/118 (various parts) and 'Parliamentary – Ministerial and Cabinet submissions', files 1985/420, parts 5, 6 and 7.
3. Arthur Tange, *Defence Policy-Making: A Close Up View, 1950–1980*, edited by Peter Edwards, ANU Press, Canberra, 2008, p. 119.
4. Tange, *Defence Policy-Making*, p. 120.
5. Sir Rupert Myers oral history transcript OH36, p. 61, interview by Sue Knights and edited by Linda Bowman and Victoria Barker, UNSW Archives.
6. 'Military Academy', *CT*, 12 May 1978, p. 2.
7. Burns to Myers, 1 June 1967, MS 239, ADFA and UNSW Canberra History Collection, Special Collections, Defence Academy Library.
8. 'Faculty of Military Studies, General File, 1973–1980', Box 363, CN 1160, UNSW Archives.
9. See Paul Kelly, *The End of Certainty*, Allen & Unwin, Sydney, 1982, pp. 40–2.
10. Wolfgang Kasper, Richard Blandy, John Freebairn, Douglas Hocking and Robert O'Neill, *Australia at the Crossroads: Our Choices to the Year 2000*, Harcourt Brace, Sydney, 1980.
11. Ken Fry, CPD (Reps), 13 September 1978.
12. Hugh Smith papers, Folder 10, MS 148, Special Collections, Defence Academy Library.
13. Papers on FAUSA contained in Hugh Smith papers, MSS 148, Folder 5, Special Collections, Defence Academy Library.
14. 'Report Relating to the Proposed Construction of a Defence Force Academy in the ACT', 22 May 1979, *Parliamentary Papers*, 1979, volume 7, paper 115.
15. Standing Committee on Public Works, 'Report on Defence Force Academy, ACT, 1979–1980', NAA file A1209, 1979/871.
16. MINDEF Press Release 105/79 'Casey University', Killen papers, folder 7, CN 1166, UNSW Archives.
17. Tange, *Defence Policy-Making*, p. 122.
18. Frank Cranston, 'Defence Forces Academy: Backbench anxiety over Casey plans', *CT*, 9 June 1979, p. 3.

19 The preparation to the Government's response can be tracked in the Killen papers, 'ADFA – Case against PWC Reports, 1979', folder 9, CN 1166, UNSW Archives.
20 Defence file, 'ADFA', 18/2 Part 6, folios 3 and 19.
21 Cabinet decision no. 10380 of November 1979.
22 MINDEF Press Release 253/79, 'Tri-service Academy'.
23 Cabinet decision No. 10380 dated 27 November 1979.
24 NAA: A12909, 3572, 23 October 1979.
25 Hugh Smith, 'The case for a tri-service academy: clarifying some of the issues', *CT*, 18 December 1979, p. 2.
26 Cabinet submission no. 3895 dated 2 March 1980; Cabinet decision no. 10967 dated 25 March 1980.
27 Mel H Bungey, 'Defence Force Academy: PWC Comments on Analysis by Department of Defence', April 1980, p. 1.
28 Cabinet submission no. 3977 dated 28 April 1980.
29 Defence file, 'ADFA', 18/2 Part 6, folios 51 and 52.
30 Hugh Smith, 'Educating the Guardians: the politics of the Australian Defence Force Academy', *Politics*, vol. 19, no. 1, May 1984, p. 31.
31 Hugh Smith, 'The Education of Officers: Academy or University', *Australian Quarterly*, vol. 46, no. 3, September 1974, pp. 20–34.
32 FMS Faculty Meeting, 6 June 1980, GV Wilson (Dean).
33 Myers interview with Damien McCoy, 9 August 1996, transcript, UNSW Archives.

9   From affiliation to Academy, 1981–1986

1 Photograph of Prime Minister Malcolm Fraser unveiling plaque, NAA file A6180, 19/2/81/4.
2 See 'Contract Administration (non-file) records relating to Defence Force Academy construction project, 1980–1989', NAA file A8381; 'General Administration (non-file) records relating to Defence Force Academy construction project, 1980–1989', NAA file A8382; and 'Financial Administration (non-file) records relating to the Defence Force Academy construction project, 1980–1989', NAA file A8383.
3 See file 1985/429 Part 1, 'Policy – Negotiations with UNSW, 1980', and file 1985/428 Part 2, 'Policy – Negotiations with UNSW, 1980–81', NAA Series A7996.
4 Julia Horne (ed.), *Not an Ivory Tower: The Making of an Australian Vice Chancellor*, UNSW Archives, Kensington, 1997.
5 Dovers died in 2007. His professional career was assessed at: <www.navyhistory.org.au/obituary-rear-admiral-w-j-dovers-cbe-dsc-ran-retd>
6 A full set of minutes for the Academy Interim Council are available in the Ray Golding papers, Box 15, CN 825.
7 See Sir Edward Woodward, *One Brief Interval: A Memoir of Sir Edward Woodward*, Miegunyah Press, Melbourne, 2005.
8 Bennett to Woodward, letter dated 18 February 1982, CGS 129/1982, Woodward papers, Melbourne University Archives.
9 Bennett to Woodward, letter 29 June 1982, 658/1982, Woodward papers, Melbourne University Archives.
10 File 1985/422, 'Australian Defence Force Academy Charter, 1975–85', NAA Series 7996.

11 I am indebted to my colleague Hugh Smith for pointing out this very significant principle.
12 The force and effect of the agreement was considered by UNSW officials; see Golding papers, CN 825, Box 17, folders 1–3.
13 Private letter from Hugh Smith to Professor Colin Pask, 19 June 1996, copy made available to the author by A/Professor Smith.
14 Interview with Rob Tonkin, 29 July 2015.
15 Major Klaus Felsche, 'A Political Compromise: A Study of the Origins, Structure and Performance of the Australian Defence Force Academy', unpublished Masters dissertation, Humanities & Social Sciences, UNSW Canberra, 1991, p. 1.
16 Felsche, 'A Political Compromise', p. 61.
17 Felsche, 'A Political Compromise', p. 73.
18 MS 239, ADFA and UNSW Canberra History Collection, Box 11, Folder 46, 15 December 1982.
19 Executive Committee of the RMC Interim Council, memo dated 24 August 1981.
20 'Man says he was "victim" at Duntroon', *SMH*, 11 April 1974, p. 3; 'Court told of Duntroon life', *Courier Mail*, 11 April 1974, p. 10.
21 The whole episode is described in length and critically interpreted by James Connor, 'The military, masculinity and the media: the 1983 Duntroon bastardisation scandal', The Australian Sociological Association, 2013, see <www.tasa.org.au/wp-content/uploads/2013/11/Connor.pdf> accessed 7 May 2017. See also 'Managing ADFA – Study/Disciplinary Problems', Folder 120, Box 33, MS 239, ADFA and UNSW Canberra History Collection, Special Collections, Defence Academy Library.
22 Greg Pemberton to Wilson, 2 May 1983.
23 Wilson to Golding, 10 May 1983, D138/83.
24 Coates to Golding PVC, 29 September 1983.
25 UNSW Canberra Collection, Box 16, Folder 67.
26 O'Farrell to Ray Golding, undated but late April 1982, RANC Historical Collection.
27 Memorandum compiled by Patrick O'Farrell of a telephone conversation with Ray Golding, 2 May 1982.
28 O'Farrell to Director of Studies, RANC, dated 3 June 1982.
29 CO RANC to Chief of Naval Personnel, 31 October 1984.
30 Director of Naval Education to CO RANC, dated 21 June 1984, letter headed 'Options for achieving an alternative to the Creswell Course'.
31 Interview with Professor Geoff Wilson, 13 August 2013.
32 Interview with Professor John Burns, 28 July 2015.
33 Email from Dr John Sneddon, 11 February 2017.
34 Interview with Professor Geoff Wilson, 13 August 2015.
35 Professor Golding, 'Structure of the Proposed University College' draft paper dated 25 October 1983. The proposal went to University Council in Kensington for its approval on 19 July 1984.
36 CDFS Administrative Instruction 1/1984 headed 'Formation of ADFA'.
37 Email to the author, 10 March 2017.
38 Notes on Staff Administration Briefing, 20 September 1983.
39 Interview with Professor Jeremy Davis, 2 March 2015.

40 Norman Attwood, interview with author, 27 April 2015.
41 *ADFA Newsletter*, vol. 1, no. 2, October 1985.
42 See the minutes of the RANC Academic Advisory Council, 1982–1986, file 1-7-30, RANC Historical Collection.
43 Tim Griggs, 'Top Navy Alumnus – UNSW/Military Links Crucial', *Alumni Papers*, vol. 2, no. 3, November 1985, p. 3.
44 'Pro VC Golding – It is one of our success stories', *Alumni Papers*, vol. 2, no. 3, p. 4.

10 Reviews, reforms and restructures, 1986–1995

1   Australian Defence Force Academy, *Yearbook 1986*, produced internally. Copy held by author.
2   Email from Major General Peter Day, 14 January 2017.
3   Interview with Tony Ayers, 16 July 2015.
4   Jarlath Ronayne to Peter Baldwin, letter dated 22 January 1991, Woodward Papers, Melbourne University Archives.
5   Chapters 8 and 9 of Professor Wilson's memoir, *An Academic Journey*, reflect on his time at Duntroon and as the Foundation Rector of the University College. It was privately published and is available as an e-book. Copy held by the author.
6   Harry Heseltine to Michael Birt, 3 December 1991, 'Correspondence Academy (Military)', Box 2012/19, Special Collections, Defence Academy Library.
7   Interview with John Niland, 4 April 2016.
8   Harry Heseltine, UNSW, oral history transcript, 14 December 1998.
9   *Acadfinitis*, vol. 7, no. 3, September 1991.
10 Richard Brabin-Smith to Richard Bomball, 4 August 1992, copy held by author.
11 Bruce Moore, *A Lexicon of Cadet Language, Royal Military College, Duntroon in the Period 1983 to 1985*, ADB Centre, ANU, Canberra, 1992, p. i.
12 John Burns and Bob Bradford, 'Enhancement of the Performance of Officer Cadets', Report of the Working Party Established by the Commandant and Rector, October 1990. The report contains nine chapters and a conclusion with recommendations spread throughout the text. Copy held by author.
13 Inspector-General Division, Department of Defence, *Program Evaluation: Report of the Australian Defence Force Academy (Component 1.3.5)*, AGPS, Canberra, 23 December 1992 [hereafter IG's Report].
14 Interview with Professor Heseltine, 13 March 1996.
15 IG's Report, p. 1.
16 Professor John Richards, email to author, 25 January 2016.
17 Honourable Ian Sinclair, email to author, 9 April 2015.
18 Group Captain Ken Given, email to author, 2 September 2015.
19 Interview with Group Captain Ken Given, 5 August 2015.
20 Roger Price, written responses to questions from the author dated 22 April 2015.
21 Joint Standing Committee on Foreign Affairs, Defence and Trade, 'Officer Education: The Military After Next' (hereafter 'Price report'), tabled October 1995, Australian Government Publishing Service, Canberra, 1995.
22 Price Report, p. 164.
23 Roger Price, written responses to questions from the author dated 22 April 2015.
24 Verona Burgess, 'Defending the ADFA realm', *CT*, 25 October 1995, p. 21.

25 Bruce Juddery, 'Price not right about ADFA', *CT*, 1 November 1995, p. 17.
26 Anthony Lavers, 'What Price civilian defence training?', *CT*, 28 October 1995.
27 Hugh Smith and Anthony Bergin, 'Educating for the Profession of Arms in Australia', *ASPI Special Report*, Issue 48, August 2012, p. 20.
28 See also Hugh Smith, 'Officer education and the 'Military After Next': A Response to the Price Report', *ADSC Working Paper*, No. 33, December 1995.
29 James Warn, Paul Tranter and Glenn Fulford, 'The Value of the Tertiary Education and Military Training Provided by the Australian Defence Force Academy', *ADSC Working Paper*, 1996.
30 Warn et al., 'The Value of the Tertiary Education and Military Training', p. 3.
31 Warn et al., 'The Value of the Tertiary Education and Military Training', p. 8.
32 Brief to Commandant by Wing Commander PA Lavelle, 1 November 1995.
33 Hickling to Woodward (ref: Comdt 445/95) letter dated 2 November 1995, Woodward Papers, Melbourne University Archives.
34 Senator Conroy, Question without Notice, CPD (Senate), 29 May 1996.

11 Controversy and consolidation, 1996–1998

1 Report of the University College Postgraduate Student Council, *Acadfinitas*, Vol. 12, no. 3, September 1996.
2 *Acadfinitis*, Vol. 12, no. 2, June 1996.
3 Interview with Professor John Richards, 24 February 2016.
4 The Report of the Defence Efficiency Review team was entitled 'Future Directions for the Management of Australia's Defence' and presented to the Minister of Defence on 10 March 1997. A 'Fact Sheet Summary' was also produced. Copies held by author.
5 'Future Directions for the Management of Australia's Defence', p. 4.
6 'Future Directions for the Management of Australia's Defence', p. 49.
7 'Future Directions for the Management of Australia's Defence', p. 49.
8 'Future Directions for the Management of Australia's Defence', p. 50.
9 'Future Directions for the Management of Australia's Defence', p. 51.
10 'Message from the Rector', *Acadfinitis*, vol. 13, no. 3, September 1997.
11 'Farewell Comment', *Acadfinitis*, vol. 13, no. 4, December 1997.
12 Woodward to Richard King (Cordiner King Hever executive recruiters), 25 April 1998, Woodward Papers, Melbourne University Archives.
13 John Richards, statement to Council, 22 Oct 1998, RAC Minutes.
14 Stuart Martin, 'ADFA cadets are isolated', *CT*, 13 November 1995.
15 T Walford, 'Officers better trained', *CT*, 20 November 1995.
16 Price Report, p. 164.
17 Editorial, 'Defence Academy', *CT*, 12 May 1978.
18 Frank Cranston, 'Defence academy cadets claim harassment', *CT*, 12 February 1986, p. 1.
19 Professor Jane Morrison, email to author, 5 July 2016.
20 'Report to the Vice-Chancellor on the implementation of equity policies, requirements and opportunities at University College, ADFA', June 1998.
21 Professor Jane Morrison, email to author, 29 June 2016.
22 Professor Jane Morrison, email to author, 5 July 2016.

23 The correspondence relating to this matter can be found in Defence file 97/37674 'UNSWR Relocation – Facilities Aspects'.
24 CDF368/1997, John Baker to John Niland, September 1997.
25 Rear Admiral Brian Adams, email to author, 15 February 2017.
26 Chris Alder, 'Opportunities Lost; The Australian Defence Force Academy and the Canberra Community, 1986–1998', BA Hons dissertation, UNSW Canberra, October 1999, p. 1.
27 Alder, 'Opportunities Lost', p. 2.
28 John Leonard, 'From Transformation to Transcendence: Cultural Change at the Australian Defence Force Academy 1989–1998', *Australian Defence Force Journal*, no. 146, January–February 2001, pp. 5–9.

12 Commercial contracts and institutional partnerships, 1999–2004

1 'The New University of Technology', editorial, *SMH*, 25 February 1950; 'A Challenge', editorial, *SMH*, 3 April 1996.
2 March 1999, Rector's Advisory Committee (RAC) Minutes and attached correspondence.
3 September 1999, RAC minutes and attached correspondence.
4 Email from Professor Bruce McKern, 14 August 2014.
5 Interview with Colonel Clive Badelow, 12 September 2014.
6 Zimmer Report, Part II, p. 105.
7 Zimmer Report, Part II, p. 9.
8 Zimmer Report, Part II, p. 73.
9 Zimmer Report, Part II, p. 75.
10 The 'Group of Eight' (Go8) Universities was formed in 1999 as a non-profit consortia consisting of the following universities: the Australian National University, Sydney University, Melbourne University, Adelaide University, the University of Western Australia, the University of Queensland, Monash University and UNSW.
11 Zimmer Report, Part II, p. 99.
12 Zimmer Report, Part II, p. 105.
13 Zimmer Report, Part II, p. 105.
14 Interview with Laurie Olive, 24 September 2016.
15 The relevant files are 'Project Sandstone – Pre-Sandstone agreements with UNSW at Australian Defence Force Academy, 1981–2001', Defence file 2003/51474; 'UNSW – Contract Management (Non-financial), 2003–2009', Defence file 2003/47228, and 'ADFA – Contractual Arrangements between Defence and UNSW, 1998–2001', Defence file 98/20269, Defence Records Centre, Queanbeyan.
16 Dennis O'Connor, email to author, 5 September 2016.
17 PREP Report, letter from Niland to King dated 2 August 2001 tabled at the August 2001 RAC meeting.
18 Defence Media Release MECC 307/22, 'Defence and UNSW sign new agreement', 25 June 2002.
19 Zimmer Report, Part II, p. 98.
20 Zimmer Report, Part II, p. 98.
21 Quoted in 'Oz strategists: 50 years of SDSC', Australian Strategic Policy Institute <www.aspistrategist.org.au/oz-strategists-50-years-sdsc/>
22 RAC meeting minutes dated 18 May 2001.

13 Stability and sustainment, 2005–2017

1. Matthew Thomson, 'Anger at uni chief's sudden secret exit', *Sydney Morning Herald*, 9 April 2004.
2. Professor Rory Hume, email to the author, 17 May 2016.
3. Professor Rory Hume, email to the author, 17 May 2016.
4. Linda Doherty and Matthew Thompson, 'On his uni highway, Hume became roadkill', *Sydney Morning Herald*, 10 April 2004.
5. Interview with Professor John Baird, 2 May 2016.
6. Business Manager ADFA to Rector, letter dated 19 November 2001, Defence Corporate File 2001/33293/3.
7. Minutes, RAC meeting, 7 May 1999.
8. Senate References Committee, Employment Workplace Relations, Small Business and Education (EWRSBE), Monday 13 August 2001, transcript of evidence, p. 1270.
9. EWRSBE transcript, p. 1287.
10. EWRSBE transcript, pp. 1275–76.
11. The memorandum was not tabled or shown to Defence's representatives but quoted during the Committee's hearing.
12. GSI Discussion Paper No. 3, presented to the RAC in April 2002.
13. 'Contract Management, UNSW', Defence file 2005/1092418, Defence Records Centre, Queanbeyan.
14. Commodore James Goldrick, Haul Down report, February 2006, copy provided by RADM Goldrick.
15. Much of the detail in this chapter came from an interview with Professor Hilmer, 21 August 2015.
16. 'ADFA, 2003 Agreement with UNSW – Contract Negotiations for 10 year extension to 2023', Defence file 2008/1122572, Defence Records Centre, Queanbeyan.
17. '12-301, Review of the UNSW & ADFA Agreement – Performance Management Framework', Defence file 2011/1211880, Defence Records Centre, Queanbeyan.
18. Cameron Stewart, 'ADFA chief slams defence minister Stephen Smith', *The Australian*, 15 August 2013.
19. Steve Cannane, 'Investigation into ADFA calls for cultural change', *Lateline*, ABC <www.abc.net.au/lateline/content/2011/s3355732.htm?source=rss>
20. See 'Shut the ADFA up', ABC News <www.abc.net.au/news/2013-09-13/shut-the-adfa-up/4957612>
21. David Low and Kate Wilson, 'Persistent Gender Gaps in First Year Physics Assessment Questions', *ACSME Proceedings*, Curtin University, September–October 2015, pp. 118–24. The notion that both enrolments and results in the first-year Physics course were influenced by the Skype scandal was not raised by the authors of this paper. This is entirely my suggestion.
22. Australian Human Rights Commission, 'University sexual assault and sexual harassment project' <www.humanrights.gov.au/our-work/sex-discrimination/projects/university-sexual-assault-and-sexual-harassment-project>
23. Special Meeting of the Interim Council of RMC, 27 April 1977.
24. Chairman of the Canberra Division of the Institution of Engineers (SD Hardy) to Scholes, 29 March 1984, Woodward Papers, Melbourne University Archives.
25. Sinclair to ACPOL (Defence), 17 December 1985, Woodward Papers, Melbourne University Archives.

26 Reported in the *Australian*, 14 May 1986.
27 Ryan to Beazley, 4 July 1986, Woodward Papers, Melbourne University Archives.
28 Defence file 1985/1141, Part 1, 'Education, ACT – Engineering, 1986-87', NAA Series A 7996.
29 Email from Rector to all UNSW staff, 11 November 2015. Copy held by the author.
30 For a positive appraisal of the UNSW Canberra Business degrees, see Department of Defence file 2011/1208237, Defence Records Centre, Queanbeyan.

14 Observations and conclusions

1 John Stone, email to author, 26 April 2015.
2 Hugh Smith, 'UNSW–Defence Relations: A Personal Perspective', unpublished manuscript dated 24 August 2015, prepared for the author.
3 Chris Field, 'To Lead, to Excel', *Australian Army Journal*, vol. IV, no. 1, pp. 175–83.
4 David Lovell, 'Reflections on Defence and Education', *Australian Defence Force Journal*, Issue 181, 2010, p. 30.
5 Lovell, 'Reflections on Defence and Education', p. 32.
6 Lovell, 'Reflections on Defence and Education', p. 33.
7 Lovell, 'Reflections on Defence and Education', p. 33.
8 Hugh Smith and Anthony Bergin, 'Educating for the Profession of Arms in Australia', *ASPI Special Report*, no. 48, August 2012, p. 19.
9 See David Blaazer and Richard Henry, 'UNSW Canberra: A University within a University', in Richard Henry, Stephen Marshall and Prem Ramburuth (eds.), *Improving Assessment in Higher Education: A Whole Institution Approach*, UNSW Press, Sydney 2013, pp. 263–279.

Index

1967 Agreement ix, 13, 36, 84–85, 92–94, 96–97, 141, 163, 228
   extension to 158, 163, 204–205
1981 Agreement
   'Heads of Agreement' 408–409
   new agreement to replace 400–417
   'Project Sandstone' 402, 409
   section 3.1 262
   Zimmer review and 392, 395–396
2003 Agreement 416, 439
   Consultative Council 438–439
   disruptive tendering 440–441
   'evergreening provision' 440
   'transition out' plan 432–434
   Triennium Review 430–434
2009 Agreement 441–442

A
abuses *see* bastardisation scandals; 'Skype scandal'
Academy *see* Australian Defence Force Academy
Academy Military Education and Training (AMET) program 443–444
ACT School of Engineering 449–450
Adams, Brian 376–384, 474
ADC Strategic Studies program 426–427
ADF Language School 422
aeronautical engineering 196, 320, 449
agreements 472–474, 477–478 *see also* names of specific agreements, eg, 1981 Agreement
air force *see* Department of Air; RAAF Academy; Royal Australian Air Force
Alder, Chris 3, 82
Anderson, John 121
Andrews, Jason 96–97
Annapolis, US Naval Academy at 179, 181, 297, 458
anti-intellectualism 7, 275, 455, 479
ANU *see* Australian National University
Army *see also* Department of Army
   officer education 37–45

university partnership canvassed 23–24
Arnold, Allan 423
Attwood, Norm 241–242, 251, 294
Australia at the Crossroads 240
Australian Centre for the Study of Armed Conflict and Society 414
Australian College of Defence and Strategic Studies, Weston Creek 421–422, 424
Australian Command and Staff College (ACSC) 424, 427, 434–435
Australian Communist Party 63–64, 80
Australian Defence College (ADC) 363–364, 416, 424, 435
Australian Defence Force Academy (ADFA) *see also* University College; UNSW Canberra
   ADFA–UNSW Canberra common attributes 443–444
   corporate branding 325–326, 381–382, 476
   cost questions 349–351, 392, 398–399, 405–411
   Defence support for 330
   'enemy from within' 353–384
   Functional Brief 212, 218
   international comparative visits 297–298
   legal proceedings 367–368
   open tender environment 360, 399, 404, 440–441, 479
   organised opposition to 171
   plans for autonomy 215–216
   public relations 382–383
   university status 155–156, 159, 167–168, 208, 210–212, 215–216, 218–221, 224–225, 227, 244
Australian Defence Force Academy (ADFA), committees and groups
   Academy Advisory Council 294
   Academy Development Committee 154, 160, 164, 261
   Academy Development Council 169, 203–206, 208, 211, 214–215, 217–218, 437–438

Accreditation Committee 211–212
ADFA Users' Committee 318
ADFA Working Group 438–439
Interim Council 261, 287–289
Australian Defence Force Academy (ADFA), development of
  1970–1980, joint educational enterprise 178–199
  1974–1976, Academy commitment 153–177
  1978–1981, political involvement in 230–259
  1981–1986, from affiliation to Academy 260–304
  1996–1998, controversy and consolidation 353–384
  ADFA Master Plan 434
  agreement to establish 257–259
  architectural design 213
  better use of 359–360
  challenges to 156–157
  cultural reform 445–448, 474
  draft Bill, comments on 222–223
  library 292, 306, 320
  location 293, 295
  name of 171, 219–220, 271–272
Australian Defence Force Academy (ADFA), reviews and reforms *see also* Grey Review; Price Report
  1986–1995, reviews, reforms and restructures 305–352
  Management Audit Report 333
Australian Defence Force Academy (ADFA), staff, students and subjects
  civilian students *see* civilian students
  Commandants 356–357
  curriculum questions 400
  discrimination, bullying and intimidation at 272–277, 365–369, 444–448
  enrolment numbers 278, 353
  Executive Officer 200–204
  female staff 291, 316, 371, 374–376
  female students 206, 212, 244, 303, 327, 367, 384, 445–448
  feminist perspectives and gender studies 371, 374–376
  first intake 303–304
  integration of military and academic components 369–371, 379–381
  Master of Defence Studies 354
  military education and training (MET) 262–269
  postgraduate education *see* postgraduate education
  proposed subjects 207, 212–213
  research *see* research
  seen as a 'military nunnery' 338, 342, 344, 348, 366
  staff 206–207
  standards expected 447–448
  Visiting Military Fellow Scheme 296–297, 327, 398
  whether to separate officers' mess 213–214
Australian Defence Force (ADF) 322–325
  see also Department of Defence
  change in 8
  course recognition 420–422, 426, 435
  domestic affairs 10–11
  formal entity in 1976 125
Australian Defence: Report of the Reorganisation of the Defence Group of Departments 125
Australian Defence Studies Centre (ADSC) 412–414
Australian Human Rights Commission 447
Australian Joint Warfare Establishment 125
Australian National University (ANU) 15–16, 43–44, 147–148, 215, 217, 221, 226, 228
  ACT School of Engineering 449–450
  ADFA planning and 243, 247, 294–295
  future prospects 479
  pay and conditions 290
  RMC Duntroon and 60–65, 67–68, 83, 90
  School of General Studies 61–62, 115
  Strategic and Defence Studies Centre (SDSC) 90–92, 147, 412, 414, 435
  student activism 90–92
Australian Universities Commission 19
Australian Vice-Chancellors' Committee (AVCC) 14–15, 118, 171, 209, 218, 220–228, 245, 253
aviation research and training 50, 197
Ayers, Tony 306–307, 332, 360
Ayres, Philip 133
Ayscough, Fred 72–73, 78–79

B
Badelow, Clive 393, 402, 407, 409
Baird, John 320, 419–420, 430–431, 433

Index

retirement 442–443
Baker, John 342, 378–379
Bannon, John 89–90
Barff, Henry 26
Barnard, Lance 111, 126, 139–140, 142, 144–145, 149, 151
   ADFA planning 153–155, 158, 162, 168–169
Barrie, Chris 379–380, 383, 389–390, 408–409
bastardisation scandals 272–277, 365–366
   see also 'Skype scandal'
Basten, Henry 170, 203–204, 208
Baxter, Philip 17–18, 20–22, 66, 68–70, 80–84, 86–89, 93–94
   academic responses to 78
   establishing a tradition 98–99, 452–453, 466–467
   feedback to 89–90
   goodwill from collaboration 96
Bearman, Richard 142–143
Beazley, Kim (Junior) 315, 449
Beazley, Kim (Senior) 171
Beck, Gary 356–357, 362–364
Beddie, Brian 104, 120–121, 143–149, 162, 165–167
Bennett, Bruce 355
Bennett, Phillip 261, 273, 276
Bergin, Anthony 337, 344–345, 412
Birt, Michael 260–261, 286, 311
Bishop, Bronwyn 368
Blamey, Thomas 40
Bland, Henry 20, 125
Bligh, William 10
Bomball, Richard 312–313, 318–319, 356–357
Bowden, Eric 27
Brabin-Smith, Richard 318–319
Braddon Clause, Australian Constitution 22
Bradford, Bob 326
Bridges, William 38
Britannia Royal Naval College, Dartmouth 25, 30–31, 33, 174, 321 *see also* Royal Navy
British Army 23 *see also* Royal Military Academy at Sandhurst
   recruitment 175–177
Broderick, Elizabeth 445–446
bullying *see* bastardisation scandals; 'Skype scandal'
Bungey, Mel 230, 246–247, 254

Burgess, Trish xv
Burns, John 82, 104, 140, 144, 148–149, 326
   ADFA planning 157, 159
   autonomy discussions 201–202, 207, 210, 216–217, 220
   political involvement 240–242, 257–258
Burton, Joe 60, 62–64

C
cadet misconduct *see* bastardisation scandals; 'Skype scandal'
cadets, UNSW campus time for 71–73, 80–81, 83, 87–88, 90, 94, 104–105, 107, 110, 117, 185–195
Cairns, Jim 111–112
Calder, Sam 246
Callinan, Ian 438
Campbell, David 321
Campbell, Ian 45
Canada *see* Royal Military College Kingston, Canada
Canberra campus *see* Australian Defence Force Academy; University College; UNSW Canberra
Canberra College of Advanced Education (CCAE) 449–450
Canberra Times 238–243, 343–344, 366–367, 372
Canberra University College (CUC) 59–60, 115
   ANU and 60
   University of Melbourne and 60
Cannane, Steve 445–446
Capp, Alan 360
Carmody, John 373
Carrick, John 226, 248–249, 254
Carter, Alan 279–280
Carwardine, AM 'Gerry' 315, 357
Casey, Baron Richard 219–220
Casey University 219–220, 222–228, 243–246, 249–250, 256
   name dropped 272
Casey University–ADFA Bill 238
   Section 5(i)(a) 244
Chaney, Fred 36, 64
Chapman, Tom 242
Cheeseman, Graeme 87, 371
Chief of the Defence Force Students (CDFS) Programs 436
China 86

501

Chipp, Don 36, 68
Chubb, Ian 307–308
citizen forces 38, 199
civilian students 157, 453–455, 468
Clark, John 98
Clark, Manning 63–64
Coates, John 272–274, 276–278, 412
Coburn, Bill 296
Cold War 18, 177
  standing peacetime force 58
collaboration, model for 12
colleges of advanced education 19, 119, 128, 163, 171, 220
Colston, Mal 254
Combes, Bertrand 40
commercial contracts, 1999–2004 385–417
Committee on Australian Universities 19
Commonwealth
  responsibility for defence 22–23, 57, 62
  role in universities 15, 18–20, 62–64, 70
Commonwealth Public Service 74, 100
Commonwealth Reconstruction Training Scheme (CRTS) 15
Communist party 63–64, 80
Connor, James 97
Conroy, Stephen 349–350
Constitution (Australia), Braddon Clause in 22
Cook, Joseph 38
Corbett, Arthur 82, 104
Cranfield University 298–299, 433 *see also* Royal Military College of Defence Science at Shrivenham
Cranwell, Britain, RAF College 46, 175
Creswell, William 24
Current Affairs Bulletin 85, 90
Curtis, Rodney 323
Cutler, Charles 69

D
Daly, Thomas 83, 135–136, 257
Dartmouth *see* Britannia Royal Naval College, Dartmouth; Royal Navy
David, Edgeworth 27
Davies, George 223–224
Davis, Edwin 94
Daw, David 163
Day, Peter 356
Deane, William 364–365
Dedman, John 15

Defence *see also* Australian Defence Force; Department of Defence; UNSW–Defence relationship
  'Academic Statement of Requirement' (ASOR) 395
  'jointery' 122, 126, 140, 229, 248, 265, 338–339, 475
  leadership changes 464–465
  universities and 17–18, 22, 123–152, 451
Defence Academy *see* Australian Defence Force Academy
Defence Act (1903)
  Section 29 23
  Section 32C 288
Defence Administration Committee 49
Defence and Security Applications Research Centre 437
Defence Department *see* Department of Defence
The Defence Force and the Community 339–340
Dempsey, Boyd 163
Denning, Arthur 17–18, 20–21
Dennis, Peter 273, 321
Department of Air 50, 124, 129
Department of Army 121, 124, 161
Department of Defence 123–126
Department of Education 140–141, 166
Department of Navy 94, 134
Department of Supply 124
Derham, David 52, 149–150, 168–169, 245–246
Dibb, Paul 414
Diro, Ted 147
Dixon, Norman 6–8
Dovers, Bill 31–33, 170, 200–201, 203, 207, 210–211, 214, 217–218, 261
Downes, Cathy 175–177
Drake-Brockman, Tom 132
Drakeford, Arthur 47
Dunbar, 'Noel' 95, 226–227
Dunstan, Donald 243
Dunstan, Graeme 88–89
Duntroon *see* Royal Military College
Dykes, Ewart 95, 181–182

E
education and training 3–6, 9, 231, 421, 425–426, 460–461 *see also* names of educational institutions, eg, UNSW Canberra

Index

balanced and liberal education xviii, 13, 395, 425, 432, 455, 464–466, 481
   for uniformed leaders 1–2
   indoctrination 10–11
   national culture and 9
   opinions on 467–469
   proposed voucher system 340, 343, 349
   vocational learning 421–422, 426, 429
Education Department (Commonwealth) 140–141, 166
Edwards, Peter 137–138, 248
Electrical and Mechanical Engineers Corps (RAEME) 21
Empire Air Training Scheme 46
engineering
   ADFA 212–213, 226, 234, 315–316, 320, 325, 330, 354, 415, 436–437
   aeronautical engineering 196, 320, 449
   air force students 46–47, 50, 196, 320, 334
   ANU 64, 221, 226, 247
   army students 37–39, 43–45, 78, 80–81, 104–108, 206, 271, 286
   civilian students 209, 212, 448–450
   competencies required 354–355
   naval students 25, 30, 34, 69, 174, 181–183, 185, 192–193
   UNSW 66
English, Cyril 127–128, 173–174
ethical training 9–10
Ewing, James Alfred 24

F

Faculty of Military Studies (FMC), RMC Duntroon 81–82, 93, 96–97, 141–143, 146, 294 *see also* Royal Military College
   1967–1972, establishing a tradition 98–122
   1967 agreement extended 158, 163, 204–205
   academic requirements 100–101
   access to official documents 146–147
   ADFA planning and 162, 202–205, 235, 241, 256–257, 301
   approval to establish 98
   computing facilities 102
   military work an integral part 101
   research 217, 462
   staff 98, 101–104, 214, 456
   status of degrees from 111–112
Fairbairn, David 137, 139
Fairhall, Allen 36, 68, 94, 125

Faulkner, John 336
Federation of Australian University Staff Associations (FAUSA) 244–246
Felsche, Klaus 265–269, 472–473
female staff 291, 316, 371, 374–376
female students 206, 212, 244, 298, 303, 327, 367, 384, 445–448
feminist perspectives and gender studies 371, 374–376
Field, Chris 468–469
Finlay, CH 'Basil' 61, 63–64, 83–84, 99–100
Fitzgibbon, Joel 441
Forbes, Jim 65
Forde, Frank 42
Foster, Henry 37
Fox, Mr Justice (Russell Walter) 111, 113
   on Interim Council 102
Fox Report 111, 113, 276
Fraser, Colin 108–109, 112
Fraser, Malcolm 68–70, 73, 84–85, 93–94, 122, 464–465, 467
   ADFA planning 171–172, 260
   autonomy discussions 199, 228
   establishing a tradition 99, 103
   political involvement 238–239, 242, 247–248, 254
   UNSW as unifying agent in Defence 125–126, 132–138
Frater, Michael 443
Frederick, Wilfred 48
Frost, Roy 51, 198–199
Fry, Ken 208, 240–241
Fulford, Glenn 345–346

G

Gannon, John 292
Garnaut, Ross 90–91
Gates, Raydon 427, 441–442
gender studies 371, 374–376
George, Don 253
George, Paul 72, 78, 117
Gerrity, George 162–163
Gilbert, Alan 287, 426–427
Given, Ken 334, 340–342, 393, 405–406
Golding, Ray 242, 274–275, 278, 282, 286–287, 301–302
Goldrick, James 439, 444
Gorton, John 67–68, 70, 125, 133–137
governance, UNSW–Defence relationship 437–438

503

Gration, Peter 332, 336
Gravell, Bill 197
Green, Harry 163, 203, 241
Grey, Bronwyn 368–372
Grey, Jeffrey xiv
Grey Review 368–372, 377
  cultural reform agenda 382, 384
  equity and diversity program by CIT 376
  UNSW response to 372–373
Gullet, Jo 119

H
Hackett, John xviii
Hall, Bob 87
Hall, Peter 386
Hammer, Gordon 76–77, 83
Hard, Lynn 292
Hardy, Walter 196–197
Harrington, WH 'Arch' 33
Hartley, Frank 299
Harverson, Dennis 300–301
Hassett, Frank 172
Hawke Labor Government 271
Hay, Bob 142, 160
Hayden, Bill 209, 220, 271
Heffernan, Paddy 46
Heffron, Bob 16
Heseltine, Harry 302, 310–311, 313–314, 322–325, 421–422, 474
  response to Price report 346–347
  retirement 355
Hickling, Frank 342, 347, 357
Hicks, Edwin 125
higher degrees *see* postgraduate education
Hill, Alec 109
Hill, Robert 404–405, 410
Hilmer, Fred 439–441
Hirschhorn, Jeremiah 72–73
Histed, George 180–181
historical approach ix–xi
HMAS *Creswell* 95, 179 *see also* Royal Australian Navy College
  'Creswell Course' 190
  HMAS *Creswell* as site for oceanography 279
HMAS *Watson*, South Head, Sydney 178, 187–188, 191
Hodges, Alan 264
Hodges, Sid 119, 163–164
Holt Government 127

Hope, AD 63–64
Horner, David 296–297
Hosking, Richard 46
Howard Coalition Government 349–350, 358
  Defence Efficiency Review 358–360
  Defence Reform Program (DRP) 360–361
Howard–English Report 127–128, 173–174
Howson, Peter 52, 129, 133
humanities 137–138, 271, 302, 311, 437, 463
Hume, Len xiv, 161, 170–171, 199, 217
Hume, Rory 418–419
Hunter, Chris 332
Hurley, David 440–441
Huxley, Leonard 60, 63–64

I
Imperial defence 22–23, 28, 37
incompetence, military 6–8
indoctrination 10–11
Indonesia 86
Institution of Engineers Australia 105–107, 225, 270–271, 354–355, 449
institutional partnerships, 1999–2004 385–417

J
Jervis Bay *see* HMAS *Creswell*; Royal Australian Naval College
Johnson, Keith 233–234
Johnson, Lyndon 89
Johnston, Grahame 104
Joint Intelligence Organisation 125
Joint Services Staff College, Weston Creek 125, 333, 336, 341–343, 424–425
'jointery' 122, 126, 140, 229, 248, 265, 338–339, 475
Jordan, JC 'Sam' 236–237, 261

K
Kafer, Bruce 444–445
Karmel, Peter 226
Kasper, Wolfgang 239–240, 243, 270–271, 282, 358
Kemp, Rod 349
Kensington campus *see* UNSW
Killen, Jim 132, 172, 202, 208
  autonomy discussions 210, 216, 228
  political involvement 247–253, 255

Index

King, Robert 388–390, 401, 406–407, 415–416, 418, 428
Kitchener, Lord 'Horatio' 38, 40
Knights, Robert 61
Korean War 17, 20

L
language studies 422, 429
Leonard, John 383–384
Lever, Susan 390
Livermore, Mr 161
Lockhart, Greg 87
Lovell, David 419, 470–471
Low, Anthony 221–223, 225–226, 242
Lowe, Ian 86–88
Luxton, Lewis 102
Lynch, Phillip 111–113

M
Macandie, George 26
Macauley, Godfrey 21, 142, 257
MacDonald, Arthur 218–219, 231, 242–243
Mahler, Ferdinand 159
Manadon, Royal Naval Engineering College at 174, 181–182, 193
marine sciences 279–280
Martin Committee 5, 64–65, 94, 128–129, 135, 169, 195, 266
Martin, Leslie 19, 48, 64, 94, 127–128
   1967–1972, establishing a tradition 99, 102–104, 106, 109
   Dean at RMC 82
Martin, Ray 222, 224
Masland, John 115
McAllister, Ian 321
McBride, Philip 49
McCallum, Doug 254, 301
McCauley, John 48
McDonald, Neil 182–183
McKeegan, Jim 179, 302
McKern, Bruce 393–394
McLachlan, Ian (RAAF officer) 49, 129
McLachlan, Ian (Defence Minister) 358, 370
McLeod, Ron 313
McMahon, William 135–136
McNeill, Ian 87
media, 'Skype scandal' and 444
Melbourne University *see* University of Melbourne
Melbourne University Private, ADC Strategic Studies program 426–427
Mench, Paul 145–148, 160–161
Menzies, Robert 18–19, 64–65, 102, 123–124
military education *see* education and training
military incompetence 6–8
Mills Committee 18
Milner, Kit 77
misconduct *see* bastardisation scandals; 'Skype scandal'
Moore, Bruce 322–325
Moore, Darren 110
Moore, John 384, 390–392
Moorhouse, Charles 119
moral education 9–10
Mordike, John 87
Morrison Report 373–376, 386
Morshead Review 123–125
Munro, Crawford 79
Murray Report 19–20, 30–31, 33, 123
Myers, Rupert 66–67, 71, 73, 75, 81–83, 88–89, 465
   foreword xvii–xix
   1964–1967, firm foundations and the faculty solution 96
   1967–1972, establishing a tradition 98–99, 110–111, 121–122
   1967–1975, university as a unifying agent in Defence 139–142, 144–145, 148–152
   1974–1976, Academy commitment 154–155, 158–159, 163–165, 168–171
   1975–1978, autonomy discussions 200–203, 210–211, 215–216, 219–222, 225, 227–228
   1978–1981, political involvement 232–236, 240–241, 246, 249–258
   1981–1986, from affiliation to Academy 260

N
national character 8–9
National Service scheme 58, 139
National Union of Australian University Students 89–90
Naval College *see* Royal Australian Naval College
naval history 321–322
navy *see* Department of Navy; Royal Australian Navy
New South Wales University of Technology 16–17, 19, 21

Newman, Jocelyn 349–350
Newton, Charles 390
Newton, Maxwell 109
Niland, John 311–312, 337, 378–379, 418–419
  institutional partnerships and 385–390, 406
  response to Price report 347–348

O
oceanography 279–280
O'Connor, Dennis 402
O'Farrell, Patrick 3, 5, 17, 280–283
officer education *see* education and training
official documents, access to 147–148
Olive, Laurie 390–391, 401
On the Psychology of Military Incompetence (Dixon) 6–8
O'Neill, Robert 236, 250
Operation Sovereign Borders 10
Osborne, Fred 49
Overall, John 102

P
Paltridge, Shane 36, 65, 67, 124
Papua New Guinea 136, 143–148
Parliamentary Standing Committee on Public Works (PWC) 230–238, 241, 243–249, 251–256
  report tabled 246–247
Parsonage, Phillip 378–379
Paton, George 49
peacekeeping missions 334
peacetime armies 8, 41
Peacock, Andrew 112
Pearson, Sandy 113, 161
Pemberton, Gregory 274–275
Pepper, Barbara 393
Pfennigwerth, Ian x–ix
philosophy 137
Playford, Thomas 24
Point Cook, Melbourne *see* RAAF Academy
political involvement *see* Parliamentary Standing Committee on Public Works
Pollard, Reginald 61–62
postgraduate education 146, 215–217, 423, 427–428, 430–432, 436–437, 453 *see also* research
  ADFA 207, 264, 341–342, 353–354
  air force students 50

army students 114–116, 146
future prospects 478, 480
naval students 190, 283–284
University College 315–318, 325, 398, 410–411, 423–424, 427–428, 430–432, 436–437
UNSW Canberra 448
Price Report 5, 333–334, 336–351, 468
  new Coalition Government's response to 349
  on cadet maturity 366
  Zimmer review and 392–393, 395
Pritchett, WB 'Bill' 121
profession of arms 8–9, 59
project management 424
protests *see* student activism
Public Service, Commonwealth 74, 100
Public Works Committee *see* Parliamentary Standing Committee on Public Works
Purcell, Gerry 35
PWC *see* Parliamentary Standing Committee on Public Works

R
RAAF *see* Royal Australian Air Force
RAAF Academy, Point Cook 49–53, 149–150, 195–199 *see also* RAAF College
  Accreditation Committee 211–212
  ADFA planning 50–53, 158–159, 168–169, 291–292
  conduct of cadets 196
  higher degrees 50
  tri-Service academy concept and 67, 127, 129–130, 133, 196–199
RAAF College, Point Cook 46–49, 198 *see also* RAAF Academy
  reconstituted as RAAF Academy 49–50
  review of 48–49
RAAF New South Wales University Squadron 199
Radway, Laurence 115
RAFC *see* Royal Air Force College at Cranwell, Britain
RAN *see* Royal Australian Navy
RAN College *see* Royal Australian Naval College
Ratcliffe, Jack 279–280
Ray, Robert 336, 347–348
recruitment 175–177, 180–181, 206, 430–431
  academic requirements 101

Index

'Skype scandal' and 447
tertiary education as a recruiting tool 22
Reeve, John 321–322
Reith, Peter 392, 404–405, 410
religious affiliations 77
research 115–116, 306–307, 432, 462 *see also* postgraduate education
  ADFA 306–307
  leadership areas 437
  RAAF Academy 50, 52–53
  RMC Duntroon 103
  University College 387
Richards, John 384
  as Rector 310, 355–357, 362, 364, 381–382, 422–423
  on UC acronym, as Rector 305
  reviews and reforms 332, 337, 360–361, 368
RMC... *see* Royal Military College...
RMIT, RAAF and 334
Roach, Terry 288
Robertson, Horace 40
Robertson, John 119–120
Robin, Mr Q de Q 33
Ronayne, Jarlath 308–309
Ross, Peter 284–285
Rowell, Sydney 42, 249
Royal Air Force College at Cranwell 46, 175
Royal Australia Air Force Academy (RAAFA) *see* RAAF Academy
Royal Australian Air Force (RAAF) 96 *see also* RAAF Academy
  association with RMIT 334
  at RMC Duntroon 45–46
  flying training 46–48
  officer education 45–53
  tri-Service academy and 129
  university partnership canvassed 24
Royal Australian Corps of Engineers 104
Royal Australian Naval College (RANC) 24, 32–36, 69–70
  Academic Standing Committee 95
  Accreditation Committee 211–212
  ADFA and 301
  'Creswell Course' 190
  education at 25–28, 30–31
  future of 279–280
  personnel system 190
  relocated to Westernport, Victoria 28–29
  return to Jervis Bay 31
  site leased as a holiday resort 28
  student failure and dissatisfaction 185–195
  tri-Service academy and 127, 291–292
  UNE and 35
  UNSW and 69–70, 94–96, 178–179, 184–189, 195, 283–286
Royal Australian Navy (RAN)
  Academic Advisory Committee 34–35
  arguments in favour of tertiary education 31–32
  entry age for naval cadets 27–29, 35–36
  officer education 24–37
  types of entry 28–30, 33, 194–195
  university partnership canvassed 23
Royal Melbourne Technical College 45
Royal Military Academy at Sandhurst 38, 172–173
Royal Military College Kingston, Canada 38, 108, 114, 165, 298
Royal Military College of Defence Science, Shrivenham 114, 173, 236, 297–299, 318, 432–433
  female students 298
Royal Military College (RMC Duntroon), committees and groups
  Interim Council 4, 83, 99–102, 106, 112–113, 117–118, 120, 301
  Standing Academic Committee 59, 65, 74
Royal Military College (RMC Duntroon), development of *see also* Australian Defence Force Academy; Faculty of Military Studies
  accommodation 114, 117
  ADFA planning and 154, 245–246
  affiliation with UNSW 65–97, 158, 163, 204–205, 249–256, 258
  ANU and 60–65
  Australian Defence Studies Centre 412–414
  facilities 106
  library 205, 207
  opened 1911 38–43
  plans for autonomy 64, 117–122, 141–143, 150–152, 154, 163, 168, 215–216
  tri-Service academy and 26–27, 67, 126–127, 130, 138–142
Royal Military College (RMC Duntroon), reviews and reforms 40–43
Royal Military College (RMC Duntroon),

507

staff, students and subjects
  abuse allegations 89, 108–112, 272–277
  cadet class system 274–275
  campus time *see* cadets, campus time for
  cultural change 97
  influences on cadets 86–87
  lexicon of cadet slang 322–325
  pay, superannuation and conditions 74, 84, 92, 100, 157, 214–215, 221–222
  PNG seminar controversy 143–145, 166
  postgraduate education *see* postgraduate education
  RAAF training at 45–46
  RAN College compared 25
  RMC Scholarship Scheme 108
  security clearances for faculty 74, 92
  staff appointments 81–82, 100
  student activism and 85–91
  Student Union 110
  tension between the military and academic staffs 97
  validation and accreditation 59–60
Royal Naval Engineering College, Manadon 174, 181–182, 193
Royal Navy 28, 30–31, 34–35, 174–175
  see also Britannia Royal Naval College, Dartmouth
  advice and assistance from 23
Rudd, Kevin 441
Rum Rebellion 10
Russell, Roger 227
Ryan, Susan 449

S
Sandhurst, Royal Military Academy 38, 172–173
Sargeant, Brendan 425–426
scandals *see* bastardisation scandals; 'Skype scandal'
Scherger, Frederick 48
scholarships 468
Scholes, Gordon 271, 274, 276, 449
school, RANC acting as 25
science and technology 68, 195, 197, 278–279
  RAAF College 48, 50–52
  role of university 17–18, 20
Scott, John 223
Second World War, education and 40, 57
Service colleges 3–4, 393 *see also* RAAF Academy; Royal Australian Navy College;

Royal Military College
Service rivalry 124, 135
sexual abuse and harassment 447–448
  see also bastardisation scandals; 'Skype scandal'
Shimmin, Ted 179, 283–284
Short, Trevor 291–292
Shrivenham *see* Royal Military College of Defence Science at Shrivenham
Sinclair, Ian 333–334, 350
Sinclair, Peter 288–290, 297–298, 300, 304, 340, 356, 449
single Service colleges 3–4, 393 *see also* RAAF Academy; Royal Australian Navy College; Royal Military College
'Skype scandal' 444–448
Smith, Hugh 121, 146, 155–156, 167–168, 250–251, 300, 337, 412
  ADFA planning 263–264
  on Price report 344–345
Smith, Stephen 444
Smith, Victor 126, 136
Sneddon, John 206, 286
Snow, John 194
socialist society (New Left) 85
Soldiers and Scholars 115
Spurling, Kathryn 445–446
Stephen Report 307–308
Stevens, Jack 102
Stone, John 239, 458
Stranks, Donald 225
student activism
  affiliation agreement 85–92
  anti-military sentiment 22, 65–66, 85
Supply Department 124
Sutherland, Traill 84, 112
Sweetman, John 177
Sydney Technical College, Ultimo 16
Sydney University *see* University of Sydney
Synnot, Anthony 218–219, 243, 465

T
Tange, Arthur 125–126, 137–138, 148, 170, 203, 226, 228, 230–231, 258, 465
  ADFA fellow 315
  ADFA planning 260
  autonomy discussions 218, 221
  political involvement 247–248
Tardif, John 163
Taylor, Bill 333–334

Index

Technical and Further Education (TAFE) 420–421
Technical Education and University of NSW Act 1949-61 71, 102, 263
technology see engineering; science and technology
tertiary education see universities
Tertiary Education Commission (TEC) 226
Tertiary Education (Services' Cadet Colleges) Committee 127, 130
Tharunka 88–89
Thompson, Roger 155, 296–297
Tocqueville, Alexis de 8
Tonkin, Rob 264–265, 294
Townley, Athol 49
training see education and training
Trakman, Leon 414
Tranter, Paul 345–346
tri-Service education 12–13, 172 see also Australian Defence Force Academy
  aims to establish 67, 122
  cost-effective solution 130, 132–133
  critical juncture 53–55
  Fraser and 84–85
  recommendations for 49, 124, 126, 130–142, 199–229
  responses to 153–168
Turner, Len 82, 94, 102–104, 110–111

U
UNE 35
uniformed leaders, education for 1–2, 56–58, 317
United States Military Academy at West Point 38, 41–42
United States Naval Academy at Annapolis 179, 181, 297, 458
United States Navy 179–180
universities see also colleges of advanced education; education and training; names of specific universities, eg, UNSW
  arguments for tertiary education 31–32
  as State entities 57
  as unifying agent in Defence 123–152
  development in Australia 14
  education at a civilian campus 47
  external affiliations 77
  faculty solution 56–97
  little flexibility from 54–55
  service officer education and 14–55
  sexual abuse and harassment 447–448
  vocational learning 79, 421–422, 426, 429
Universities Australia see Australian Vice-Chancellors Committee
Universities Commission 15, 214–215
University College 260–304, 327–328 see also Australian Defence Force Academy; UNSW Canberra
  academic results and retention rates 326, 337
  annual lectures 460
  Asia Pacific language focus 361–362
  budgets 313–314
  cadet slang controversy 322–325
  departments 287–288, 415–416
  enrolment patterns 315–316, 325
  equity and diversity 374–376
  established 257–259
  External Affairs Committee 318–319
  Graduate School 343, 423
  Graduate Studies Institute 428–429
  Master of Defence Studies 318
  Media Liaison Unit 314–315
  name of 272, 305–306, 416–417, 476
  Open Days 314–315
  pay and conditions 410–412
  Rector 263–265, 287
  research see research
  response to Price report 346–347
  staff 316–317, 410–411
  Steering Group 373–376
  strategic studies and management courses 370
  UNSW and 293–294, 386
University of Canberra 427
  ACT School of Engineering 449–450
University of Melbourne see also Canberra University College
  ADFA planning 164
  RAAF Academy and 49–53, 96, 149–150, 195–199
  RAAF College and 47–49
  RMC and 38–39, 44–45
  School of Naval Science proposed 24–25
University of New England (UNE), agreement with Naval College 35
University of New South Wales see UNSW
University of NSW Act see *Technical Education and University of NSW Act 1949–61*

University of Queensland, RMC and 38
University of Sydney 16, 196
   Department of Military Studies 37
   RAAF Academy and 50
   RANC cadets and 26
   RMC and 38–39, 44
University of Technology, Kensington 16–17, 21
   renamed 'The University of New South Wales' 19–20
UNSW, affiliations *see also* names of specific agreements, eg, 1981 Agreement; University College; UNSW Canberra; UNSW–Defence relationship
   1965 defence negotiations with 36
   ADFA planning 164
   benefits from collaboration 96
   funding from Defence 402–404
   NIDA at 66
   operating in ACT 69, 92
   RANC and 69–70, 94–96, 185–195
   reporting to Defence 331
   RMC affiliation 45, 65–97, 249–256, 258
   RMC autonomy plans 141–143, 150–152, 154, 163, 168, 215–216
   tri-Service academy and 126
   unifying agent in Defence, 1967–1975 123–152
   Zimmer review and *see* Zimmer review
UNSW, committees and groups
   Professorial Board 70–71, 73–75, 79, 82–83, 88, 93, 202–203, 257
   University Council 80
UNSW, development *see also* Wollongong campus of UNSW
   Act of Incorporation 77
   corporate branding 325–326, 381–382
   militia unit/regiment proposed for 20–22
   named 19–20
   reputation 395–397
   size of 397
   strategic plan 312–313
UNSW, reviews and reforms 386–387
UNSW, staff, students and subjects
   'Academic Statement of Requirement' 395
   campus time *see* cadets, campus time for
   distance education 422–423
   Faculty of Military Studies *see* Faculty of Military Studies

Military Communication Training to Support AMET 422
   Staff Association 110–111, 113
   student activism 65–66, 85–91
   Students Union 110–111, 113
   Tharunka 88–89
UNSW–Defence relationship *see also* names of specific agreements, eg, 1981 Agreement
   introduction 1–13
   1901–1966: universities and service officer education 14–55
   1964–1967: firm foundations and the faculty solution 56–97
   1967–1972: establishing a tradition 98–122
   1967–1975: UNSW as unifying agent in Defence 123–152
   1970–1980: joint educational enterprise 178–199
   1974–1976: Academy commitment 153–177
   1975–1978: from autonomy to uncertainty 200–229
   1978–1981: Parliamentary works and political will 230–259
   1981–1986: from affiliation to academy 260–304
   1986–1995: reviews, reforms and restructures 305–352
   1996–1998: controversy and consolidation 353–384
   1999–2004: commercial contracts and institutional partnerships 385–417
   2005–2017: stability and sustainment 418–451
   difficulties 460–461, 467, 470–471
   future prospects 477–478
   governance 437–438
   liaison 461–462
   observations and conclusions 452–481
   success of 453–454, 456–457
   vision for education 461–464
UNSW Canberra 421–422 *see also* Australian Defence Force Academy; University College
   academic results and retention rates 445
   ADFA–UNSW Canberra common attributes. 443–444
   civilian students 448–450
   Learning and Teaching Group 442

Index

names of 417
performance metrics 431, 434
postgraduates *see* postgraduate education
priority activities 420
professional short courses 442
recognising professional military training 421
revenue 436–437, 442
role 420
School of Engineering and Information Technology 450–451
'silence' in 'Skype scandal' 446–447
specialist contracts 430
Strategy 2025 480
student numbers 430, 436–437, 450, 454
Student Progress Rate 431
success of 454–456
UNSW PREP (Performance, Reporting, Evaluation and Planning) feedback process 386
UNSW Regiment 21–22, 378–379

V
Van der Werf, Paul 179–181
Vasey, Alan 40–42
Victoria Barracks, Sydney 39–40
Vietnam 64–66, 85–89, 91, 124–125, 378
  anti-Vietnam protest movement 65–66
  ending of involvement in 139
  RMC graduates in 106–107
vocational learning 79, 421–422, 426, 429
Vowels, Rex 61–62, 66, 71, 73, 75, 80–81, 83, 257, 465
  appointed Pro Vice-Chancellor 98–99

W
Wade, Ronald 124
Wainwright, Mark 419
Walsh, Gerry 108–114
Walsh, Richard 85, 90
Warn, James 345–346
Weeden, Jock 33, 99
Weeden Report 33–35

West, Francis 224–225
West Point, US Military Academy at 38, 41–42, 297, 458
Weston Creek, Canberra, ACDSS at 421–422, 424
White, Bruce 65, 83–84, 103, 258
White, Brudenell 26
Whitlam, Gough 92, 126, 139
Wicken, Tony 386
Willett, John 223
Williams, Bruce 220
Willis, Albert 'Al' 61–62, 79–82, 103, 110, 113, 117, 148, 465
  ADFA planning 158–160, 162–163, 211, 242
  appointed Pro Vice-Chancellor 98–99
Wills, Howard Arthur 116
Wilson, Brian 327–329
Wilson, Geoff 164–165, 170, 236, 309–311, 449
  ADFA planning and 274–275
  as Rector of University College 286–287, 289–290, 297–298, 300
  autonomy discussions and 202–203, 211, 217, 222
  political involvement 242, 246, 251–252
Wilton, John 45, 59–61, 125, 127, 465
Windeyer, Victor 21, 102
Wollongong campus of UNSW 17, 22, 36, 69, 102–103, 121, 177–178
Wood, John 76, 78, 83
Woodman, Stewart 428–429, 434–435
Woodward, Edward 261, 308–309, 347–348, 364, 438
World War II, education and 40, 57
Wright, Graham 126
Wrixon, Henry 24

Y
Young, Ian 386

Z
Zimmer review 391–402, 413, 415

511